Intelligent Systems Reference Library

Volume 65

Series editors

For further volumes:
http://www.springer.com/series/8578

About this Series

The aim of this series is to publish a Reference Library, including novel advances and developments in all aspects of Intelligent Systems in an easily accessible and well structured form. The series includes reference works, handbooks, compendia, textbooks, well-structured monographs, dictionaries, and encyclopedias. It contains well integrated knowledge and current information in the field of Intelligent Systems. The series covers the theory, applications, and design methods of Intelligent Systems. Virtually all disciplines such as engineering, computer science, avionics, business, e-commerce, environment, healthcare, physics and life science are included.

Mrutyunjaya Panda · Satchidananda Dehuri
Gi-Nam Wang
Editors

Social Networking

Mining, Visualization, and Security

Editors
Mrutyunjaya Panda
Gandhi Institute for
Technological Advancement
Bhubaneswar
India

Satchidananda Dehuri
Department of Systems Engineering
Ajou University
Korea
Republic of (South Korea)

Gi-Nam Wang
Department of Industrial
Engineering
Ajou University
Korea
Republic of (South Korea)

ISSN 1868-4394 ISSN 1868-4408 (electronic)
ISBN 978-3-319-34413-3 ISBN 978-3-319-05164-2 (eBook)
DOI 10.1007/978-3-319-05164-2
Springer Cham Heidelberg New York Dordrecht London

Softcover reprint of the hardcover 1st edition 2014

Printed on acid-free paper

Springer is part of Springer Science+Business Media (www.springer.com)

To Wife and Kids
Mrutyunjaya Panda
To Father: Kunja Bihari Dehuri and Mother: Kuntala Dehuri
Satchidananda Dehuri
To Wife
Gi-Nam Wang

Preface

The purpose of this volume is to attract a wide range of researchers and readers from computer science, network science, social sciences, mathematical sciences, medical and biological sciences, financial, management, political sciences, industrial and systems engineering by providing them the state-of-the-art, novel, and worthy contributed works towards various aspects of social networking from prospective authors for sharpening their way of thinking.

We solicit empirical and theoretical work on social network analysis based on a wide range of techniques from the disciplines like data mining, social sciences, mathematics, statistics, physics, network science, machine learning with visualization techniques, and security. Such a collection intends to illustrate the potential of multi-disciplinary techniques in various real life problems and motivate researchers of social network analysis to design effective tool for dredging actionable knowledge.

A social network is an umbrella with nodes of individuals, groups, organisations, and related systems that tie in one or more types of interdependencies. In other words it is intended to understand the social structure, which subsists amongst entities in an organisation. Social network analysis is focused on uncovering the patterning of people's interaction. Further, it is based on the intuition that these patterns are important features of the lives of the individuals who display them. The social network approach has primarily involved in two important aspects such as: (1) it is guided by formal theory organised in mathematical terms, and; (2) it is grounded in the systematic analysis of empirical data.

In chapter 1, Louni and Subbalakshmi have described several models in details to describe the mechanism of spread of information in social networks. Thereafter, they study the importance of "influencers" (nodes that have a higher influence on information spread in a network) and discuss the spread of both truthful and mis-information in a network. Methods to control the spread of mis-information through a social network are also discussed. Further, they have discussed the inverse problem of discovering the single and multi-source of any given piece of information.

With the exponential rise in popularity of Online Social Networks (OSNs) in recent years, there have been a number of studies which measure the topological properties of such networks. Several network evolution models have also been proposed to explain the emergence of these properties, such as those based on preferential attachment, heterogeneity of nodes, and triadic closure. Ghosh and Ganguly have made a survey of these studies in chapter 2 to describe in detail a preferential attachment based model to analyze the evolution of OSNs in the presence of restrictions on node-degree that are presently being imposed in all popular OSNs.

The importance of machine learning for social network analysis is realized as an inevitable tool in forthcoming years. This is due to the unprecedented growth of social-related data, boosted by the proliferation of social media websites and the embedded heterogeneity and complexity. Alongside the machine learning derives much effort from psychologists to build computational model for solving tasks like recognition, prediction, planning and analysis even in uncertain situations. In chapter 3, De and Dehuri have presented different network analysis concepts such as implication of machine learning for network data preparation and different learning techniques for descriptive and predictive analysis. Finally, they have presented some machine learning based findings in the area of community detection, prediction, spatial-temporal, and fuzzy analysis.

Social networks of various kinds demonstrate a strong community effect. Actors in a network tend to form closely-knit groups; those groups are also called communities or clusters. Detecting such groups in a social network (i.e., community detection) remains a core problem in social network analysis. Among the challenges that face the researchers to come up with advanced community detection methods, there is a key challenge, which is the validation and evaluation of their methods. The limited benchmark data available, the lack of ground truth for many of the available network datasets, and the nature of the social behavior factor in the problem, turned the evaluation process to be very hard. Accordingly, understanding such challenges may help in designing good community detection methods. Hafez et al., in chapter 4 presents testing strategies for community detection approaches and explores a number of datasets that could be used in the testing process as well as stating some characteristics of those datasets.

In the present scenario the social networks are more dynamic than static. The introduction of societal networks by Fiksel is an evolutionary step in the study of social networks which originated the concept of dynamic social networks. The term societal network was used by Fiksel in order to distinguish it from the diversified ways in which the term 'social network' was used prior to the 1980s. There seems to be little work done following his approach. In chapter 5 Tripathy et al., have introduced the social structures which form the basis of Fiksel's concept and also present his results.

Most social networks are dynamic and connections between people change naturally overtime. Researchers have begun to consider the problem of tracking the formation of groups of users in social network. The situation is complicated by the fact that subgroups may split or merge, so that cohesiveness is not necessarily a property of a single subgroup, but may sometimes relate to a family of one or more related subgroups. However, in general, cohesive families of subgroups at one time period should be similar to corresponding subgroups at a different time period. In chapter 6, Sharma and Purohit presents a state of the art survey of the works done on community tracking in social network. Additionally, their goal is to provide a road map for researchers working on different measures for tracking communities in Social Network.

In chapter 7, Choudhuri and Jana have made an empirical attempt for studying caste, class, and social support in Rural Jharkhand and West Bengal of the nation India through Social Network Analysis Approach.

Interaction across social networking sites leads to different kinds of ideas, concepts, choices, and sharing or some nourishing effects, which might influence others to believe or trust. Seldom may this cause some malicious effect for members and their peers only. As social network is the domain for sharing opinion and comments, subsequently it can also propagate malicious signature as well. Security and privacy is essential component to protect user profile from this kind of malicious program, which basically evolves from any close acquaintances, that also belongs to same vector plane. The degree of malicious attack of a social network depends on the number of flow links from one user to another with forward operations. It is true that the probability of malicious attack evolves from friend's community is of greater attack prone magnitude than the degree of attack from unknown members. Sarkar et al., in Chapter 8 focuses the different verticals of such possibilities of attack under social network processes and also tries to investigate the rudimentary precautionary measure pertaining security algorithm behind it.

The chapter 9 contributed by Basu aims to bring to the reader an overview of the work done since the 9/11 terrorist attack, in the field of Social Network Analysis as a tool for understanding the underlying pattern /dynamics of terrorism and terrorist networks. SNA is particularly suitable for analyzing terrorist networks as it takes relationships into account rather than merely attributes, which are difficult to obtain for covert networks. Using graph theoretic methods and measures and open source data it has been possible to map terrorist networks and examine roles of different actors, as well as identify groups and structures within the network. The methodology is illustrated by reviewing two case studies: the 9/11 terrorist network study by Krebs, that used data from a single terrorist attack, and a study by Basu that uses data from about 200 terrorist incidents in India to create a network of terrorist organizations for predictive purposes.

As the Internet continues to grow, the proliferation of online social networks raises many privacy concerns. The users of these OSNs are divulging endless details about their lives online. This personal information can be used by attackers to perpetrate significant privacy breaches and carry out attacks such as identify theft and credit card fraud. The privacy concerns arise from not just the users posting their personal information online, but also from OSNs publishing this information for analysis. Driven by Web 2.0 applications, more and more social network has been made publicly available. Preserving the privacy of individuals in this published data is an important concern. Although privacy preservation in data publishing has been studied extensively and several important models such as k-anonymity and l-diversity as well as many efficient algorithms have been proposed, most of the existing studies deal with relational data only. Those methods cannot be applied to social network data straightforwardly. Anonymization of social network data is a much more challenging task than anonymizing relational data. Firstly, in relational databases, attacks come from identifying individuals from quasi-identifiers. But in social networks, information such as neighbourhood graphs can be used to identify individuals. Secondly, tuples can be anonymized in relational data without affecting other tuples. But in social networks, adding edges or vertices affects the neighbourhoods of other vertices in the graph as well. In this chapter 10, Tripathy et al., give a brief overview of the privacy concerns in online social networks and provide a detailed description of our algorithm, GASNA, a greedy algorithm for social network anonymization. This algorithm provides structural anonymity and sensitive attribute protection by achieving k-anonymity and l-diversity in social network data. We also discuss the challenges faced by the existing algorithms/models for social network data privacy and suggest techniques to counter these challenges. The issues discussed are the high cost of achieving k-anonymity when the value of k is fixed and the need for a better anonymity model which suits the current scenario of social networks. Further, they also propose a new model called partial anonymity which can help reduce the number of edges added for anonymization when the value d of d-neighbourhood is greater than 1.

Social Network Analysis is a non-conventional Data Mining technique which analyzes social networks on web. The technique is used frequently for studying network behaviors using centrality measures viz. Degree, Betweenness, Closeness and Eigenvector. Hence has also led to the concept of Terrorist Network Mining which aims at detection of the terrorist group, studying the hierarchy they follow for the communication (using SNA) and then finally destabilizing of the network activities. The chapter 11 contributed by Chaurasia and Tiwari focuses on an approach under SNA known as Brokerage which finds brokers who serve as the leading nodes in the network. Brokerage is expected to be beneficial in case of estimating the terrorist groups where different subgroups of terrorist organization coordinate to fulfill their awful deeds. The brokerage on whole estimates the influential roles as it would be done by individually calculating the centrality measures, with much more useful information aiding to amend terrorist network analysis.

Bhubaneswar, India,
Suwon South Korea,
Suwon, South Korea
December, 2013

Mrutyunjaya Panda
Satchidananda Dehuri
Gi-Nam Wang
Editors

Acknowledgements

This project was partially supported by Defense Acquisition Program Administration and Agency for Defense Development under the contract UD110006MD and the Industrial Strategic Technology Development Program, 10047046, funded by the Ministry of Science, ICT & Future Planning (MSIP), Korea. We would like to thank these funding bodies for their financial support.

All the chapters included in this book are self-content and addresses the different issues of social networking along with remedies and applications to many real life societal problems. We would like to express our sincere appreciation for their contributions. We are deeply grateful to Springer International Publishing AG for their support starting from acceptance of proposal to final production of this book.

We would also like to thank Mr. Sagar S. De for his help towards compilation of this project in latex.

Finally, we would like to offer our deepest and most heartfelt thanks to our wives and children for their support, encouragement, and patience. In particular, Dr. Satchidananda Dehuri would like to thanks his children RISHNA and KHUSYANSEI for their support of creating a noiseless environment.

Contents

List of Contributors

Alireza Louni · K.P. Subbalakshimi
Stevens Institute of Technology, Hoboken, NJ, USA
e-mail: {alouni,ksubbala}@stevens.edu

Saptarshi Ghosh
Department of Computer Science and Technology,
Bengal Engineering and Science University Shibpur, Howrah – 711103, India
e-mail: sghosh@cs.becs.ac.in

Niloy Ganguly
Department of Computer Science and Engineering,
Indian Institute of Technology Kharagpur, Kharagpur – 721302, India
e-mail: niloy@cse.iitkgp.ernet.in

Sagar S. De
S.N. Bose National Centre for Basic Sciences, Block-JD, Sector-III, Salt Lake, Kolkata-98, India
e-mail: sagar.s.de@gmail.com

Satchidananda Dehuri
Department of Systems Engineering, Ajou University, San 5, Woncheon-dong, Yeongtong-gu, Suwon 443-749, Republic of Korea
e-mail: satchi.lapa@gmail.com

Ahmed Ibrahem Hafez
Faculty of Computer and Information, Minia University, Minia - Egypt,
Scientific Research Group in Egypt (SRGE)
e-mail: ah.hafez@gmail.com

Aboul Ella Hassanien
Faculty of Computers and Information, Cairo University, Cairo - Egypt,
Scientific Research Group in Egypt (SRGE)
e-mail: aboitcairo@gmail.com

Aly A. Fahmy
Faculty of Computers and Information, Cairo University, Cairo - Egypt
e-mail: aly.fahmy@gmail.com

B.K. Tripathy · M.S. Sishodia · Sumeet Jain
SCSE, V.I.T. University, Vellore, Tamil Nadu. 632 014, India
e-mail: tripathybk@rediffmail.com,
{mayanksshishodia, sumeet.jain44}@gmail.com

Sanjiv Sharma · G.N. Purohit
Banasthali Vidyapith, Bansthali, Rajasthan, India
e-mail: er.sanjiv@gmail.com, gn_purohitjaipur@yahoo.co.in

Anil Kumar Choudhuri
Sociological Research Unit, Indian Statistical Institute, Giridih, Jharkhand, India
e-mail: anil@isical.ac.in

Rabindranath Jana
Sociological Research Unit, Indian Statistical Institute, Kolkata, West Bengal, India
e-mail: rabindranathjana65@gmail.com

Manash Sarkar · Soumya Banerjee
Department of Computer Science, Birla Institute of Technology, Mesra, India
e-mail: manashsarkar53@gmail.com,
soumyabanerjee@bitmesra.ac.in

Aparna Basu
CSIR-National Institute of Science Technology and Development Studies,
Dr. K.S. Krishnan Marg, Pusa Gate, New Delhi 110012, India
e-mail: aparnabasu.dr@gmail.com

Anirban Mitra
Dept. of CSE, M.I.T.S., Rayagada, Odisha-765017, India
e-mail: mitra.anirban@gmail.com

Nisha Chaurasia · Akhilesh Tiwari
Department of CSE & IT, MITS Gwalior, India
e-mail: {chaurasianisha21, atiwari.mits}@gmail.com

Acronyms

WoM	Word of Mouth
Vloggers	Video Bloggers
SIR	Susceptible-Infected-Recovered
SIS	Susceptible-Infected-Susceptible
MLE	Maximum Likelihood Estimator
MAP	Maximum a Posteriori Probability
ML	Machine Learning
MDL	Minimum Description Length
OSN	Online Social Network
Triadic Closure	Triangle Closing
GEO	P-Geo-Protean
SNA	Social Network Analysis
GML	Graph Modeling Language
GTL	Graph Template Library
WWW	World Wide Web
SSSP	Single Source Shortest Path Problem
APSP	All Pair Shortest Paths Problem
DFS	Depth First Search
BFS	Breadth First Search
MRF	Markov Random Fields
ERGM	Exponential Random Graphical Models
OSN	Online Social Network
CMS	Content Management Systems
MC	Clean Model
MN	Noise Model
MR	Corruption Matrix
PCA	Principal Component Analysis
CIM	Collective Inference Models
nBC	Network only Bayesian Classifier
SSL	Semi-Supervised Learning
SVM	Support Vector Machine

ANN	Artificial Neural Network
SNCF	Social Network Collaborative Filtering
EM	Expectation-Maximization
ODF	Out-Degree Fraction
NMI	Normalized Mutual Information
GN	Girvan and Newman
START	Study of Terrorism and Responses to Terrorism
GTD	Global Terrorism Database
TOPs	Terrorist Organization Profiles
TKB	Terrorist Knowledge Base
MIPT	Memorial Institute for the Prevention of Terrorism
TPDRC	Terrorism and Preparedness Data Resource Centre
ICPSR	Inter-University Consortium for Political and Social Research
CASOS	Computational Analysis of Social and Organizational Systems
IDSA	Institute of Defense Studies and Analysis
MDS	Multidimensional Scaling
ULFA	United Liberation Front of Asam
LTTE	Liberation Tigers of Tamil Eelam
ISI	Inter Service Intelligence
Bipartite	Bigraph
IDM	Investigative Data Mining
IDS	Intrusion Detection System
TP	True Positive
FP	False Positive
DC	Dependance Centrality
GASNA	Greedy Algorithm for Social Network Anonymization
AEN	Edge Addition to Existing Nodes
AFN	Anonymization of Fake Nodes
NLP	Natural Language Processing
SNL	Social Network List
SSQ	Social Support Questionnaire
SCAN	Social Cohesion Analysis of Networks
DISSECT	Data Intensive Socially Similar Evolving Community Tracker
CC	Caste Community
SL	Source of Livelihood
Sigraphs	Sidigraph

Diffusion of Information in Social Networks

Alireza Louni and K.P. Subbalakshmi

Abstract. Social networks are a growing phenomenon in today's Internet media consumption. Social networks are used to not only stay in touch with friends and family, but also to seek and receive information on specific products/services as well as social activism. Understanding and quantifying the information flow within these networks is, therefore, of great interest to individuals, groups and businesses. Several models have been proposed to describe the mechanism of spread of information. We describe these models in detail in this chapter. We then study the importance of "influencers" (nodes that have a higher influence on information spread in a network) and discuss the spread of both truthful and mis-information in a network. Methods to control the spread of mis-information through a social network is also discussed. We then discuss the inverse problem of discovering the source of any given piece of information. Both single and multi-source problems are considered.

Keywords: Social networks, information diffusion, influencers, diffusion source estimation.

1 Introduction

Social networks, such as Facebook, Twitter, etc, have enabled individuals to influence each other's decisions through a network of friendships and subscriptions. The Internet and the physical world are no more separate, unrelated entities; events in the real world influence the activities on the Internet and vice versa [1]. About 23% of people in the United States use the Internet as their primary source of information in decision making [2] on dietary habits [3], shopping [4] and professional issues [5].

Alireza Louni · K.P. Subbalakshmi
Stevens Institute of Technology, Hoboken, NJ, USA
e-mail: {alouni@stevens.edu, ksubbala}@stevens.edu

M. Panda, S. Dehuri, and G.-N. Wang (eds.), *Social Networking*
Intelligent Systems Reference Library 65,
DOI: 10.1007/978-3-319-05164-2_1,

To illustrate the connection between real life events and the Internet media, we collected tweets related to "Wikileaks" and the number of searches on "Wikileaks", from Google-trends [6] for the year 2010-2011 (Fig. 1). It can be seen that the number of searches and tweets peaked on three occasions: around April 2010 (when the founder of Wikileaks was first wanted on criminal charges); around Nov. 2010 (when Wikileaks released around 400K files related to the war in Iraq) and in Dec. 2010 (because of the Wikileaks controversy).

Since information on the Internet plays a vital role in real life, it is essential to understand the dynamics of information flow over the Internet. It is well known that an individual who adopts a new technology is often influenced by their friends or subscribers. Two reasons have been postulated [7] to describe why people imitate the behavior of their friends. One reason ("informational effects") is that the decisions made by friends or subscribers provides indirect information about the technology in question. For example, people observe the decisions of their friends about a new phone and potentially describes features that these phones carry. That, and the individual's trust on the friend's decision, makes the individual also buy the same model of phone. The second reason is "direct benefits". This refers to direct payoffs

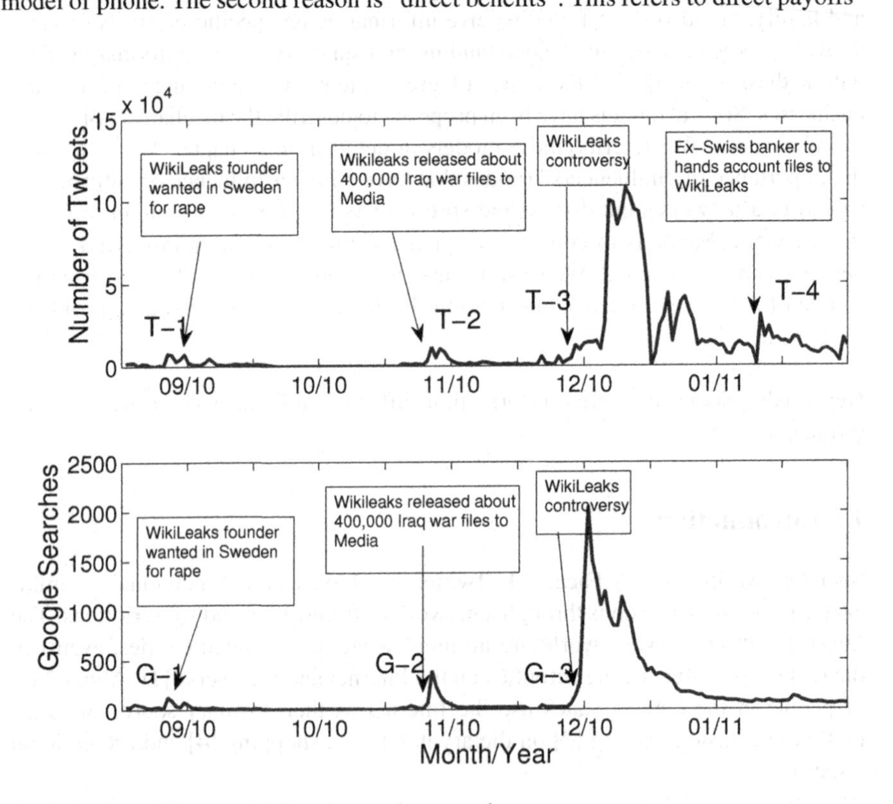

Fig. 1 Number of Tweets and Google searches over time

that result from copying the decision of a friend or subscriber. For example, one is more likely to download a piece of software or an App that is compatible with one's friend's version in order to better stay in touch or share a common experience.

This social mechanism of adopting a friend's decision can have a cascading effect on the spread of a new behavior or technology. This gives rise to the phenomenon, particularly in YouTube, of the technology "going viral". This sociological feature of humans has been used by the industry as a quick and cheap marketing strategy. In fact Word of Mouth (WoM) in social media is a popular marketing concept [8]. It was shown [8], that WoM referrals can be even more efficient than traditional marketing strategies. For example, the Ford Motor Company released their Fiesta model through social media networks using the top 100 video bloggers (Vloggers) in 2009, which resulted in 3000 reservations for the model by March 2010, by first time Ford buyers [9].

However, not all disseminators of information on social media use this social mechanism for non-malicious intent. Social networks can also be used to spread malicious gossip, untruthful information or outright lies. For example, in Summer 2012, false rumors about impending attacks on northeast Indian migrant workers all over India, caused wide spread panic and mass migration of these workers back to their home states in northeastern India [10]. Similarly, following a recent fake tweet about an explosion in the White House, the Dow Jones industrial average dropped 152 points within seconds [11].

In light of this strong influence that social networks have on real lives, it is critical to understand how information spreads rapidly through social networks. Similarly identifying the most influential disseminators of information can prove to be useful in either pro-actively spreading information or in quickly zeroing in on disseminators of false/malicious information. Suppose that the goal is to promote a trend or market a new technology within a social network. In other to do so, we target a set of individuals, known as influencers, for seeding a piece of information about the trend or the new technology to help propagate the information quickly. Similarly, the influencers also can act as a counter campaign to a group of people who spread malicious rumors or gossip in a social network.

A social network can be modeled as a graph where the nodes represent individuals or groups that have a social network presence. Nodes are connected by links if there is a direct relationship between these nodes. For example, a Facebook friendship network can be represented as an undirected graph where the nodes represent people and the undirected links between them indicates friendship. Whereas, the Twitter network is better modeled as a directed graph where a directed link connects a user to their follower. This general framework is used to study most problems in social networks.

Several researchers have studied the problem of modeling the diffusion of information *given* the knowledge of the source [12, 13]. By comparison, the inverse problem of localizing the source of information is harder and not yet fully studied. We will discuss each of these two important problems in the following sections of this chapter.

2 Models of Information Diffusion in Social Networks

Some of the popular models for information diffusion in social networks come from other areas like epidimiology, sociology and economics. Three of the most popular models are listed below [12].

1. **The Contagion Model**: This model is motivated by the fact that information or "trend" may potentially spread through the social network in much the same way as contagious diseases spread through a population [14, 15].
2. **The Social Influence Model**: This model also assumes that information or trends spread through neighbors in a network. In contrast to the contagion model, however, this model assumes that a specific node will not adopt a trend (or believe a piece of information) unless the number of neighbors who subscribe to this trend or information exceeds a certain threshold.
3. **The Social Learning Model**: In contrast to the first two models, the nodes in this model are assumed to be rational decision makers. They observe the outcomes of prior adopters among their neighbors and then decide whether or not to believe a piece of information.

2.1 Mathematical Modeling of Information Diffusion

2.1.1 The Contagion Model

In the contagion model of information propagation, the information is modeled as contagions in diseases in a population. These contagion of information spread in a population can be modeled either assuming an underlying structure of the diffusion network, as in the case of the susceptible-infected-recovered (SIR)[16, 17] and the susceptible-infected-susceptible (SIS)[15] models, or they can be modeled without considering the structure of the underlying social network, like in the case of the Bass model [12, 18, 19, 16].

Of these the Bass model is more widely used and studies contagion using two parameters: internal rate of contagion (λ_{int}) and external rate of contagion (λ_{ext}). λ_{int} is the rate at which a node that has not adopted a piece of information, hears about it from the prior adopters within the friendship or follower network. λ_{ext} is the rate at which a such node hears about a new technology from sources outside the friendship network. The external sources of contagion, for example, can be mass media advertisements of a new technology in television, radio, and billboards, etc. Let $f(t)$ be the fraction of nodes that have adopted a given technology at time t, the diffusion of the information (contagion) can be formulated as an ordinary differential equation:

$$\frac{df(t)}{dt} = (\lambda_{\text{int}} f(t) + \lambda_{ext})(1 - f(t)) \tag{1}$$

where $f(t)$ is given by

$$f(t) = (1 - c\lambda_{ext} e^{-(\lambda_{\text{int}}+\lambda_{ext})t})(1 + c\lambda_{\text{int}} e^{-(\lambda_{\text{int}}+\lambda_{ext})t}). \tag{2}$$

λ_{int} and λ_{ext} are non negative values that can be estimated from fitting the model to the empirical data. The diffusion curve $f(t)$ forecasts information diffusion over time. Many studies have found diffusion curves that are S-shaped where diffusion begins slowly, then accelerates, and then eventually slows down [16]. The best known example of such studies are [20, 21], which demonstrated S-shaped diffusion curve in case of adoption of hybrid corn among Iowa farmers [1]. The network structure is not involved in analysis of diffusion in the Bass model and therefore, many basic properties of information flow over social networks remain unseen in this model.

In contrast to the Bass model, the SIR model considers structure of the network in that, it uses the degree distribution of the diffusion network to calculate the flow. In this model information is modeled as spreading among prior adopters (infected nodes) and susceptible nodes. Once an individual adopts a piece of information or technology they are considered no longer susceptible. One might interpret the SIR model as follows: a node who has received and believed a rumor on the Twitter is infected and those that have not heard are considered susceptible. When the user decides to retweet the rumor, the rumor diffuses in the user's Twitter followers. Later, if the user deletes the rumor from her page on Twitter, that is akin to a recovery from the "infection" [22].

The mathematical analysis of SIR and SIS has drawn on tools from percolation theory. Percolation theory is a useful tool to study systems that exhibit abrupt change in behavior beyond a critical point. The nodes in a social network can be interpreted as holes in the porous membrane, which are connected by links to each other. The general question in percolation theory is whether a liquid can flow from one side of a porous membrane to the other [23]. The same question can be posed in the context of the spread of rumor spread in Twitter: can a rumor spread across the entire network when not all nodes are infected and some of the infected nodes recover from the infection.

Let $P(d)$ be the degree distribution of the diffusion network and t be the probability that an adopter will affect a susceptible node successfully. Let $P_t(d)$ be the degree distribution of the network of non adopters. Then $P_t(d)$ can be written in terms of $P(d)$ and t as follows [16, 17]:

$$P_t(d) = \sum_{k>d} P(k) \binom{k}{d} t^d (1-t)^{k-d}. \tag{3}$$

Assuming independent diffusion across different edges in the social network, it can be shown that if the probability t exceeds some threshold, t_c, then the information will spread over the entire network. The threshold t_c often takes small values, and so diffusion can occur even with high fraction of immune nodes. In

[1] Although the references quoted here dealt with a network of farmers in Iowa, it is possible to see the potential application of these theories and techniques to information diffusion in social networks.

general, t_c is dependent on the degree distribution of the diffusion network and therefore differs for different types of network

$$t_c = \frac{<d>}{<d^2> - <d>} \tag{4}$$

where $<d>$ and $<d^2>$ are the first and the second moments of the degree distribution $P_t(d)$. For a scale-free network, where degree distribution $P_t(d)$ is $d^{-\gamma}$, $<d^2>$ diverges for $\gamma < 3$ and therefore the threshold t_c is zero [24].

The SIS model differs from the SIR model in that nodes can be susceptible to further infections even after recovery. The SIS model is commonly used to study the spread of computer virus. Each computer either is healthy or infected. Infected computers can be cleaned after using anti-virus software, but since the computer viruses mutate, it is possible for an infected computer to be susceptible again [25]. Let each susceptible node be infected with rate v and become susceptible again with rate w. Then effective spreading rate is defined as $\lambda = \frac{v}{w}$. The minimum value of effective spreading rate at which a disease spreads to the entire network is known as the epidemic threshold (λ_T). The epidemic threshold is generally a function of the average degree and structure of the network [14]

$$\lambda_T = \frac{<d>}{<d^2>}. \tag{5}$$

Using mean field approximations, a non-zero epidemic threshold is derived for the random graph model of a complex network [17, 26]. In contrast to random graphs, it is shown in [27] that the epidemic threshold is zero for computer virus propagation in scale-free networks (since $<d^2>$ diverges and correspondingly $\lambda_T = 0$). Although the authors consider the propagation of viruses in a network of computers, this analysis can be also used to model information spread in social networks. Wang *et al.* [28] proved that the epidemic threshold is closely related to the largest Eigen value of the adjacency matrix of the network.

2.1.2 Social Influence

Social influence between nodes will obviously affect the spread of information within a network. It is possible to categorize the ties between nodes as strong and weak. Nodes who interact frequently with each other and whose actions influence each other more strongly share strong ties, whereas nodes that do not interact often with each other, share weak ties. Although weak ties provide access to information [29, 7]; they can potentially preclude the information from being adopted inside a dense cluster of strong ties [30, 7]. Note that having access to information and adopting that piece of information are two different things. For example, an individual can hear about a new iPhone, but may decide not to buy it. Effect of weak ties on the spread of information can be described by introducing two basic definitions: "*density of cluster*" and "*social influence threshold*".

Definition 0.1. *The density of a cluster* is k if at least a k fraction of each node's neighbors are placed inside the cluster [7]. For example, for an isolated cluster, all of each node's neighbors are placed inside the cluster, and therefore the density of cluster is 1.

Definition 0.2. A node adopts the information if the fraction of neighbors who have already adopted the information is greater than *the social influence threshold* r.

In Figure 2, we show how information spreading through a set of nodes with weak ties may stop spreading through a network with strong ties. In this figure, a piece of information is starting at node i spreads to a dense cluster with the density k, via the node v. The only neighbors of v that had adopted the piece of information are outside of the cluster and their number is less than a fraction (r) of v's neighbors. Therefore, the required social influence threshold is not met and the information is unable to spread further. It can be concluded that if the density of a cluster k is greater than $1-r$, the cluster precludes the information spread [7].

Most studies consider static values of strength of ties. However, in reality, the temporal behavior of nodes can affect the strength of social ties. This dynamically varying strength of ties is investigated in [31]. A SIR model is used in which a susceptible node becomes infected with probability λ. This problem is mapped to a percolation model where the transmissibility, the probability that information transmits between any pair of nodes, describes the dynamic strength of ties. The transmissibility is modeled as a function of λ and the number of communication events within a given time interval. It was shown that this dynamic variation in the strengths favors the spread of information in a network for small λ and hinders the spread of information for large λ.

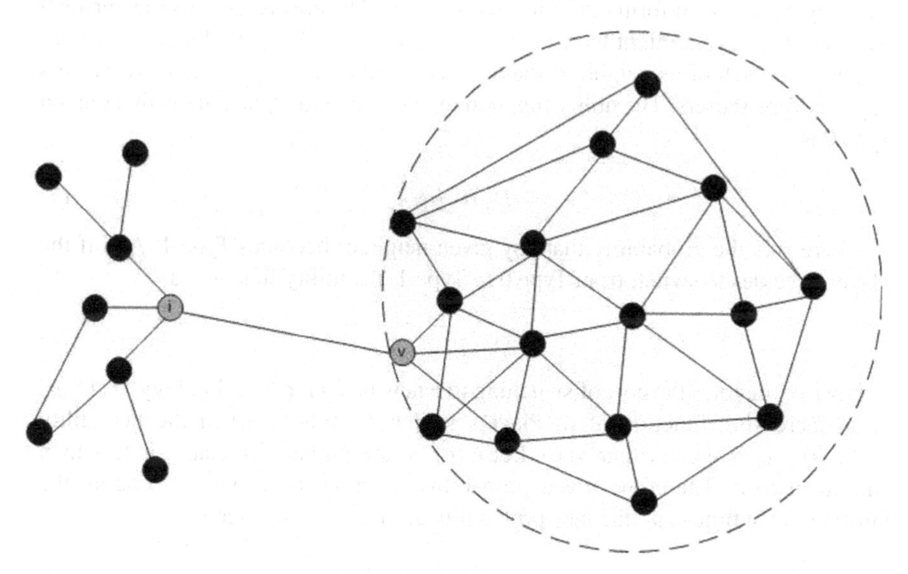

Fig. 2 A schematic representation of a sub-network shows how the source of diffusion (i) is connected to a dense cluster via a weak tie (the edge between i and v). This weak tie precludes information from diffusing into the cluster.

2.1.3 Social Learning

In real social networks, people tend to observe the outcomes of prior information on early adopters for a period and then decide to adopt the technology or not based on these observations. The previous models do not allow for this sort of rational behavior among the nodes. Unlike the models discussed earlier for information spread in social media, the nodes, in this model, are considered to be rational agents that want to maximize some utility for themselves. The De-Groot Model [13] accounts for mutual trust between any pair of neighbors. Let $p(0)$ be initial belief vector (an $n \times 1$ vector where n is the number of nodes in the network) that represents the probability that the nodes will adopt a given piece of information. Let T be the $n \times n$ matrix such that $T_{i,j}$ denotes the mutual trust between nodes i and j. Trust values are constant over time, and the nodes monitor their neighbors and update the belief values as follows:

$$p(n) = T\,p(n-1) = T^n p(0). \tag{6}$$

In this model, the external source of information can be modeled as a node i that does not change the initial belief $p_i(0)$. The Google page rank method is analogous to the De-Groot model with $T_{i,j} = \frac{1}{d_i}$, where d_i is the number of direct links that page i has to other pages [16].

Social learning can be studied from a game theoretic standpoint. A coordination game is considered in [32, 16] in which at every time slot, each node chooses to either adopt (Type 1) or not adopt (Type 0) a piece of information. The question is which of these two types takes over the network? There is a cost associated with adopting a piece of information for Type 1 nodes. The cost represents the personal bias that the players might have on any technology, product, etc. Player i receives a payoff for each of its neighbors that selects the same type as player i. All players begin as Type 0 users. The utility function for a player with d_i neighbors that remain Type 0 is

$$U_{d_i}(0,p) \tag{7}$$

where p is the probability that any given neighbor becomes Type 1. And if the player i decides to switch from Type 0 to Type 1, the utility function is

$$U_{d_i}(1,p) - c_i \tag{8}$$

where c_i denotes the cost of switching to a new behavior or technology. $U_{d_i}(1,p)$ is an increasing function of p. Players switch to Type 1 when the net utility $U_{d_i}(1,p) - c_i$ is greater than zero. Let $F(c)$ be the probability that c_i is less than and equal to c. Therefore, given probability of choosing behavior to adopt the information at time t, p_t, the new probability at time $t+1$ is given by

$$p_{t+1} = \phi(p_t) = \sum_d P(d) F(U_d(1,p_t)). \tag{9}$$

Equilibrium can be described mathematically as a fixed point, i.e. a point p such that $\phi(p) = p$. If p is close to one at equilibrium, Type 1 behavior takes over the network and vice versa. The equilibrium has been studied for different degree distributions including power law, Poisson and uniform. It is demonstrated that

$$\phi^{power}(p) \geq \phi^{Poisson}(p) \geq \phi^{regular}(p) \quad (10)$$

for all p when the utility function is either increasing with d or convex. [16]. Therefore, change from a Poisson degree distribution to a power-law degree distribution will lead to more diffusion of Type 1 behavior and a change from a Poisson degree distribution to a uniform degree distribution will lead to overall dominance of Type 0 behavior.

2.2 *Influencers in Social Networks*

Influencers are defined as nodes in a social network that have a big impact on the spread of information. Identifying key influencers in a social network can help in trying to effect a fast spread of information. For example, the Ford Company used the top 100 influencers in Vlogger to market their new model. Each vlogger was given a newly released Fiesta and were invited to talk about their experiences on their social media networks. Surprisingly, by March 2010, Ford received over 4 million tweets and 6.2 million YouTube views. Fiesta received 100,000 hand-raisers[2] and 3,000 reservations by people who had not bought a car from the Ford Company before [9, 33, 34].

2.2.1 Measures of Centrality

Network centrality is a powerful tool in the analysis of large scale networks. It can determine the relative importance of a node in the network. Given the network graph, we can assign centrality scores to the nodes based upon their structural characteristic in the network graph [35, 36, 37]. In this sub-section, we provide definitions of widely used measures of the network centrality in the analysis of social networks: degree centrality, betweenness, closeness, and Page-Rank centrality.

Definition 0.3. *Degree centrality* for an undirected network refers to the number of ties a node has to other nodes. For a directed network, two different measures, in-degree centrality and out-degree centrality, are usually used. In-degree centrality refers to the number of ties ending at a node and out-degree centrality refers to the number of ties starting from a node.

Definition 0.4. *Betweeness* refers to the number of times that a node lies on the shortest paths between any pair of nodes in the network graph.

Definition 0.5. *Closeness* refers to the inverse of the sum of the shortest distances between each node and every other node in the network graph.

[2] Individuals that sign up to receive the latest deals and brochures from the company.

Definition 0.6. *Page-Rank* was introduced by Google to quantify the importance of a particular web-page [38]. Given web-pages as nodes and hyperlinks as edges, the Page-Rank for a particular web-page can be quantified as follows

$$PR(p_i) = \sum_{p_j \in \psi_{p_i}} \frac{PR(p_j)}{L(p_j)} \tag{11}$$

where $PR(p_i)$ is the Page-Rank value for page p_i, ψ_{p_i} is the set of web-pages link to p_i, and $L(p_j)$ is the number of outbound links on p_j. The above recursive equation captures the fact that importance of a web-page (the Page-Rank value) is dependent on importance of all web-pages that link to it. Although this centrality measure is originally designed for web-pages, it is widely used to quantify influence of a node in social networks [39, 40].

2.2.2 Identifying Influencers in Social Networks

Studying influencers has proved challenging in most social networks, because the networks themselves are not completely observable. Twitter, however, proves to be an excellent platform for this type of studies because of two reasons: (a) the original intent of Twitter was to spread information [41] and (b) any repost on Twitter in the form of retweets can be traced back to prior retweeters, thereby making it possible to observe the influencers.

Several researchers have used heuristics like different notions of network centrality to identify influencers. A topic based method to quantify influence in Twitter was proposed in [39]. In another example [40], nodes were ranked and compared based on three different measures: number of followers, Page-Rank of the follower network, and the number of retweets. It was observed that the number of followers and the Page-Rank showed similar ranking, but there was a gap between these measures and the number of retweets.

The problem of identifying influencers can also be considered as a resource allocation issue where the aim is optimally choose the initial active nodes so as to maximize the range of information diffusion. For instance, the problem of viral marketing wherein recommendation by the top k influential nodes (k is given) resulting in the largest cascade is studied in [42]. They show that their method signicantly out-performs heuristic methods such as degree centrality and closeness centrality.

It is believed that the most influential node has the highest value of network centrality such as betweenness centrality, degree centrality, etc. However, in many cases, the most influential node is not the highly connected node. For example, a highly connected node at the periphery of a network, will be less influential than a less connected node that is located in the core of the network [43]. Kitsak *et al.* [43] used the k-core (or k-shell) decomposition of the network [44, 45] to differentiate nodes based on their position in the network. They assign an integer index or coreness, k, to each node, representing its position according to successive layers (k shells) in the network. This is shown in Figure 3. In this figure, the pink node has a higher coreness (k=3) than the blue node (k=1) under the same degree and therefore is more influential.

In many cases, individuals who have already adopted an innovation are more likely to become friends because of their the common interests (homophily) [46]. Therefore, distinguishing influence driven diffusion from homophily driven diffusion in social networks is of great importance in the analysis of influencers. In fact, it is an instance of the general problem of distinguishing correlation from causality [47, 48]. When source of adoption is homophily, the timing of adoption for each friend should be independent of the timing of other friends. Employing this fact, authors in [47] proposed a test based on time series of user actions to identify the source of correlation.

2.3 *Influencers and Controlling Mis-information Spread in Social Networks*

Spread of incorrect news on the social media network can have disruptive effects in real life. For instance, after the recent fake tweet about an explosion in the white house, the Dow Jones industrial average dropped 152 points within seconds [11]. Similarly, following the earthquake in Japan in 2011, a fake twitter account spread mis-information about nuclear reactors, stating that a nuclear meltdown was in progress killing 80% of Japanese and 50% of the population in the West Coast of the United States and Canada [49]. More interestingly, new studies [50] demonstrates that Twitter is one million times more effective than a person in spreading incorrect medical information. Fortunately, it is possible to limit the disruptive effect of mis-information by a judicious use of influencers in a counter campaign.

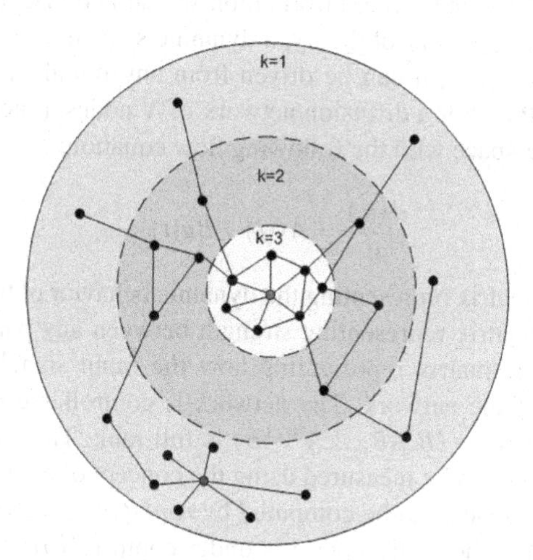

Fig. 3 A network under k-shell decomposition [43]. The coreness value (k) indicates which node is more influential under the same degree. For example, the pink node (k=3) is more influential than the blue node (k=1).

Early solutions to control the spread of misinformation relied on ideas borrowed from biology. Immunization was one popular mechanism [51, 52, 53] where a limited number of influencer nodes are removed from the network to stop the spread of the misinformation. Degree centrality can be used [16] to select nodes for the immunization. It was seen that selecting even only a handful of nodes with higher degree of centrality for immunization helped control the spread of misinformation quickly, in comparison to choosing a random set of nodes for immunization. Although immunization methods work very well in static networks, the network topology is not necessarily constant and therefore immunization is not likely to be as efficient in practice at controlling the spread of misinformation.

In practice, the problem of mis-information spread can be modeled as a cascade of two pieces of information, one of truthful information and one of misinformation, evolving simultaneously in a social network. The spread of truthful information can be controlled by targeting a set of influencers to help the spread. In this group of methods, it is assumed that if the mis-information and truthful information reach a node at the same step, the nodes will choose to believe the truth. Moreover, once a node becomes active in one campaign, it never changes campaign. The set of influencers can be identified using centrality measures such as degree and Page-Rank centrality. While all the techniques referred to before used heuristics to determine influencers, it is also possible to formulate this as an optimization problem. For example, the problem of limiting the spread of mis-information in social networks using optimization tools is addressed in [54]. The main idea is that one strategy to deal with a misinformation campaign is to limit the number of users who are willing to accept and spread it. In this regard, influencers are selected so as to minimize the number of nodes that end up believing the mis-information. Although the optimization problem is NP-hard, a greedy algorithm yields a sub-optimal solution.

Another approach to controlling information spread is to use ideas from control theory [55]. According to control theory, a dynamic system is controllable if, with a suitable choice of inputs, it can be driven from any initial state to any desired final state in finite time. For a diffusion network of N nodes, time evolution can be considered as a state-space with the following flow equation:

$$\frac{dx(t)}{dt} = Ax(t) + Bu(t) \tag{12}$$

where $x(t)$ is the matrix representing the dynamic behavior of the whole network over time. A is the matrix representing strength between any pair of nodes in the network and B is the matrix representing how the input signal reflects into the dynamic behavior of the network. The network is controllable if and only if the controllability matrix $C = (B, AB, ..., A^{N-1}B)$ is full rank. The ability of any node to control the network can be measured using the concept of control centrality. The control centrality of node i can be computed by $rank(C^{(i)})$, where $C^{(i)}$ denotes the controllability matrix when only node i is under control. If $rank(C^{(i)}) = N$, node i alone can control the whole network. This concept can be applied to controlling information spread in social networks, where the aim is to find the minimum sets of influencers in the diffusion network, whose time-dependent control can guide the whole network to any desired final state [55]. In our case, the influencers can be fed truthful information to limit the spread of misinformation. It was shown [55] that the number of influencers needed to send the network to the desired state is determined mainly by degree distribution of the network. In general, it is easier to control information spread in a dense homogeneous network using only a few influencers, whereas information spread in a sparse non-homogenous network is much more difficult to control. Note that nodes with high degree of control centrality may just as easily be controlled by mis-information spreaders [56]. Targeted attacks based on degree of centrality can cause greater harm in terms of spread of misinformation than random attacks [57].

2.4 Multiple Sources of Diffusion and the Problem of Information Confusion

In practice, a network may have multiple sources of information diffusion. People also often opt to receive this piece of information from multiple sources especially in the case of high risk activities or unproven technology before making a decision on whether to adopt that piece of information [29]. This problem is modeled as a complex contagion problem.

Traditional studies consider independent behavior of multiple contagious sources. However, it is conceivable that individual sources of information may not only be active simultaneously but also interact with each other. Karrer and Newman [58] investigate a network with two sources of diseases which spread concurrently and compete with each other, once a node is infected by the disease, it becomes immune to the other disease. The SIR model discussed previously in Section 2.1 is used to represent the diffusion of the diseases. Using percolation theory, they show that the relative rate of diffusion of sources determines the diffusion range for each source. Although the authors consider the propagation of diseases in a real network of people, it is possible to see the potential application of these theories to information diffusion in social networks. It is also of interest to study the effect of collaboration or competition between several sources of information. A recent study on Twitter [59] studies this interaction where they model Twitter users are nodes that are exposed to different contagions (tweets from other users they follow). It is found that competing contagions (tweets with un-related contents) decrease each other's probability of spreading, whereas co-operating contagions (tweets with related contents) help each other. Diffusion of the same content by multiple sources is studied in [60] and [43]. Adoption of politically controversial hashtags on Twitter is investigated in [60]. These studies show that repeated exposures (via multiple information providers) to the same content have significant effect on its adoption. Kitsak *et al.* in [43] discussed how spread of the same content can be made much more efficient by a predetermined distance of sources from each other on the social network graph.

Until now, we have assumed that nodes have an unlimited capacity to receive and retransmit social signals. However, generally in social networks, cognitive and perceptual factors can limit a person's capacity to process incoming messages, referred to as *limited attention* [61, 62]. For example, nodes in Twitter can retweet a few tweets for a given period of time. This concept is examined for the case of Twitter in [63] where it was demonstrated that predictions on heterogeneity in the popularity and persistence of hashtags are consistent with empirical data from Twitter. In a social network of multiple sources of diffusion, limited attention leads to indirect competition among sources.

Information consumers receive different content on the same topic from various primary information providers (e.g. reputable news agencies) and the other information providers (e.g. friends in Facebook). This can result in what is known as *information confusion.* The amount of confusion can be modeled as a function of trust between any pair of information consumer and information provider, functionality of the information provider and characteristics of the information consumer [64].

Each information provider transmits information with different intensities (e.g. different advertisement volumes). This network is modeled as a communication system [64] in which each information provider and information consumer acts like a transmitter and a receiver respectively. Intensity of information may either be amplified or reduced, dependent on trust between the information provider and consumer which is modeled as the channel gain. On the other hand, the amount of information which contradicts the information provided by the primary information provider is modeled as interference. Let p_i denote that intensity of information transmission by the primary information provider and h_{ii} represents trust between the ith consumer and the primary information provider. Therefore, the effective information received by the ith consumer is $p_i h_{ii}$. Let N_i be the natural confusion experienced by the ith consumer. This models the consumer's ability to process information. For example, even if the ith consumer only listens to the primary information source, he still suffers from an amount of confusion.

The total confusion level suffered by the ith consumer is $\sum\limits_{j \neq i} p_j h_{ji} + N_i$. Similar to the signal to interference and noise ratio (SINR) in wireless communication, information to confusion and noise ratio is considered as an objective function. For example, the SINR for the ith consumer while the ith provider is assumed as the primary information provider is given by

$$\frac{p_i h_{ii}}{\sum\limits_{j \neq i} p_j h_{ji} + N_i}. \tag{13}$$

Maximizing the proposed objective function gives the optimal intensity of the information transmission (p_i) or equally optimal strategy for the information providers. This problem is modeled as a power control problem and can be solved using a non-cooperative game theory model. The set of players are the set of information providers and the strategy set is the values of the intensities, p_i. It is demonstrated that the confusion level is the highest in a social network where all information sources are trusted equally.

3 Locating the Source of Diffusion

Imagine a rumor has spread through the Twitter network, and that the only available data about the diffusion is the list of nodes that are affected and a list of nodes that have been communicating with each other. In this setting, it is a non-trivial task to automatically and reliably identify the source/sources of the rumor in a large scale social network. Whereas the modeling of the information spread in social networks has attracted a significant research, the reverse problem of identifying the source of diffusion has been addressed only recently.

Although this problem is admittedly complex, it is reasonable to assume that the source of information is likely to be a node with a high degree of centrality, where centrality is appropriately defined. For example, in [65], closeness, betweenness, and eigenvector centralities are used to locate the origin of the information. They assume that information is propagated along a breadth-first-search (BFS) tree, which is not necessarily under all circumstances.

Another line of approaches uses information from a snapshot of the infected nodes, rather than the entire network. This approach can be used to identify a single source of diffusion [66, 67, 68, 69, 70, 71] or multiple sources of diffusion [72, 73].

- **Single source of diffusion and full observation**

The Susceptible-Infected (SI) diffusion model, a special case of the general SIR model, is used in [66, 67, 68] to study the problem of locating the diffusion source. The question of interest is to estimate the source node, based on the noisy states observed at the nodes. A maximum likelihood estimator (MLE) is employed to identify the source using a snapshot of n infected nodes:

$$\overset{*}{s} = \arg\max_{s \in G_n} P(G_n|s) \tag{14}$$

where G_n is the observed graph of infected nodes. In general, the MLE suffers from computational complexity. However, it is demonstrated that the MLE is equivalent to a combinatorial problem in a regular tree and can be approximated by *rumor centrality* $R(s, G_n)$. Rumor centrality is the number of paths available for the spread of information in a tree rooted at node $s \in G_n$. Therefore, the estimator becomes

$$\overset{*}{s} = \arg\max_{s \in G_n} P_G(G_n|s) = \arg\max_{s \in G_n} R(s, G_n)\,. \tag{15}$$

It can be shown that the node with the maximum closeness centrality is the solution of the MLE on regular-trees. Simulations performed on small scale-free networks such as the U.S. electric power grid, demonstrates that this estimator finds the target nodes to within a few hops [66]. The computational complexity is $O(N^2)$ for generic graphs with N nodes and $O(N)$ for trees.

The MLE is a good tool to use when nothing is known about the propensity of a node to spread rumors. In real life, though, it is possible that some nodes are more likely to spread rumors than others and it is also possible to have some a priori knowledge about these nodes. The problem of identifying the source of rumors, when there is some a priori knowledge of suspect nodes is studied [69] under SI model and a snapshot of the infected nodes. In cases where such a priori knowledge is available, it is possible to design a maximum a posteriori probability (MAP) estimator to identify the source:

$$\overset{*}{s} = \arg\max_{s \in \{S \cap G_n\}} P_G(G_n|s) \tag{16}$$

where $P_G(G_n|s)$ is the probability of observing G_n (the graph of the infected nodes), given s as a rumor source. S is the set of suspect nodes. Similar to the previous example, the MAP estimator is equivalent to a combinatorial problem in a regular tree. So, rumor centrality in [66, 67, 68] can be leveraged to overcome the computational complexity. Therefore, the optimal estimator will be

$$\overset{*}{s} = \arg\max_{s \in \{S \cap G_n\}} P_G(G_n|s) = \arg\max_{s \in \{S \cap G_n\}} R(s, G_n) \tag{17}$$

where $R(s,G_n)$ is the rumor centrality for the node s in the observed snapshot G_n. This problem can be solved easily for a tree. For a general graph, the number of permitted permutations which begin with the source (spanning graphs) is too large. To reduce complexity further, a BFS tree is assumed. The rationale for this assumption is that the rumor spreads along the minimum length path. So, the estimator for a general graph is given by

$$\overset{*}{s} = \arg\max_{s \in \{S \cap G_n\}} R(s, \tau_{bfs}(s)) \tag{18}$$

where $\tau_{bfs}(s)$ is the BFS tree associated with the diffusion tree rooted at the node s.

The Susceptible-Infected-Recovered (SIR) model can also be used [70] to study the problem of locating the diffusion source. In contrast to the SI model, any nodes in the SIR model is a potential candidate since the susceptible nodes cannot be distinguished from the recovered nodes. A sample path based method is proposed where the identified source belongs to the sample paths that most likely lead to the observed snapshot of the diffusion graph. Let $X_v(t)$ be the state of node v in time slot t which can take three possible values: susceptible (S), infected (I), or recovered (R). The infection process at time slot t is modeled as a discrete time Markov chain $X(t) = \{X_v(t), v \in \upsilon\}$ where υ is the set of all nodes in the network. However, $X(t)$ is not fully observable since we cannot distinguish susceptible nodes and recovered ones. So, an observable vector $Y = \{Y_v, v \in \upsilon\}$ is defined, where Y_v is 1 if the node v is observed in the state I, otherwise it is 0. Let $X[0,t] = \{X(\tau) : 0 < \tau < t\}$ be a sample path of the infection process from 0 to t. The source node associated with the sample path that most likely lead to Y is identified as the source of diffusion:

$$X^*[0,t^*] = \arg\max_{t, X[0,t] \in \chi(t)} P(X[0,t]) \tag{19}$$

where $\chi(t)$ is the set of all paths leading to Y. Defining the infection eccentricity of a node to be the largest distance from the node to every other infected node, it can be shown that the source node of the optimal sample path is the node with the minimum infection eccentricity. The sample path based estimator is closer to the actual source than the nodes with the maximum infection closeness [66, 67, 68]. It should be noted that since all nodes, whether infected or not, must be observed the number of possible path is very large and therefore makes this setting impractical in large scale social networks.

- **Multiple sources of diffusion and full observation**

The problem of identifying multiple diffusion sources and their infected regions is investigated in [72]. The SI model is used to represent information diffusion. Let G_n be the graph of the infected nodes and A_n be the union of their infected regions. Since there is no prior knowledge of G_n and A_n, we define a maximum likelihood (ML) estimator:

$$\max_S P(S, A_n | G_n) \tag{20}$$

where S is the initial set of infected nodes. Maximizing Eqn.20 for a general G_n is computationally hard. A sub-optimal estimator is designed through splitting Eqn.20 into two maximum likelihood problems

$$\overset{*}{S} = \arg\max_S P(G_n | S) \tag{21}$$

$$\overset{*}{A}_n(\overset{*}{S}) = \arg\max_A P(A | \overset{*}{S}, G_n). \tag{22}$$

It is shown that if the diffusion graph is a tree with at most two sources, the probability of accurately estimating the source goes to one.

The Minimum Description Length (MDL) principle is a type of hypothesis testing which is usually used in coding theory and learning theory. The main idea is that the best model to describe a limited set of observed data, is the one that permits the greatest compression of the data [74]. Let M be a set of candidate models. The best model $m \in M$ to describe the data D is the one which minimizes $L(m) + L(D|m)$, where $L(m)$ is the length, in bits, of the description of M, and $L(D|M)$ is the length, in bits, of the description of the data when encoded with m. For the problem of locating the source of diffusion, the observed graph of infected nodes, G_n, is the data D we want to describe succinctly using our models M which are all infected nodes.

The MDL is used in [73] to identify the best set of sources of information diffusion for the case when the number of sources is not known.

- **Single source of diffusion and partial observation**

Up to this point all previous methods were based on the knowledge of the state of the entire network. While this is a good approach for small networks, it becomes quickly impractical as the size of the network increases because gathering information about the state of a larger network starts to become cumbersome. One method to deal with this problem is to place a small set of sensor nodes within the network. Using this approach an SI model was constructed to study information spread [71]. The source of diffusion (s^*) is modeled as a uniformly distributed random variable over the set of all nodes. The unknown source of information, s^*, starts to disseminate the information at some unknown time t^*. The sensors (o) measure from which neighbors (v) and at what time ($t_{v,o}$) the information arrives at sensors. Sensors use these measurements to calculate the most likely source of the information

$$\overset{*}{s} = \arg\max_{s} p((o, v, t_{v,o})|s). \tag{23}$$

The computational complexity of this estimator increases exponentially with the number of nodes. Hence, a reduced complexity estimator based on inter arrival times, d, (which is a sufficient statistic for estimating diffusion) is also proposed. Propagation delays are Gaussian distributed. Therefore, d is a Gaussian vector and the new estimator becomes

$$\overset{*}{s} = \arg\max_{s} f(d|s) \tag{24}$$

where $f(d|s)$ is the pdf of d, given s is the source. It is proved that the proposed estimator is optimal for a tree. For a general graph, the number of spanning trees can be exponentially large. Therefore, it is assumed that the diffusion graph is a BFS tree which results in a sub-optimal solution for the estimator. This method was tested on a real network for the spread of diseases. In methods that use sensors for additional monitoring of the network, the key question is the positioning of these sensors. Placing sensors at nodes with high degree resulted in smaller error in localization than randomly placed sensors [71].

4 Summary

In this chapter, some of the popular models for information diffusion were presented: the contagion model, the social influence model, and the social learning model. The first two models assumed more of a passive participation from the nodes in the network, in contrast to the third model where the nodes in the social network are assumed to be rational actors, trying to maximize a payoff for themselves. The contagion model and the social influence model are used to study the static behavior of information spread, and the dynamic behavior of information spread was considered in the social learning model. We then discussed the effect of structural properties of the network on the extent of information spreads. Strength of ties between nodes was assumed to be constant in these models, but in reality, this value can fluctuate over time due to potentially waxing and waning trust between nodes with time. We then discussed the role of influencers in facilitation the rapid dissemination of information in the network. Two methods were discussed to identify the set of influencers, employing measures of network centrality. We then look at multiple sources of information and consider both the cases where the sources collaborate or compete with each other. Finally we look at the problem of locating the source of diffusion in a social networks. Both, techniques that assume full knowledge of the network and those that do not are presented. Several challenges still remain in this area of study including optimal placement of sensors in the networks to aid in zeroing in on the sources of diffusion, finding the source of diffusion when the diffusion graph is not completely known, etc.

References

1. Kakihara, M.: Understanding real-virtual perspective of human behavior from the active stream perspective. In: Proc., Pacific Asia Conf. on Info. Sys. (PACIS 2010) (2010)
2. McRoberts, B.: Internet plays integral role in decision making: Study. China Daily (2010), http://www.chinadaily.com.cn/business/2010-09/03/content_11253060.htm
3. Impact of food advertising on childhood obesity. American Psychological Association, http://www.apa.org/topics/kids-media/food.aspx
4. Senecal, S., Kalczynski, P.J., Nantel, J.: Consumers' decision making process and their online shopping behaviors: A Clickstream analysis. Elsevier Jl. of Business Research 58(11), 1599–1608 (2005)
5. Bullas, J.: 12 key findings on social media's impact on business and decision makings by CEOs and managers. Social Media Networks, Social Media and the CEO, http://www.jeffbullas.com/2010/04/05/12-key-findings-on-social-media%E2%80%99s-impact-on-business-and-decision-making-by-ceos-and-managers/
6. http://www.google.com/trends
7. Easley, D., Kleinberg, J.: Networks, Crowds, and Markets: Reasoning About a Highly Connected World. Cambridge University Press, New York (2010)
8. Trusov, M., Bucklin, R.E., Pauwels, K.: Effects of word-of-mouth versus traditional marketing: Findings from an internet social networking site. Journal of Marketing 73(5), 90–102 (2009)
9. Barry, K.: Ford bets the fiesta on social networking (2009), http://www.wired.com/autopia/2009/04/how-the-fiesta/
10. Northeast rumours: 254 websites blocked, http://m.ibnlive.com/news/northeast-rumours-254-websites-blocked/283906-3.html
11. Strauss, G., Shell, A., Yu, R., Acohido, B.: Sec, fbi probe fake tweet that rocked stocks (2013), http://www.usatoday.com/story/news/nation/2013/04/23/hack-attack-on-associated-press-shows-vulnerable-media/2106985/
12. Young, H.P.: Innovation diffusion in heterogeneous populations: Contagion, social influence, and social learning. American Economic Review 99(5), 1899–1924 (2009)
13. DeGroot, M.H.: Reaching a consensus. Journal of the American Statistical Association 69(345), 118–121 (1974)

14. Bailey, N.: The Mathematical Theory of Infectious Diseases and its applications. Hafner Press (1975)
15. Kermack, W.O., McKendrick, A.G.: A contribution to the mathematical theory of epidemics. Proceedings of the Royal Society of London. Series A 115(772), 700–721 (1927)
16. Jackson, M.O.: Social and Economic Networks. Princeton University Press (2008)
17. Newman, M.E.J.: The structure and function of complex networks. SIAM Rev. 45, 167–256 (2003)
18. Bass, F.M.: A new product growth for model consumer durables. Manage. Sci. 15(5), 215–227 (1969)
19. Bass, F.M.: The relationship between diffusion rates, experience curves, and demand elasticities for consumer durable technological innovations. Journal of Business 53(3), 551–567 (1980)
20. Ryan, B., Gross, N.C.: The diffusion of hybrid seed corn in two iowa communities. Rural Sociology 8, 15–24 (1943)
21. Griliches, Z.: Hybrid corn: An exploration in the economics of technological change. Econometrica 25, 501–522 (1957)
22. Nekovee, M., Moreno, Y., Bianconi, G., Marsili, M.: Theory of rumour spreading in complex social networks. Physica A: Statistical Mechanics and its Applications 374(1), 457–470 (2007)
23. Stauffer, D., Aharony, A.: Introduction to Percolation Theory. Taylor and Francis (1992)
24. Cohen, R., Erez, K., ben Avraham, D., Havlin, S.: Resilience of the internet to random breakdowns. Phys. Rev. Lett. 85, 4626–4628 (2000)
25. Pastor-Satorras, R., Vespignani, A.: Epidemic dynamics and endemic states in complex networks. Phys. Rev. E 63, 066117 (2001)
26. Boguñá, M., Pastor-Satorras, R., Vespignani, A.: Absence of epidemic threshold in scale-free networks with degree correlations. Phys. Rev. Lett. 90, 028701 (2003)
27. Pastor-Satorras, R., Vespignani, A.: Epidemic spreading in scale-free networks. Phys. Rev. Lett. 86, 3200–3203 (2001)
28. Wang, Y., Chakrabarti, D., Wang, C., Faloutsos, C.: Epidemic spreading in real networks: An eigenvalue viewpoint. In: Proceedings of the 22nd International Symposium on Reliable Distributed Systems, pp. 25–34 (2003)
29. Centola, D., Macy, M.: Complex contagions and the weakness of long ties. American Journal of Sociology 113, 702–734 (2007)
30. Granovetter, M.: The strength of weak ties. The American Journal of Sociology 78(6), 1360–1380 (1973)
31. Miritello, G., Moro, E., Lara, R.: Dynamical strength of social ties in information spreading. Phys. Rev. E 83, 045102 (2011)
32. Jackson, M., Yariv, L.: The diffusion of behavior and equilibrium properties in network games. American Economic Review 97, 92–98 (2007)
33. Greenberg, K.: Ford fiesta movement shifts into high gear (2010), `http://www.mediapost.com/publications/article/123503/`
34. Yoganarasimhan, H.: Impact of social network structure on content propagation: A study using youtube data. Quantitative Marketing and Economics 10(1), 111–150 (2012)
35. Bonacich, P.: Power and centrality: A family of measures. American Journal of Sociology 92(5), 1170–1182 (1987)
36. Sabidussi, G.: The centrality index of a graph. Psychometrika 31(4), 581–603 (1966)
37. Friedkin, N.E.: Theoretical foundations for centrality measures. American Journal of Sociology 96(6), 1478–1504 (1991)

38. Page, L., Brin, S., Motwani, R., Winograd, T.: The pagerank citation ranking: Bringing order to the web. Technical Report 1999-66 (1999)
39. Weng, J., Lim, E.P., Jiang, J., He, Q.: Twitterrank: finding topic-sensitive influential twitterers. In: Proceedings of the Third ACM International Conference on Web Search and Data Mining, WSDM 2010, pp. 261–270 (2010)
40. Kwak, H., Lee, C., Park, H., Moon, S.: What is twitter, a social network or a news media? In: Proceedings of the 19th International Conference on World Wide Web, WWW 2010, pp. 591–600 (2010)
41. Bakshy, E., Hofman, J.M., Mason, W.A., Watts, D.J.: Everyone's an influencer: quantifying influence on twitter. In: Proceedings of the Fourth ACM International Conference on Web Search and Data Mining, WSDM 2011, pp. 65–74 (2011)
42. Kempe, D., Kleinberg, J., Tardos, E.: Maximizing the spread of influence through a social network. In: Proceedings of the Ninth ACM SIGKDD International Conference on Knowledge Discovery and Data Mining, KDD 2003, pp. 137–146 (2003)
43. Kitsak, M., Gallos, L., Havlin, S., Liljeros, F., Muchnik, L., Stanley, H., Makse, H.: Identification of influential spreaders in complex networks. Nature Physics 6(11), 888–893 (2010)
44. Carmi, S., Havlin, S., Kirkpatrick, S., Shavitt, Y., Shir, E.: From the cover: A model of internet topology using k-shell decomposition. Proceedings of the National Academy of Sciences 104(27), 11,150–11,154 (2007)
45. Serrano, M.A., Boguñá, M.: Clustering in complex networks. ii. percolation properties. Phys. Rev. E 74, 056115 (2006)
46. McPherson, M., Smith-Lovin, L., Cook, J.M.: Birds of a feather: Homophily in social networks. Annual Review of Sociology 27(1), 415–444 (2001)
47. Anagnostopoulos, A., Kumar, R., Mahdian, M.: Influence and correlation in social networks. In: Proceedings of the 14th ACM SIGKDD International Conference on Knowledge Discovery and Data Mining, KDD 2008, pp. 7–15 (2008)
48. Aral, S., Muchnik, L., Sundararajan, A.: Distinguishing influence-based contagion from homophily-driven diffusion in dynamic networks. Proceedings of the National Academy of Sciences 106(51), 21,544–21,549 (2009)
49. Fraser, W.: Social media and the spread of misinformation, `http://www.worldoffemale.com/social-media-and-the-spread-of-misinformation/`
50. Miller, T.: Medical misinformation spreads quickly on twitter, sometimes reaching millions: study (2010), `http://www.nydailynews.com/life-style/health/medical-misinformation-spreads-quickly-twitter-reaching-millions-study-article-1.173304`
51. Tong, H., Prakash, B., Tsourakakis, C., Eliassi-Rad, T., Faloutsos, C., Chau, D.: On the vulnerability of large graphs. In: 2010 IEEE 10th International Conference on Data Mining (ICDM), pp. 1091–1096 (2010)
52. Briesemeister, L., Lincoln, P., Porras, P.: Epidemic profiles and defense of scale-free networks. In: Proceedings of the 2003 ACM Workshop on Rapid Malcode, WORM 2003, pp. 67–75 (2003)
53. Prakash, B.A., Tong, H., Valler, N., Faloutsos, M., Faloutsos, C.: Virus propagation on time-varying networks: theory and immunization algorithms. In: Balcázar, J.L., Bonchi, F., Gionis, A., Sebag, M. (eds.) ECML PKDD 2010, Part III. LNCS, vol. 6323, pp. 99–114. Springer, Heidelberg (2010)
54. Budak, C., Agrawal, D., El Abbadi, A.: Limiting the spread of misinformation in social networks. In: Proceedings of the 20th International Conference on World Wide Web, WWW 2011, pp. 665–674 (2011)

55. Pósfai, M., Liu, Y.Y., Slotine, J.J., Barabasi, A.L.: Effect of correlations on network controllability. Scientific Reports 3(1067), 1–7 (2013)
56. Liu, Y.Y., Slotine, J.J., Barabasi, A.L.: Control centrality and hierarchical structure in complex networks. Plos One 7 (2012)
57. Pu, C.L., Pei, W.J., Michaelson, A.: Robustness analysis of network controllability. Physica A: Statistical Mechanics and its Applications 391(18), 4420–4425 (2012)
58. Karrer, B., Newman, M.E.J.: Competing epidemics on complex networks. Phys. Rev. E 84, 036106 (2011)
59. Myers, S.A., Leskovec, J.: Clash of the contagions: Cooperation and competition in information diffusion. In: ICDM 2012, pp. 539–548 (2012)
60. Romero, D.M., Meeder, B., Kleinberg, J.: Differences in the mechanics of information diffusion across topics: idioms, political hashtags, and complex contagion on twitter. In: Proceedings of the 20th International Conference on World Wide Web, WWW 2011, pp. 695–704 (2011)
61. Hodas, O., Lerman, K.: How limited visibility and divided attention constrain social contagion. In: ASE/IEEE International Conference on Social Computing (2012)
62. Lerman, K., Jain, P., Ghosh, R., Kang, J.H., Kumaraguru, P.: Limited attention and centrality in social networks. In: Proceedings of International Conference on Social Intelligence and Technology (SOCIETY) (2013)
63. Weng, L., Flammini, A., Vespignani, A., Menczer, F.: Competition among memes in a world with limited attention. Scientific Reports 2(335) (2012)
64. Anand, S., Subbalakshmi, K.P., Chandramouli, R.: A quantitative model and analysis of information confusion in social networks. IEEE Transactions on Multimedia 15(1), 207–223 (2013)
65. Comin, C.H., da Fontoura Costa, L.: Identifying the starting point of a spreading process in complex networks. Phys. Rev. E 84, 056105 (2011)
66. Shah, D., Zaman, T.: Detecting sources of computer viruses in networks: theory and experiment. In: Proceedings of the ACM SIGMETRICS International Conference on Measurement and Modeling of Computer Systems, SIGMETRICS 2010, pp. 203–214 (2010)
67. Shah, D., Zaman, T.: Rumors in a network: Who's the culprit? IEEE Transactions on Information Theory 57(8), 5163–5181 (2011)
68. Shah, D., Zaman, T.: Rumor centrality: a universal source detector. In: Proceedings of the 12th ACM SIGMETRICS/PERFORMANCE Joint International Conference on Measurement and Modeling of Computer Systems, SIGMETRICS 2012, pp. 199–210 (2012)
69. Dong, W., Zhang, W., Tan, C.W.: Rooting out the rumor culprit from suspects. arXiv:1301.6312 [cs.SI] (2013)
70. Zhu, K., Ying, L.: Information source detection in the sir model: A sample path based approach. In: Proc. of Inform. Theory and Applications Workshop, ITA 2013 (2013)
71. Pinto, P.C., Thiran, P., Vetterli, M.: Locating the source of diffusion in large-scale networks. Phys. Rev. Lett. 109, 068702 (2012)
72. Luo, W., Tay, W.P.: Identifying multiple infection sources in a network. In: 2012 Conference Record of the Forty Sixth Asilomar Conference on Signals, Systems and Computers (ASILOMAR), pp. 1483–1489 (2012)
73. Prakash, B., Vreeken, J., Faloutsos, C.: Spotting culprits in epidemics: How many and which ones? In: 2012 IEEE 12th International Conference on Data Mining (ICDM), pp. 11–20 (2012)
74. Grünwald, P.D.: The Minimum Description Length Principle (Adaptive Computation and Machine Learning). The MIT Press (2007)

Structure and Evolution of Online Social Networks

Saptarshi Ghosh and Niloy Ganguly

Abstract. Social networks are complex systems which evolve through interactions among a growing set of actors or users. A popular methodology of studying such systems is to use tools of complex network theory to analyze the evolution of the networks, and the topological properties that emerge through the process of evolution. With the exponential rise in popularity of Online Social Networks (OSNs) in recent years, there have been a number of studies which measure the topological properties of such networks. Several network evolution models have also been proposed to explain the emergence of these properties, such as those based on preferential attachment, heterogeneity of nodes, and triadic closure. We survey some of these studies in this chapter. We also describe in detail a preferential attachment based model to analyze the evolution of OSNs in the presence of restrictions on node-degree that are presently being imposed in all popular OSNs.

Keywords: Online social networks, topology, evolution, complex network theory, degree cutoffs, preferential attachment.

1 Introduction

Online Social Networks (OSNs), such as Facebook, Twitter and LinkedIn, are some of the most popular sites on the Web at present, each having hundreds of millions of users. These OSNs differ among themselves in the character of the

Saptarshi Ghosh
Department of Computer Science and Technology, Bengal Engineering and Science University Shibpur, Howrah – 711103, India
e-mail: sghosh@cs.becs.ac.in

Niloy Ganguly
Department of Computer Science and Engineering, Indian Institute of Technology Kharagpur, Kharagpur – 721302, India
e-mail: niloy@cse.iitkgp.ernet.in

M. Panda, S. Dehuri, and G.-N. Wang (eds.), *Social Networking*
Intelligent Systems Reference Library 65,
DOI: 10.1007/978-3-319-05164-2_2,

service they provide to users. For instance, Facebook (www.facebook.com) can be seen mainly as a platform for keeping in touch with close friends, Twitter (twitter.com) is mostly used to propagate and receive current news, while LinkedIn (www.linkedin.com) is primarily for the maintenance of professional contacts. In spite of these differences in functionality, all these platforms share some common ingredients, e.g., the users are connected to each other by an underlying social network that helps them to interact with others.

Social networks are typical examples of complex systems which evolve through the interactions among a growing set of actors / users. A popular methodology of studying such systems is to model them as dynamic networks where the nodes represent the actors / users and the links represent the interactions among the users.[1] Tools of complex network theory [48, 3] can then be used to analyze the evolution of the networks and the topological properties that emerge through the process of evolution.

Social networks in the off-line world – friendship networks among people having a common affiliation (e.g., among students in a school) – have long been studied to gain insights into their evolution and emerging topological properties [34, 51, 50]. However, studies on off-line social networks were typically limited to few hundreds of individuals, since data on social relationships had to be collected using painstaking methods such as asking the individuals to fill survey-forms. With the advent of OSNs, detailed data of the activities of millions of users are now available, which help test the theories of social network evolution over populations of a size that is unimaginable in the off-line world.

There have been a number of studies in recent years which measure the emerging topological properties of OSNs, and several network evolution models have also been proposed to explain these properties. Some of these studies are discussed in Section 2 and Section 3 respectively. Finally, section 4 describes a complex network theoretic model to analyze the evolution of OSNs in the presence of restrictions on node-degree that are presently being imposed in most popular OSNs.

2 Topological Properties of Social Networks

Social networks, both in the off-line and online worlds, have been found to show rich topological properties that are very different from those of random Erdos-Renyi networks [20, 21]. We first introduce some of the topological properties by which large networks are characterized, and then discuss the properties of OSNs.

[1] Apart from this representation, some social systems are modeled using other types of networks. For instance, users' interactions with online resources are represented as bipartite user-resource networks [7], while social bookmarking sites or folksonomies (e.g., Flickr) are represented as tripartite hypergraphs [16]. However, this chapter focuses on the most common representation of social networks, where nodes represent users and the links represent social interactions among users.

2.1 Topological Properties Used to Characterize Large Networks

Degree Distribution: The most fundamental characteristic of a network is its degree distribution $P(k)$ which measures the fraction of nodes having a certain degree k, for various values of k. The degree distribution of social networks are usually broad, with a heavy tail, i.e., there is a large variation in the number of friends of different users. A large majority of users have small number of friends (low degree) while a few have very large number of friends (high degree).

Clustering and Community Structure: The clustering coefficient measures the probability that two nodes sharing a common neighbor are connected between themselves. This property can be quantified by the local clustering coefficient c_i of a node i which is defined as:

$$c_i = \frac{\text{number of closed triples centered on node } i}{\text{number of triples centered on node } i}$$

where a triple centered on node i is a set of 3 nodes $\{i, j, k\}$ having the links (i, j) and (i, k), and a *closed* triple centered on node i is such a set of nodes having all the three links (i, j), (i, k) and (j, k). The global clustering coefficient of a network can be obtained by averaging the local clustering coefficient over all nodes in the network. An alternative definition of the global clustering coefficient C of a network is

$$C = \frac{\text{number of closed connected triples in the network}}{\text{number of connected triples in the network}}$$

where a connected triple is a set of 3 nodes which have at least 2 links between them, and a closed triple is such a set of nodes having 3 links between them.

Social networks, both in the online and off-line worlds, exhibit high clustering coefficients since friends of an individual tend to have friendship links among themselves as well [50]. Topologically, this leads to the existence of a large number of triangles in such networks. This is a well-known phenomenon which leads to the emergence of community structure in social networks [23, 30], where a community is a group of nodes which are more densely connected among themselves than with the rest of the network. A large number of community detection algorithms have been proposed to identify such groups of nodes from the network topology; see [27] for a survey.

Assortativity and Homophily: Another common feature of social networks is that users who are similar to each other tend to be connected. In sociology, this property is known as *homophily* or 'birds of a feather flock together' phenomenon. For instance, people belonging to the same country are frequently found to be linked in the Facebook OSN [59]. In fact, homophily is the underlying cause of the community structure in networks, whereby similar nodes form densely connected communities.

A specific form of homophily in networks is that nodes having similar *degrees* tend to be connected to each other. This property is quantified by the *degree assortativity* coefficient, which, in simple terms, can be described as the Pearson correlation coefficient of the degrees of nodes which are connected to each other [47]. In an assortative network (having assortativity $\in [0, 1]$), high-degree nodes tend to connect to other high-degree nodes while low-degree nodes are mostly connected to other low-degree nodes.

Shortest Path Distances: The distances between nodes is another interesting macroscopic property of a complex network. This is quantified by the average shortest path distance between any pair of nodes in the network. For an undirected network of n nodes, the average shortest path distance between nodes is

$$l = \frac{1}{\frac{n(n+1)}{2}} \sum_{i \geq j} d_{ij}$$

where d_{ij} is the shortest distance between nodes i and j.

It is well known that in social networks, the average shortest path distance is almost always low even when the network may consist of millions of nodes. In fact, the average shortest path distance in social networks has often been found to be within 6; this property is popularly known as the "six degrees of separation" observed in Milgram's famous experiments [58]. Recent studies show that in the online world as well, the average shortest path distance in OSNs is very small. For instance, the degree of separation between a large majority of nodes in the Facebook OSN has been found to be close to *four* [5].

Small-world Characteristics: A network is said to exhibit small-world characteristics [60] if it has (i) high local clustering coefficient, (ii) low density, and (iii) low mean shortest-path distance among nodes – specifically, the average distance between two randomly chosen nodes grows no faster than the logarithm of the number of nodes in the network. Effectively, in a small-world network, most nodes are not neighbors of one another, but most nodes can be reached from every other node by a small number of hops. Social networks are known to demonstrate small-world characteristics, primarily due to the presence of many densely connected groups of nodes (the communities) as well as many high-degree nodes or hubs which serve as the common connections mediating the shortest path distances between other nodes.

2.2 Topological Properties of OSNs

We now describe the topology of OSNs in terms of the properties described above. There have been several large-scale measurement studies on the topological characteristics of OSNs, considering snapshots of the networks containing millions of users and billions of social links. For instance, Mislove *et al.* [42, 43] studied the four OSNs Orkut, Youtube, Flickr and LiveJournal, Ahn *et al.* [2] studied the Cyworld, MySpace and Orkut OSNs, while Kwak *et al.* [37] and Cha *et al.* [17] studied the Twitter social network. Ferrara *et al.* [24] proposed models to capture the structure of popular OSNs such as Arxiv, Facebook, Wikipedia and YouTube.

Recently Ugander *et al.* [59] studied a snapshot of the Facebook OSN, consisting of more than 720 million active Facebook users and more than 68 billion social links.

These empirical studies have found that OSNs in general demonstrate a set of common topological characteristics. Degree distributions in OSNs have been found to be broad, usually with a heavy tail, with possibly a cut-off at some degree of the order of few thousands (which is typically imposed by the OSN authorities [59, 29]). Note that while some OSNs have power-law degree distributions [29], some other have been found to exhibit *non* power-law degree distributions [59].

Almost all users in an OSN are connected in a single largest connected component, while the other components are much smaller in size [36, 59]. Some studies also point out that OSNs contain one or more densely connected 'cores' consisting of groups of high-degree nodes that hold the network together and connect low-degree nodes at the fringes [42]. The presence of such cores provides short paths connecting the nodes in different parts of the network, thus reducing the average shortest-path distances. Hence OSNs generally exhibit small-world characteristics [2, 42, 59]. OSNs also demonstrate homophily [59] and community structure [23].

While the above mentioned topological properties of OSNs are similar to those of social networks in the off-line world [50], there also exist certain differences between online and off-line social networks. For instance, node-degrees (i.e., the number of social links of a given user) in off-line social networks are almost always limited to at most a few hundred. This is consistent with the theory in sociology that there exists a cognitive limit to the number of *meaningful* social relationships that an individual can maintain; this limit is commonly known as the Dunbar number which is proposed to lie between 100 and 300. On the other hand, it is very common for users in OSNs to have thousands of social links, which is attributed to the very low cost of maintaining a social link in such virtual networks as compared to that in the real human society [32]. In fact, there have been studies which attempt to identify the 'meaningful' parts of OSNs, by focusing on only those links where there has been at least a certain amount of interaction or activity between the two users forming the link. For instance, Huberman *et al.* [32] studied a sub-graph of the Twitter social network, comprising of only those social links along which there has been at least a certain number of directed tweets. They found this sub-graph is considerably sparser than the complete social network, which indicates that users typically interact with only a small subset of their declared social links.

Again, while off-line social networks are almost always assortative [50, 14], i.e., they have positive degree-degree correlations, OSNs have been found to exhibit both assortative [59] and disassortative structure [31, 2]. In fact, there are examples of OSNs [31] which underwent a transition from being assortative in the initial stages to disassortative in later stages. Again, Kwak *et al.* [37] observed that the Twitter OSN exhibits a non power-law indegree distribution, short effective diameter and low reciprocity, all of which deviate from the known characteristics of off-line social networks [50].

3 Evolution of Social Networks

A large number of evolution models have been proposed from a complex network [48] perspective, in order to explain the emergence of the topological properties of large networks (which were discussed above). This section surveys some of these network evolution models, some of which have been applied to off-line social networks and OSNs. We also discuss some of the evolution models proposed specifically for OSNs.

3.1 Evolution Models for Complex Networks

The study of evolution of networks using concepts of probability theory was pioneered by Erdös & Rényi [20, 21] who studied the formation of random graphs containing a fixed set of nodes which are connected among themselves by the addition of edges with some probability. But these models are not applicable for most real-world networks, including social networks, where both nodes and edges regularly get introduced into the network over time. A number of evolution models have later been proposed for growing networks, as discussed in the rest of this section.

Barabasi-Albert Model and Its Variants: In 1999, Albert and Barabasi [8] developed the famous *preferential attachment* model (also known as the Barabasi-Albert model or BA model), which was based on the Price model of growth of citation networks [19] and the work of Herbert Simon [56]. The BA model considers a network with initially a small set of m_0 nodes, where at each time step a new node arrives and connects to m existing nodes *preferentially*, i.e., the probability of a newly entering node connecting to a node i of degree k_i is $\frac{k_i}{\sum_j k_j}$ (the denominator is the sum of the degrees of all nodes in the network). In this model, a node having larger number of connections is more likely to get new connections; this is popularly known as a 'rich-get-richer' effect. Barabasi and Albert showed that a *power-law degree distribution* $P(k) = 2m^2k^{-3}$ naturally emerges (at large time) from this process. This distribution is also termed as *heavy tailed* distribution as there is a non-zero probability of existence of nodes with very high degrees (which are known as hubs).

It can be noted that preferential attachment can occur in a network as a result of global processes as well as local processes. The BA model uses a global process where a newly entering node can get attached to any existing node in the network. In contrast, some models have used local processes, such as the random walk model [55] where a node u links to nodes selected by taking random walks starting from u, and the common neighbours model [46] where u links preferentially to nodes having many neighbors in common with itself. These models with local rules exhibit preferential attachment because high-degree nodes get selected more often for forming new links.

While the original BA model considered undirected unipartite networks, preferential attachment based growth models have also been proposed for other types of networks. Krapivsky *et al.* [35] proposed a preferential attachment based model for *directed* networks.[2] Preferential attachment based evolution models have also been proposed for bipartite networks [53, 44, 45]. Preferential attachment has also been studied over networks where the number of nodes remains fixed but the number of links grows with time (similar to the Erdös & Rényi networks discussed above). Ben-Naim *et al.* [9] recently studied an evolution model in which links are sequentially added among a fixed set of nodes; the probability of two nodes getting connected is linearly proportional to the degrees of both the nodes (preferential attachment). Interestingly, while preferential attachment leads to power-law degree distributions in growing networks (i.e., networks in which both nodes and edges increase with time), preferential attachment in this network (where number of nodes remains fixed) leads to exponential degree distributions [9].

A recent study by Podobnik *et al.* [52] generalized the BA model for inter-dependent networks or 'network of networks', i.e., a set of networks whose dynamics affect the evolution of one another. They proposed an evolution model considering two inter-dependent networks, where each network grows through the addition of new nodes, and new nodes in each network preferentially attach to both high-degree nodes within the same network, as well as to high-degree nodes in the other network. They used the model to explain the evolution of the Internet network (i.e., the network of routers or autonomous systems (AS) connected by physical links) of three countries, where the networks within a country and the across-country networks are the inter-dependent networks.

While traditional preferential models considered preferential attachment based on node-degree, some recent studies have proposed evolution models where the preferentiality is based on other metrics of nodes, such as the clustering coefficient or betweenness centrality. For instance, in the model proposed by Bagrow *et al.* [6], new nodes preferentially attach to existing nodes based on the clustering coefficient of the existing nodes. Since the event of a new node attaching to an existing node *reduces* the clustering coefficient of the existing node, this has an effect *opposite* to the rich-get-richer phenomenon. Bagrow *et al.* used this model to explain various phenomena such as bursty dynamics, aging of nodes and community formation in networks. In another recent study, Abbasi *et al.* [1] proposed a model where preferential attachment is driven by the betweenness centrality of existing nodes, to explain the evolution of research collaboration networks.

Heterogeneity Based Models: In many real systems, especially in social networks, individual nodes (users) are very diverse and have widely varying capabilities of achieving popularity (e.g., those who are celebrities in real world vs. the normal population). To consider such factors, some models incorporate heterogeneity of nodes in the form of a fitness parameter for nodes, which is typically a random variable drawn from a given distribution [10, 13]. Bianconi *et al.* [10] modeled

[2] In the later part of this chapter, we extend this model to explain the evolution of the directed Twitter social network in presence of restrictions on node-degree.

the topological properties of such systems by assuming a fitness parameter η_u associated with every node u. Here the probability that an incoming node connects to node u depends on both the current degree of u as well as η_u. In such evolution, the emerging degree distribution follows a power-law, but the power-law exponent depends on the distribution of the fitness parameter η. Medo *et al.* [40] recently proposed a fitness-based model to explain the evolution of a citation network (among papers published by the American Physical Society). They considered individual nodes to have heterogeneous fitness values that decay with time (i.e., aging of nodes), and preferential linking dynamics. They showed that, depending on the model parameters, the emerging degree distributions can have various forms, e.g., power-law, exponential or log-normal forms.

Triadic Closure (Triangle Closing) Models: Triadic closure refers to an evolution process where links are formed to close triangles between nodes, e.g., if node A is linked to node B and B is linked to C, then the formation of a link from A to C closes a triangle formed by the nodes A, B and C. Triadic closure based evolution models are often used to explain the presence of large number of triangles in certain real networks, e.g., social networks (as stated earlier). One of the earliest such models was the 'small-world' model by Watts and Strogatz [60] in which a regular network with initially high clustering had its connections *rewired* to have realistic values of both clustering coefficient and the average shortest path length. Evolution models that grow networks through triangle closures have also been proposed [49, 38] to produce networks with high clustering coefficients.

3.2 Evolution Models for OSNs

Next, we describe some of the evolution models proposed for OSNs, to explain the emergence of the various topological properties exhibited by these networks.

One of the earliest large-scale studies on the evolution of OSNs was by Kumar *et al.* [36] who analyzed the evolution of the Flickr and Yahoo! 360 OSNs. This study used extensive data where each node and each (directed) link had an associated time stamp indicating the exact moment when the particular node or link became part of the network. The social networks were found to be composed of three types of components – (i) singletons, i.e., users who joined the network but never created social links, (ii) the giant component which represents the large group of members who are connected to one another (directly or indirectly) through friendship links, and (iii) the middle region, consisting of the remaining users who form small isolated communities and interact with one another but not with the network at large. Based on these observations, Kumar *et al.* proposed an evolution model for OSNs [36] based on *biased* preferential attachment which considered a trade-off between the relative ease of finding potential online connections within the giant component, and the relative difficulty of locating potential connections in the isolated communities.

Leskovec *et al.* [38] studied the individual node arrival and edge creation processes that collectively lead to the emerging macroscopic properties of OSNs. Using a methodology based on the maximum-likelihood principle, they investigated which of several network evolution models best explains the microscopic edge-by-edge evolution of the OSNs. Studying four OSNs (Flickr, Delicious, Yahoo! Answers and LinkedIn), they observed preferential attachment in link formation in all cases. The preferential attachment was found to be produced by edge locality, whereby most of the edges / links formed are between nodes (users) who were close to each other in the social network even before the edge was created. In fact, a large fraction of the edges were found to close triangles (i.e., triadic closure). Based on these observation, Leskovec *et al.* proposed a network evolution model for OSNs, where the rate of joining of nodes and their lifetime are chosen from given distributions, and links are created based on a triangle-closing model. New nodes entering the social network connect to some node, usually chosen preferentially, and then start closing triangles with neighbors of the chosen node.

Similar observations as in the above study were reported by Mislove *et al.* [41] who studied daily snapshots of the Flickr social network – preferential attachment was observed in the link creation dynamics, and most links were found to close triangles. Romero *et al.* [54] also investigated the role of 'triadic closure' in the evolution of the Twitter social network, and observed that follow-links leading to popular users frequently close triangles formed by the popular user and two of her followers.

Bonato *et al.* [12] proposed the geo-protean (GEO-P) model, where nodes are identified with points in Euclidean space, and links are generated depending on a mixture of the relative distance of nodes and a popularity ranking function. With high probability, the GEO-P model generates networks satisfying many observed properties of OSNs, such as power-law degree distributions, the small world property, densification power law [39], and bad spectral expansion[3]. More recently, Bonato *et al.* proposed a model named *iterated local transitivity* [11], in which a newly entering node links to neighbours of an existing node. This model has also been shown to reproduce a number of topological properties that are observed in OSNs. There have also been recent studies on the evolution of *location-based* OSNs (e.g., Foursquare and Gowalla) where users share their geographic locations with their friends [4].

All the studies described above model OSN dynamics with respect to a logical clock, e.g., one or more nodes or links are introduced into the network at each logical time-step. Different from these studies, a recent work by Gaito *et al.*[28] analyze the evolution of Renren (the largest OSN in China) in the context of physical time. They studied a large timestamped dataset describing the evolution of the network, and analyzed the temporal property of burstiness in link creation. Bursts were detected by measuring the acceleration of growth of node-degrees. They observed that most users create links in bursts (which includes a burst in activity immediately after a

[3] Since social networks have a community structure having tightly-knit groups, these networks exhibit bad spectral expansion properties manifested by small gaps between the first and second eigenvalues of their adjacency matrices [22].

user joins the OSN), interspersed with phases of steady-state activity and inactivity. They also found that when users first join an OSN, they create links based on the preferential attachment mechanism, while in later bursts, they explore various regions of the network to create links.

Note that the topological evolution of OSNs have been found to be correlated with the *content* posted by users on such sites. A recent study by Singer *et al.* [57] attempted to understand the co-evolution of social and content networks in the Twitter OSN. Along with the Twitter social network (where nodes are users and edges are the follow-links), they considered as the content network, a bipartite user-tweet network where a user is linked to the tweets she has posted. They find that the characteristics of the social network have a strong influence on the content network; for instance, the number of followers of a user (indegree of the user in the social network) strongly influences how active the user is in authoring tweets (degree of the user in the content network).

Apart from the factors considered by the models described above, there are certain other factors which influence the evolution of OSNs. For instance, in almost all popular OSNs today, various types of restrictions are imposed on the node-degree, i.e., on the number of social links that a user can create. In the next section, we describe in detail an evolution model for networks in the presence of such restrictions [29].

4 Evolution of OSNs in Presence of Restrictions on Node-degree

There has been a phenomenal growth in the number and activity of users in OSNs in the recent years. As a result, the popular OSNs are facing a number of challenges, which include scalability issues (e.g., high latency) and malicious activity by spammers who typically establish social links with thousands of users and then attempt to disseminate spam over the social links. Several OSNs have adopted a common 'tool' to deal with the above issues – they have imposed limits on the number of social links that a user can create in the network, i.e., on the node-degree. For instance, the number of friends that a user can have is restricted to 1000 in Orkut and 5000 in Facebook. Such limits help in reducing the load on the OSN infrastructure – since most OSNs support real-time one-to-many communication (by which messages posted by a user are delivered to all social contacts of that user in real-time), controlling the number of social links of users is an effective method to reduce the system overhead due to message communication. Moreover, these restrictions also act as a first line of defence against spam by preventing spammers from indiscriminately linking to others.

However, with the increasing amounts of activity and engagement of users in popular OSNs, such restrictions are frequently being criticized by the socially active and popular legitimate users, as an encroachment on their freedom to have more friends [15]. Hence the OSN authorities are facing a trade-off while imposing these restrictions. In this scenario, it is necessary to develop an evolution model for OSNs in presence of restrictions on node-degree. Such a model would be useful to predict the effects of a certain restriction on the network, such as the number of users who are likely to get blocked by the restriction. This would, in turn, help the OSN authorities to design restrictions that suitably balance the two conflicting requirements of reducing system-load and minimizing dissatisfaction among users.

In view of the trade-off stated above, the Twitter OSN has imposed a more intelligent ‘soft’ cutoff which attempts to minimize the number of users who get blocked by the restriction. We first describe the restriction imposed in Twitter, and then describe an evolution model for OSNs [29], developed using the tools of complex network theory, taking the Twitter restriction as a case-study.

4.1 *The Restriction in Twitter and Its Effects*

The Twitter social network is a directed network where an edge $u \rightarrow v$ implies that user u ‘follows’ user v, i.e., u has subscribed to receive all messages posted by v. In Twitter terminology, u is a ‘follower’ of v and v is a ‘following’ of u. The out-degree (number of followings) of u is thus a measure of u’s social activity or her interest to collect information from other users. Analogously, the in-degree of u (number of followers who are interested in u’s posts) is a measure of u’s popularity in Twitter.

4.1.1 The Restriction in Twitter

The growing popularity of Twitter has not only led to high system-load due to increasing user-activity, but also to high levels of spam-activity. To reduce strain on the website and prevent indiscriminate linking by spammers, Twitter has enforced a restriction on the number of people that a user can follow (i.e., on the out-degree) [26]. Every user is allowed to follow up to 2000 others, but “once you’ve followed 2000 users, there are limits to the number of additional users you can follow: this limit is different for every user and is based on your ratio of followers to following”, as given in the Twitter Support web-pages [25]. However, Twitter does not specify the restriction fully in public (security through obscurity). This has led to several conjectures regarding the Twitter follow-limit; among these, the most widely believed one is as follows. If a user u has u_{in} number of followers (in-degree), then the maximum number of users whom u can herself follow (maximum possible out-degree), denoted by u_{out}^{max}, is:

$$u_{out}^{max} = \max\{2000,\ 1.1 \cdot u_{in}\} \tag{1}$$

This Twitter restriction, commonly known as the “10% rule”, implies that if a user wants to follow (outdegree) more than 2000 people, she needs to have at least a certain number of followers (indegree) herself.

4.1.2 Effects of the Restriction on the Network Topology

We now discuss the degree distributions of the Twitter social network, which clearly show that the imposed restriction has significantly altered the topology of the Twitter network. For the empirical measurements, we use a large dataset made publicly available by [17], which includes a complete snapshot of the Twitter social network as of August 2009. In brief, the snapshot contains more than 52 million nodes and 1.9 billion edges; details about the dataset are available in [17].

We measure the indegree distribution (distribution of number of followers) and the outdegree distribution (distribution of number of followings) of the Twitter social network, as in August 2009, and show them in Fig. 1. The indegree distribution shows a power-law decay $p_i \sim i^{-2.06}$ over a large range of indegrees[4], similar to what was found in earlier measurements on the Twitter network [33]. However, the outdegree distribution clearly shows a departure from the simple power-law nature that was observed by measurements on Twitter *before* the restriction was imposed [33]. Now, the power-law $p_j \sim j^{-1.92}$ for the outdegrees below the point of restriction is followed by a sharp spike at around outdegree $j = 2000$, followed by a rapid decay in the outdegree distribution beyond this point. This indicates that an uncharacteristically large fraction of users have outdegree near about 2000 – this is because a significant number of users are unable to increase their outdegree beyond a certain limit near 2000 as they do not have sufficient indegree (followers).[5]

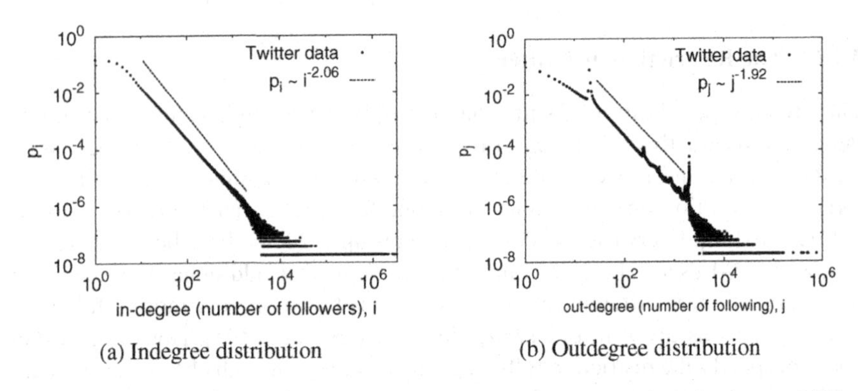

(a) Indegree distribution (b) Outdegree distribution

Fig. 1 Degree distributions of nodes (users) in Twitter social network as in August 2009, along with power-law fits. The sharp spike in the outdegree distribution at degree 2000 is produced by the restriction imposed on the number of users one can follow beyond 2000.

It can be seen that restrictions on node-degree in present-day OSNs actually result in significant changes in the topological properties of the network. These observations motivate the need for an analytical model that can be used to estimate the effects of such restrictions on the network topology. We next describe such a model.

[4] The power-law exponents are estimated by the method stated in [18].

[5] Note that the out-degree distribution also shows a peak at $x = 20$ because till 2009, Twitter used to recommend an initial set of 20 people for every newcomer to follow by a single click, and many newcomers took up this offer [37].

4.2 Evolution Model in Presence of Restrictions on Node-degree

In this section, we discuss an evolution model for networks (specifically OSNs) in presence of restrictions on node-degree [29]. The model is a customized version of the preferential attachment based growth model proposed by Krapivsky, Rodgers and Redner [35] for directed networks (henceforth referred to as the KRR model), which is modified by introducing restrictions on the outdegree of nodes, similar to the follow-limit imposed by Twitter. Here we describe the model in brief, details can be found in [29].

In this model, network growth occurs in two distinct steps. At each discrete time step, one of the following events occurs: (1) with probability p, a new node is introduced and it forms a directed out-link to an existing node, or (2) with probability $q = 1 - p$, a new directed link is created between two existing nodes.

The probability that a new node (event 1) links to an (i, j)-node (i.e., a node having in-degree i and out-degree j) is assumed to be proportional to $(i + \lambda)$, since intuitively a new user is more likely to link to (follow) a popular user having many followers (high in-degree). Analogously, the probability that a new edge (event 2) is created from a (i_1, j_1)-node to a (i_2, j_2)-node is assumed to be proportional to $(i_2 + \lambda)(j_1 + \mu)$. This again follows the intuition that a socially active user u who follows many people already is more likely to follow another user v, especially if v is popular having many followers. Here $\lambda > 0$ and $\mu > -1$ are model parameters that introduce randomness in the preferential attachment rules [35]. Lower values of these parameters indicate that the link-creation dynamics in the OSN are closer to preferential attachment.

Let $N_{ij}(t)$ be the average number of (i, j)-nodes in the network at time t. The model considers the following *rate-equations* to track how N_{ij} changes with time.

Change in N_{ij} Due to Change in Indegree of Nodes: The number N_{ij} of (i, j)-nodes increases when a new link is created leading to a $(i-1, j)$-node (of which there are $N_{i-1,j}$ in number). This can happen due to a new node linking to a $(i-1, j)$-node (with probability p) or due to the creation of a new link leading to a $(i-1, j)$-node (with probability $q = 1 - p$). With the probabilities stated above, this increase occurs with the rate $(p+q)(i-1+\lambda)N_{i-1,j}$, divided by the normalization factor $\sum_{ij}(i+\lambda)N_{ij}$. On the other hand, the number N_{ij} of (i, j)-nodes gets reduced when a new link is created leading to a (i, j)-node; this can happen due to a new node linking to a (i, j)-node (with probability p) or due to the creation of a new link leading to a (i, j)-node (with probability $q = 1 - p$). Hence, this reduction in N_{ij} occurs with the rate $(p+q)(i+\lambda)N_{ij} / \sum_{ij}(i+\lambda)N_{ij}$. Since $(p+q) = 1$, therefore the rate of change in N_{ij} due to change in indegree of nodes is as given in Eqn. 2.

$$\left.\frac{dN_{ij}}{dt}\right|_{in} = \frac{(i-1+\lambda)N_{i-1,j} - (i+\lambda)N_{ij}}{\sum_{ij}(i+\lambda)N_{ij}} \tag{2}$$

Change in N_{ij} Due to Change in Outdegree of Nodes: Since the change in outdegree of nodes is restricted (as by the Twitter restriction), we introduce the term β_{ij} in the equations below to capture the effects of the restriction on the growth dynamics. β_{ij} is defined to be 1 if and only if users having indegree i are allowed by the restriction to have outdegree j, 0 otherwise (explained in details below).

There is a gain in N_{ij} when a $(i, j-1)$-node (of which there are $N_{i,j-1}$ in number) creates a new out-link; note that only those $(i, j-1)$-nodes are allowed to do this for whom $\beta_{ij} = 1$. This event occurs with the rate $q(j-1+\mu)N_{i,j-1}\beta_{ij}$ divided by the normalization factor $\sum_{ij}(j+\mu)N_{ij}\beta_{i,j+1}$. Similarly, N_{ij} gets reduced when an (i, j)-node (which must have a sufficiently large i such that $\beta_{i,j+1} = 1$) forms a new out-link, which occurs with the rate $q(j+\mu)N_{i,j}\beta_{i,j+1} / \sum_{ij}(j+\mu)N_{ij}\beta_{i,j+1}$. These events can occur only due to creation of links among existing nodes, hence the rates have probability q as a factor. Thus the rate of change in $N_{ij}(t)$ due to change in outdegree of nodes is:

$$\left.\frac{dN_{ij}}{dt}\right|_{out} = q\left[\frac{(j-1+\mu)N_{i,j-1}\beta_{ij} - (j+\mu)N_{ij}\beta_{i,j+1}}{\sum_{ij}(j+\mu)N_{ij}\beta_{i,j+1}}\right] \tag{3}$$

Hence the total rate of change in the number $N_{ij}(t)$ of (i, j)-nodes at time t is given by Eqn. 4.

$$\frac{dN_{ij}}{dt} = \left.\frac{dN_{ij}}{dt}\right|_{in} + \left.\frac{dN_{ij}}{dt}\right|_{out} + p\delta_{i0}\delta_{j1} \tag{4}$$

where the last term accounts for the introduction of new nodes with indegree zero and outdegree one, with a probability p at every time-step. δ_{i0} (Kronecker's delta function) is 1 for $i = 0$, and 0 otherwise; similarly, δ_{j1} is 1 for $j = 1$ and 0 otherwise.

The β_{ij} Terms – Incorporating The Restriction: The role of the β_{ij} terms in Eqn. 3 is explained as follows. Due to the imposed restriction (e.g., the Twitter follow-limit), only a fraction of the existing nodes can create new out-links, and β_{ij} is defined such that it equals 1 only for this fraction of nodes. The β_{ij} terms can be defined according to the restriction that is to be modeled, thus making this model suitable to study different restriction functions. To study the Twitter follow-limit, we define β_{ij} for a generalized 'κ-% rule' starting at out-degree s ($\kappa = 10$ and $s = 2000$ in the Twitter OSN) as

$$\beta_{ij} = \begin{cases} 1 & \text{if } j \leq \max\{s, (1+\frac{1}{\kappa})i\}, \forall i \\ 0 & \text{otherwise} \end{cases}$$

4.2.1 Solving the Model to Compute Degree Distributions

At time t, let $N(t)$ be the total number of nodes in the network, and let $I(t)$ and $J(t)$ be the total in-degree and total out-degree respectively. Since at every time-step, a new edge is added, and a new node is added with probability p,

$$N(t)=\sum_{ij}N_{ij}=pt,\quad I(t)=\sum_{ij}iN_{ij}=J(t)=\sum_{ij}jN_{ij}=t \tag{5}$$

Thus parameter p controls the relative number of nodes and edges in the network. The denominator (normalizing factor) in (2) equals $(I+\lambda N)$. For the denominator in (3), we make a simplifying approximation - we assume that at a given time, the number of nodes that are actually blocked by the restriction from increasing their out-degree (i.e. number of (i,j)-nodes for which $\beta_{i,j+1}$ is 0) is negligibly small compared to the total number of nodes in the network, which implies

$$\sum_{ij}(j+\mu)N_{ij}\beta_{i,j+1}\simeq\sum_{ij}(j+\mu)N_{ij} \;=\; (J+\mu N) \tag{6}$$

By solving (4) with the above approximation for few small values of i,j, it is seen that $N_{ij}(t)$ grows linearly with time [35]; hence we can substitute

$$N_{ij}(t)=n_{ij}t \tag{7}$$

where n_{ij} is the (constant w.r.t. time) rate of increase in the number of (i,j)-nodes. Substituting (5) and (7) in (4) gives a recursion relation for n_{ij}:

$$n_{ij}=\frac{(i-1+\lambda)n_{i-1,j}-(i+\lambda)n_{ij}}{1+\lambda p}+\frac{q(j-1+\mu)n_{i,j-1}\beta_{ij}-q(j+\mu)n_{ij}\beta_{i,j+1}}{1+\mu p}+p\delta_{i0}\delta_{j1} \tag{8}$$

For brevity, we denote the first fraction on the right-hand side in (8) as A_{ij} in subsequent discussions.

To simplify the computation of the functional form of the degree distribution, we assume (as was done in the original KRR model [35]) that the power-law exponents of the in-degree and out-degree distributions are equal, which implies $\lambda=(\mu+1)/q$. Note that the exponents were actually found [33] to be equal for the Twitter OSN before the restriction was imposed.

Since we are studying restrictions only on out-degree (as in Twitter), we shall henceforth consider only the out-degree distribution. The in-degree distribution can be computed by the original KRR model [35] and will be of the form of a power-law for the entire range of in-degrees.

Let $N_j^{out}(t)=\sum_i N_{ij}(t)$ be the number of nodes with out-degree j at time t; using (7), $N_j^{out}(t)=t\sum_i n_{ij}=tg_j$, where $g_j=\sum_i n_{ij}$. Thus the out-degree distribution at j (i.e. fraction of nodes with out-degree j) can be obtained as $N_j^{out}(t)/N(t)=g_j/p$. To obtain the complete out-degree distribution, we solve (8) to get $g_j=\sum_i n_{ij}$ for all j by considering the following cases.

Case 1: $j<s$ (Before The Starting Point of Cutoff): Since there is no restriction for $j<s$, the model behaves similar to the original KRR model [35]; hence

$$g_j \;=\; G\cdot\frac{\Gamma(j+\mu)}{\Gamma(j+1+q^{-1}+\mu q^{-1})}\;\sim\; j^{-(1+q^{-1}+\mu p q^{-1})} \tag{9}$$

where $\Gamma()$ is the Euler gamma function, and G is a constant.

Case 2: $j = s$ (At The Starting Point of The Cutoff): Let α denote the fraction $\frac{1}{(1+1/\kappa)}$ in case of a κ-% rule ($\kappa = 10$ in Twitter). A node can have an out-degree $j > s$ only if it has an in-degree $i \geq \alpha j$, implying that for $\beta_{i,j+1}$(for $j \geq s$) to be 1, $i \geq \alpha(j+1)$. Hence, for $j = s$, (8) becomes

$$n_{is} = \begin{cases} A_{is} + \frac{q(s-1+\mu)n_{i,s-1}}{1+\mu p} & i < \alpha(s+1) \\ A_{is} + \frac{q(s-1+\mu)n_{i,s-1} - q(s+\mu)n_{is}}{1+\mu p} & i \geq \alpha(s+1) \end{cases} \tag{10}$$

We use a standard technique [35] to solve rate equations: summing (10) for all $i \geq 0$, the terms in A_{is} disappear (they cancel out each other, except the first term in the first equation, i.e. for the case $i = 0$, but that term is zero), and we get

$$g_s = \frac{s-1+\mu}{s+(1+\mu)q^{-1}} \cdot g_{s-1} + \frac{s+\mu}{s+(1+\mu)q^{-1}} \cdot c_s \tag{11}$$

where g_{s-1} can be computed by (9) and $c_s = (\sum_{i=0}^{\lfloor \alpha(s+1) \rfloor} n_{is})$ is the rate of increase in the number of nodes that have out-degree s but cannot increase their out-degree further (i.e. (i, j)-nodes for which $j = s$ and $\beta_{i,s+1} = 0$). Let $\lfloor \alpha(s+1) \rfloor$ be denoted by d. To compute c_s, we sum (10) in the range $0 \leq i \leq d$ to get

$$c_s = \frac{1}{1+\lambda p} \cdot \left[(s-1+\mu) \sum_{i=0}^{d} n_{i,s-1} - (d+\lambda) n_{ds} \right] \tag{12}$$

where n_{ds} can be obtained as

$$n_{ds} = \frac{(s+\mu-1)\Gamma(d+\lambda)}{\Gamma(d+\lambda(1+p)+2)} \sum_{k=0}^{d} \frac{\Gamma(k+\lambda(1+p)+1)}{\Gamma(k+\lambda)} \cdot n_{k,s-1} \tag{13}$$

from (10) after some algebraic manipulations (omitted for sake of brevity). The terms $n_{i,s-1}$ in (12) and (13) can be evaluated from the original KRR model (Eqn. 18 in [35]) since they are not affected by the restriction starting from $j = s$. Substituting c_s from (12) into (11), we can obtain a closed-form expression for the degree distribution g_s/p at $j = s$.

Case 3: $j > s$ (Beyond The Starting Point of Cutoff): In this region, (8) becomes

$$n_{ij} = \begin{cases} 0 & i < \alpha j \\ A_{ij} + \frac{q(j-1+\mu)n_{i,j-1}}{1+\mu p} & \alpha j \leq i < \alpha(j+1) \\ A_{ij} + \frac{q(j-1+\mu)n_{i,j-1} - q(j+\mu)n_{ij}}{1+\mu p} & i \geq \alpha(j+1) \end{cases} \tag{14}$$

since nodes having in-degree $i < \alpha j$ cannot have out-degree $j(> s)$, nodes having in-degree $\alpha j \leq i < \alpha(j+1)$ can have out-degree j but not $j+1$, and nodes with in-degree $i \geq \alpha(j+1)$ can increase their out-degree from j to $j+1$. Proceeding similarly as in the case $j = s$, and adding (14) over all $i \geq 0$, we get

$$g_j = \frac{j-1+\mu}{j+(1+\mu)q^{-1}} \cdot [g_{j-1} - c_{j-1}] + \frac{j+\mu}{j+(1+\mu)q^{-1}} \cdot c_j \tag{15}$$

where $c_j = \sum_{i=0}^{\lfloor \alpha(j+1) \rfloor} n_{ij}$ is the rate of increase in the number of nodes which have out-degree j but cannot increase their out-degree further, unless their in-degree increases. Proceeding from (15) in a similar way as in the case $j=s$, we can derive analytical expressions for g_j and c_j for $j>s$ iteratively using the values of g_{j-1} and c_{j-1} (e.g., g_{s+1} and c_{s+1} can be derived using g_s and c_s, and so on).

4.2.2 Fitting Empirical Twitter Data with the Proposed Model

Now we check whether the analytical model developed above can predict the degree distributions empirically observed in the Twitter social network. For this, we need to choose suitable values for the parameters in the model.

The parameter p (ratio of nodes to links) is measured as 0.028 from the empirical Twitter network data described in Section 4.1, hence p was set to this value. The parameters of the restriction function were set to $\kappa = 10$ and $s = 2000$ (as in the real Twitter) unless otherwise stated. Parameters λ and μ indicate how closely the link-creation dynamics in the OSN follow the preferential attachment model. Estimating these parameters for a real OSN is a challenging issue; hence we consider different values of these parameters. Since the model assumes $\lambda = (\mu+1)/(1-p)$ (as described above), we report results for different values of μ only.

The empirical indegree and outdegree distributions of the Twitter OSN (described in Section 4.1) are fitted with those obtained from the proposed model in Fig. 2(a) and Fig. 2(b) respectively, using suitable values of the parameters: p = 0.028, μ = 6.0, s = 2000, κ = 10. The optimal value for parameter μ is selected using a least-squares method, i.e., that value of μ is selected which minimizes the square error between the empirical and theoretical degree distributions. Both the empirical distributions show excellent fit with those from the model in general, signifying that the proposed model successfully captures the growth dynamics of the Twitter OSN.

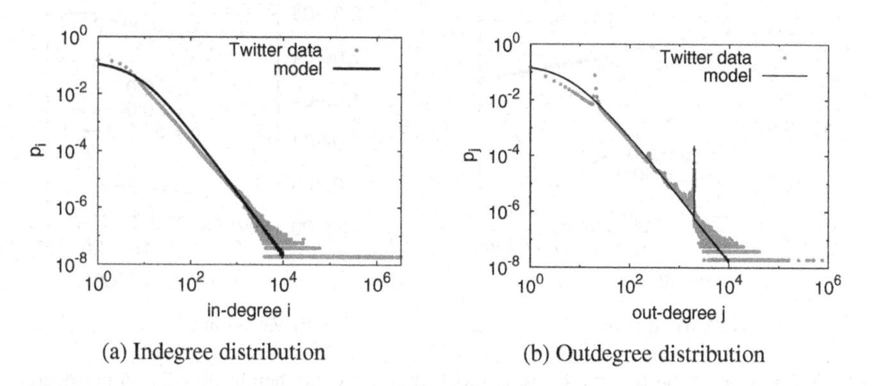

(a) Indegree distribution (b) Outdegree distribution

Fig. 2 Fitting empirical degree distributions of the Twitter social network with the distributions predicted by the proposed model (with parameters p = 0.028, μ = 6.0, s = 2000, κ = 10)

4.2.3 Utility of The Model – Estimating Number of Blocked Users

Finally we show how the analytical model can be used to estimate the number of users who are likely to get blocked (from creating new links) by an imposed restriction on node-degree.

In absence of any restriction, g_j decays as $g_j = (j-1+\mu)g_{j-1}/(j+(1+\mu)q^{-1})$ (as observed in the original KRR model [35]). Comparing this with (11), we see that due to the 'soft' cut-off at $j = s$, the fraction of nodes having out-degree s (i.e. g_s/p) includes the following additional term, which accounts for the spike in the out-degree distribution at this point:

$$\phi_s = \frac{s+\mu}{s+(1+\mu)q^{-1}} \cdot \frac{c_s}{p} \tag{16}$$

where c_s is obtained from (12). For $s \gg \mu$ and $q \simeq 1$ (for a real-world OSN, typically cut-off s is large and $p = 1-q$ is very small), $\phi_s \simeq c_s/p$ which is an estimate of the fraction of nodes (users) blocked at the point of cut-off.

Figures 3(a) and 3(b) show the variation in ϕ_s with the restriction parameters s and κ respectively; we use different values of μ to investigate varying link-creation dynamics (from highly preferential to more random). ϕ_s shows a power-law decay with increasing s (Fig. 3(a) in log-log scale); for lower values of s, a larger fraction of users get blocked leading to a greater reduction in the system-load, but at the risk of increased user-dissatisfaction. Similarly, with increase in κ, a higher in-degree becomes necessary to cross the cut-off resulting in a larger fraction of blocked users; as shown in Fig. 3(b), ϕ_s has a parabolic increase with κ.

Thus the model helps to estimate the fraction of users who are likely to get blocked if a certain restriction is imposed. Such analyses are an efficient way for the OSN authorities to make design-choices while imposing restrictions, such as fixing the values of the parameters s and κ of the restriction function.

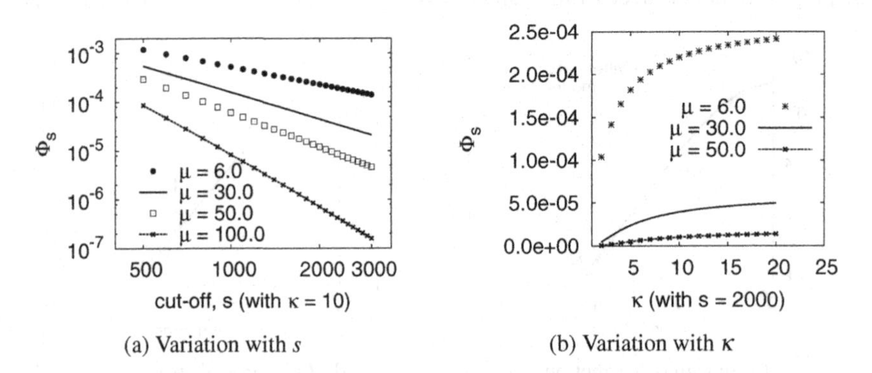

(a) Variation with s (b) Variation with κ

Fig. 3 Variation of the fraction of users blocked at $j=s$ (i.e. height of spike in out-degree distribution) (a) with s where $\kappa = 10$ (log-log plot) (b) with κ, where $s = 2000$

5 Summary

Research on online social networks has been an active field of study in recent years, primarily because the availability of data of millions of users enables large-scale studies on different aspects of user-behavior. In this chapter, we focused on the evolving structure of OSNs. We discussed the commonly observed topological characteristics of OSNs, and some of the network evolution models proposed to explain their emerging structure. We also described in detail an evolution model for networks where restrictions are imposed on the number of links that a node can create. In view of the rapid rise in the number and activity of users in OSNs and the consequent load on the infrastructure, restrictions on node-degree are being imposed in almost all popular OSNs. Considering the frequent criticism of such restrictions by the active users, it is evident that restrictions need to be designed suitably to balance the conflicting factors of reducing system-load and minimizing dissatisfaction among blocked users; this model helps to design such restrictions.

In conclusion, study of the evolution of social networks can give valuable insights into how users find other interesting users to connect to, how they behave individually as well as collectively, and so on. Though a large number of evolution models for social networks have already been proposed, we expect that this field of research will be active in the coming years, and that several more studies will be undertaken to better understand and quantitatively describe social behaviors.

References

1. Abbasi, A., Hossain, L., Leydesdorff, L.: Betweenness centrality as a driver of preferential attachment in the evolution of research collaboration networks. Journal of Informetrics 6(3), 403–412 (2012)
2. Ahn, Y.-Y., Han, S., Kwak, H., Moon, S., Jeong, H.: Analysis of topological characteristics of huge online social networking services. In: Proceedings of ACM International Conference on World Wide Web (WWW), pp. 835–844 (2007)
3. Albert, R., Barabási, A.-L.: Statistical mechanics of complex networks. Reviews of Modern Physics 74(1), 47–97 (2002)
4. Allamanis, M., Scellato, S., Mascolo, C.: Evolution of a location-based online social network: analysis and models. In: Proceedings of ACM Internet Measurement Conference (IMC) (2012)
5. Backstrom, L., Boldi, P., Rosa, M., Ugander, J., Vigna, S.: Four degrees of separation. arXiv:1111.4570 [cs.SI] (2012)
6. Bagrow, J.P., Brockmann, D.: Natural emergence of clusters and bursts in network evolution. Phys. Rev. X 3, 021016 (2013)
7. Baluja, S., Seth, R., Sivakumar, D., Jing, Y., Yagnik, J., Kumar, S., Ravichandran, D., Aly, M.: Video suggestion and discovery for Youtube: taking random walks through the view graph. In: Proceedings of ACM International Conference on World Wide Web (WWW), pp. 895–904 (2008)
8. Barabási, A.-L., Albert, R.: Emergence of scaling in random networks. Science 286(5439), 509–512 (1999)
9. Ben-Naim, E., Krapivsky, P.L.: Popularity-driven networking. Europhysics Letters 97(4), 48003 (2012)

10. Bianconi, G., Barabási, A.-L.: Competition and multiscaling in evolving networks. Europhysics Letters 54(4), 436 (2001)
11. Bonato, A., Hadi, N., Horn, P., Pralat, P., Want, C.: Models of on-line social networks. Internet Mathematics 6, 285–313 (2011)
12. Bonato, A., Janssen, J., Pralat, P.: A geometric model for on-line social networks. In: Proceedings of Workshop on Online Social Networks (WOSN) (June 2010)
13. Caldarelli, G., Capocci, A., De Los Rios, P., Muñoz, M.A.: Scale-free networks from varying vertex intrinsic fitness. Physical Review Letters 89, 258702 (2002)
14. Catanzaro, M., Caldarelli, G., Pietronero, L.: Assortative model for social networks. Physical Review E 70, 037101 (2004)
15. Catone, J.: Twitter's Follow Limit Makes Twitter Less Useful (August 2008), `http://www.sitepoint.com/twitter-follow-limit-makes-twitter-less-useful/`
16. Cattuto, C., Schmitz, C., Baldassarri, A., Servedio, V.D.P., Loreto, V., Hotho, A., Grahl, M., Gerd, S.: Network properties of folksonomies. AI Communications 20, 245–262 (2007)
17. Cha, M., Haddadi, H., Benevenuto, F., Gummadi, K.P.: Measuring user influence in Twitter: the million follower fallacy. In: Proceedings of AAAI International Conference on Weblogs and Social Media (ICWSM) (May 2010)
18. Clauset, A., Shalizi, C.R., Newman, M.E.J.: Power-law distributions in empirical data. SIAM Review 51(4), 661–703 (2009)
19. de Solla, P.: Networks of scientific papers. Science 149(3683), 510–515 (1965)
20. Erdös, P., Rényi, A.: On random graphs, I. Publicationes Mathematicae (Debrecen) 6, 290–297 (1959)
21. Erdos, P., Renyi, A.: On the strength of connectedness of a random graph. Acta Mathematica Hungarica 12, 261–267 (1961)
22. Estrada, E.: Spectral scaling and good expansion properties in complex networks. Europhysics Letters 73(4), 649 (2006)
23. Ferrara, E.: A large-scale community structure analysis in Facebook. arXiv:1106.2503 [cs.SI] (2012)
24. Ferrara, E., Fiumara, G.: Topological features of online social networks. Communications on Applied and Industrial Mathematics 2(2), 1–20 (2011)
25. Twitter help center: Following rules and best practices, `http://support.twitter.com/forums/10711/entries/68916`
26. Twitter blog: Making progress on spam (August 2008), `http://blog.twitter.com/2008/08/making-progress-on-spam.html`
27. Fortunato, S.: Community detection in graphs. Physics Reports 486(3-5), 75–174 (2010)
28. Gaito, S., Zignani, M., Rossi, G.P., Sala, A., Wang, X., Zheng, H., Zhao, B.Y.: On the bursty evolution of online social networks. arXiv:1203.6744 [cs.SI] (2012)
29. Ghosh, S., Srivastava, A., Ganguly, N.: Effects of a soft cut-off on node-degree in the Twitter social network. Computer Communications 35(7), 784–795 (2012)
30. Girvan, M., Newman, M.E.J.: Community structure in social and biological networks. Proceedings of the National Academy of Sciences of U.S.A (PNAS) 99, 7821 (2002)
31. Hu, H., Wang, X.-F.: Disassortative mixing in online social networks. Europhysics Letters 86(1), 18003 (2009)
32. Huberman, B.A., Romero, D.M., Wu, F.: Social networks that matter: Twitter under the microscope. First Monday 14(1) (January 2009)
33. Java, A., Song, X., Finin, T., Tseng, B.: Why we Twitter: understanding microblogging usage and communities. In: Proceedings of Workshop on Web Mining and Social Network Analysis (WebKDD / SNA-KDD), pp. 56–65 (2007)

34. Jin, E.M., Girvan, M., Newman, M.E.J.: Structure of growing social networks. Physical Review E 64, 046132 (2001)
35. Krapivsky, P.L., Rodgers, G.J., Redner, S.: Degree distributions of growing networks. Physical Review Letters 86(23), 5401–5404 (2001)
36. Kumar, R., Novak, J., Tomkins, A.: Structure and evolution of online social networks. In: Proceedings of ACM International Conference on Knowledge Discovery and Data mining (SIGKDD), pp. 611–617 (2006)
37. Kwak, H., Lee, C., Park, H., Moon, S.: What is Twitter, a social network or a news media? In. In: Proceedings of ACM International Conference on World Wide Web (WWW), pp. 591–600 (2010)
38. Leskovec, J., Backstrom, L., Kumar, R., Tomkins, A.: Microscopic evolution of social networks. In: Proceedings of ACM International Conference on Knowledge Discovery and Data Mining (SIGKDD), pp. 462–470 (2008)
39. Leskovec, J., Kleinberg, J., Faloutsos, C.: Graphs over time: densification laws, shrinking diameters and possible explanations. In: Proceedings of ACM International Conference on Knowledge Discovery and Data Mining (SIGKDD), pp. 177–187. ACM (2005)
40. Medo, M., Cimini, G., Gualdi, S.: Temporal effects in the growth of networks. Physical Review Letters 107(23), 238701 (2011)
41. Mislove, A., Koppula, H.S., Gummadi, K.P., Druschel, P., Bhattacharjee, B.: Growth of the Flickr social network. In: Proceedings of Workshop on Online Social Networks (WOSN), pp. 25–30 (2008)
42. Mislove, A., Marcon, M., Gummadi, K.P., Druschel, P., Bhattacharjee, B.: Measurement and analysis of online social networks. In: Proceedings of ACM SIGCOMM Conference on Internet Measurement (IMC), pp. 29–42 (2007)
43. Mislove, A.E.: Online social networks: measurement, analysis, and applications to distributed information systems. PhD thesis, Rice University (April 2009)
44. Mukherjee, A., Choudhury, M., Ganguly, N.: Understanding how both the partitions of a bipartite network affect its one-mode projection. Physica A: Statistical Mechanics and its Applications 390(20), 3602–3607 (2011)
45. Nacher, J.C., Akutsu, T.: On the degree distribution of projected networks mapped from bipartite networks. Physica A: Statistical Mechanics and its Applications 390(23-24), 4636–4651 (2011)
46. Newman, M.E.J.: Clustering and preferential attachment in growing networks. Physical Review E 64, 025102 (2001)
47. Newman, M.E.J.: Assortative mixing in networks. Physical Review Letters 89(20), 208701 (2002)
48. Newman, M.E.J.: The structure and function of complex networks. SIAM Review 45(2), 167–256 (2003)
49. Newman, M.E.J.: Random graphs with clustering. Physical Review Letters 103, 058701 (2009)
50. Newman, M.E.J., Park, J.: Why social networks are different from other types of networks. Physical Review E 68(3), 036122 (2003)
51. Newman, M.E.J., Watts, D.J., Strogatz, S.H.: Random graph models of social networks. Proceedings of the National Academy of Sciences of U.S.A (PNAS) 99, 2566–2572 (2002)
52. Podobnik, B., Horvatic, D., Dickison, M., Stanley, H.E.: Preferential attachment in the interaction between dynamically generated interdependent networks. arXiv:1209.2817 [physics.soc-ph] (2012)
53. Ramasco, J.J., Dorogovtsev, S.N., Pastor-Satorras, R.: Self-organization of collaboration networks. Physical Review E 70(3), 036106 (2004)

54. Romero, D.M., Kleinberg, J.M.: The directed closure process in hybrid social-information networks, with an analysis of link formation on Twitter. In: Proceedings of AAAI International Conference on Weblogs and Social Media (ICWSM) (May 2010)
55. Saramaki, J., Kaski, K.: Scale-free networks generated by random walkers. Physica A: Statistical Mechanics and its Applications 341, 80–86 (2004)
56. Simon, H.A.: On a class of skew distribution functions. Biometrika 42, 425–440 (1955)
57. Singer, P., Wagner, C., Strohmaier, M.: Understanding co-evolution of social and content networks on Twitter. In: Proceedings of Workshop on Making Sense of Microposts (with ACM WWW) (2012)
58. Travers, J., Milgram, S.: An experimental study of the small world problem. Sociometry 32, 425–443 (1969)
59. Ugander, J., Karrer, B., Backstrom, L., Marlow, C.: The anatomy of the Facebook social graph. arXiv:1111.4503 [cs.SI] (2011)
60. Watts, D.J., Strogatz, S.H.: Collective dynamics of 'small-world' networks. Nature 393(6684), 440–442 (1998)

Machine Learning for Auspicious Social Network Mining

Sagar S. De and Satchidananda Dehuri

Abstract. The importance of machine learning for social network analysis is realized as an inevitable tool in forthcoming years. This is due to the unprecedented growth of social-related data, boosted by the proliferation of social media websites and the embedded heterogeneity and complexity. Alongside the machine learning derives much effort from psychologists to build computational model for solving tasks like recognition, prediction, planning and analysis even in uncertain situations. In this chapter, we have presented different network analysis concepts. Then we have discussed implication of machine learning for network data preparation and different learning techniques for descriptive and predictive analysis. Finally we have presented some machine learning based findings in the area of community detection, prediction, spatial-temporal and fuzzy analysis.

Keywords: Social Network Mining Strategies, Network data collection and Preparation, Machine Learning based Network Analysis, Network Learning Methods.

1 Introduction

Social network is a linked formation of social elements in which social objects such as people, organizations, web documents etc. act as *nodes* of the network and *links* are accomplished as relationship among them. In different fields of study, the social components are named in different terminology although they behave as equivalent. For example, nodes are alternatively known as *vertices, actors, points*; and links

Sagar S. De
S. N. Bose National Centre for Basic Sciences,
Block-JD, Sector-III, Salt Lake, Kolkata-98, India
e-mail: sagar.s.de@gmail.com

Satchidananda Dehuri
Department of Systems Engineering, Ajou University, San 5, Woncheon-dong,
Yeongtong-gu, Suwon 443-749, Republic of Korea
e-mail: satchi.lapa@gmail.com

M. Panda, S. Dehuri, and G.-N. Wang (eds.), *Social Networking*
Intelligent Systems Reference Library 65,
DOI: 10.1007/978-3-319-05164-2_3,

are as *edges, ties, arcs,* or *bonds*. The theoretical formation is useful to analyse interrelation among individuals, groups, organisations, or even the society.

Over the past decades, researchers have developed several sophisticated techniques for analyzing and visualizing network data, particularly for *social network analysis (SNA)*. Social scientists have already proven the importance of network analysis for melioration of the society in the form of government, business, health care, social security, community development, political and personal benifits, and many more. Decision makers have chosen *network data mining* as one of the most crucial tool in day-to-day operations. Researchers have derived several ways of network knowledge exploration and representation using *statistical, graphical*, and *visual techniques*. They have used degree to measure immediate importance or social relationship of an individual; community to measure importance of a group; centrality to identify people or groups having capability of information flow across the network. Several tools such as *UCINET, Netminar, Pajek, Cytospace, Gephi, Mathematica, NodeXL, Graphviz, Tulip, Microsoft Automatic Graph Layout, Tom Sawyer Software, etc.* have been developed during last research. Despite of popularity of SNA and related techniques in academics, most of the social media analysts are unfamiliar with the concepts and tools that directly support these analysis. This is due to the recognition of the value of using centrality metrics to identify important individuals, clustering algorithms to identify subgroups, and network visualizations to characterize online conversions and markets.

Over the past years, people used survey as the primary source of data. Along with the formal survey techniques; the growth of the *web, web-based social network sites* (e.g. Facebook, Twitter, LinkedIn, Google+ etc.), *online social media services* (e.g. YouTube, Flickr etc.), and *large-scale information sharing communities* (e.g. Digg, Yahoo! Answers, Stack Overflow etc.) create a wealth of data that has the potential to provide new insights to marketers, social scientists, community administrators, and system developers. Most of the data falls in the form of *linked structure*: people connected through friends and follow relationships; hyperlink connects websites; shopping items connected to other shopping items by share purchasing patterns; co-authors are connected through published papers. The reliance of web and related services make it a potential mirror of our society. Traces of activities left by web users in the form of *email/IM communications, contribution of contents, surfing, commenting, rating, playing online games* can shed light on individual behaviour, social relationships and community efficiency. Tools and processes are essential to make sense of social traces for enabling researchers to study meaningful and sustainable social interaction.

Due to the unprecedented growth of social-related data, boosted by the proliferation of social media websites and embedded heterogeneity and complexity; the traditional methods of network analysis has become less effective. Therefore, it is realized that the importance of *machine learning for social network mining* can be an inevitable tool in forthcoming years. Alongside the *machine learning* (ML) derives much effort from psychologists to build computational model for solving tasks like *recognition, prediction, planning,* and *analysis* even in uncertain situations. Therefore, it is significant to study the synergy of machine learning techniques in social network analysis, focus on practical applications, and open avenues for further research.

2 Representation of Network Data

Networks are represented using graphs and matrices. *Sociogram* developed by Jacob Moreno [1] is consider as the first representation of social network, where people are depicted by points and lines represent relationships among them. The graph based view using graph theory is one of the widely used structural network representation and visual analytical process. Different *colours, shades, symbols,* and *sizes* are used to specify different actor's properties or types. To indicate a different kind of relation the actors are linked by *dotted lines, bold lines, multi-lines, arrow headed lines indicating direction* etc. along with different *colours* and *line thikness*. These helps to understand superimposed relation among actors. To reduce overlapping sometimes ties are presented using curves instead of straight lines. In Fig. 4 sociogram of *Zachary's karate club* (dataset has been downloaded from *https://wiki.gephi.org/index.php?title=Datasets*) as described by Wayne Zachary in 1977 [2] is represented after data has been clustered using *Chinese Whispers Clustering Algorithm* [3], nodes are scaled according to their degree.

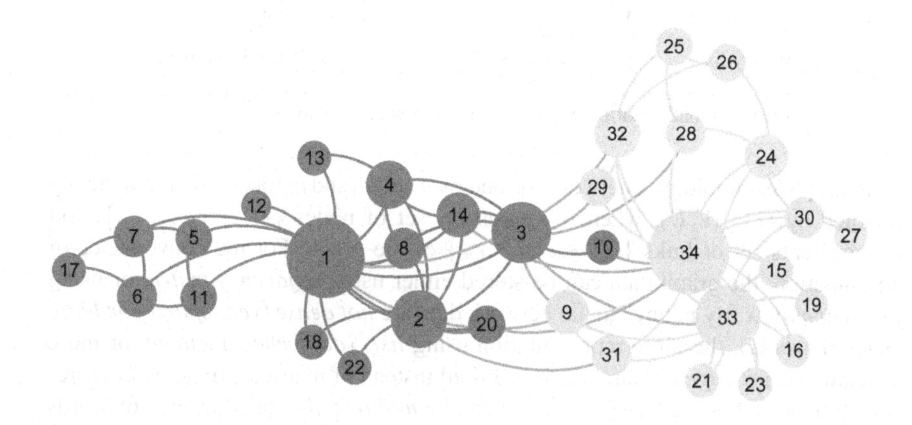

Fig. 1 Sociogram representation of *Zachary's karate club* as described in paper [2]. Data has been clustered by *Chinese Whispers Clustering Algorithm* [3], nodes are scaled according to their degree.

Several layout techniques (e.g. *force-directed layout, Spectral layout, Tree layout, Layered graph drawing, arc diagrams, Circular layout, Dominance drawing*) has been developed for efficient network visualization. [4, 5, 6] discussed several network layout techniques.

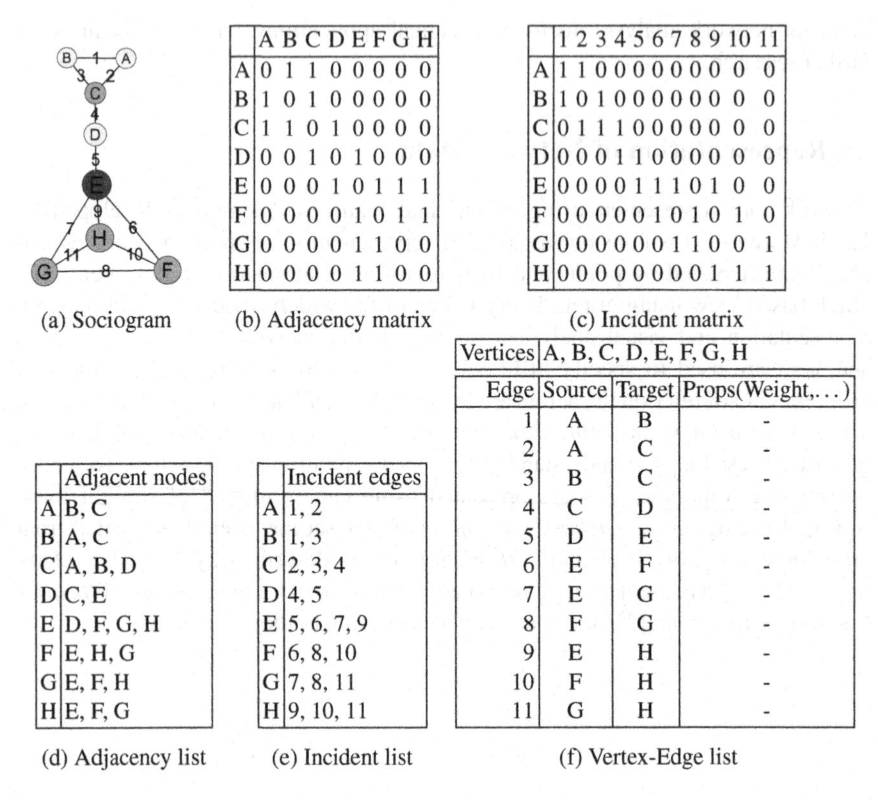

(a) Sociogram

	A	B	C	D	E	F	G	H
A	0	1	1	0	0	0	0	0
B	1	0	1	0	0	0	0	0
C	1	1	0	1	0	0	0	0
D	0	0	1	0	1	0	0	0
E	0	0	0	1	0	1	1	1
F	0	0	0	0	1	0	1	1
G	0	0	0	0	1	1	0	0
H	0	0	0	0	1	1	0	0

(b) Adjacency matrix

	1	2	3	4	5	6	7	8	9	10	11
A	1	1	0	0	0	0	0	0	0	0	0
B	1	0	1	0	0	0	0	0	0	0	0
C	0	1	1	1	0	0	0	0	0	0	0
D	0	0	0	1	1	0	0	0	0	0	0
E	0	0	0	0	1	1	1	0	1	0	0
F	0	0	0	0	0	1	0	1	0	1	0
G	0	0	0	0	0	0	1	1	0	0	1
H	0	0	0	0	0	0	0	0	1	1	1

(c) Incident matrix

	Adjacent nodes
A	B, C
B	A, C
C	A, B, D
D	C, E
E	D, F, G, H
F	E, H, G
G	E, F, H
H	E, F, G

(d) Adjacency list

	Incident edges
A	1, 2
B	1, 3
C	2, 3, 4
D	4, 5
E	5, 6, 7, 9
F	6, 8, 10
G	7, 8, 11
H	9, 10, 11

(e) Incident list

Vertices	A, B, C, D, E, F, G, H

Edge	Source	Target	Props(Weight,...)
1	A	B	-
2	A	C	-
3	B	C	-
4	C	D	-
5	D	E	-
6	E	F	-
7	E	G	-
8	F	G	-
9	E	H	-
10	F	H	-
11	G	H	-

(f) Vertex-Edge list

Fig. 2 Graphical and mathematical representations of social network

In graph terminology, a network formed by n actors and m links is mathemetically denoted by $G = (V, E)$, where V represents set of nodes $\{v_1, v_2, v_3, \ldots, v_n\}$, and E represents set of links $\{e_1, e_2, e_3, \ldots, e_m\}$ of the graph. For the convenience of computation, the graph data can be stored either using *adjacency matrix* or using *incident matrix*. By nature, social network data are *not dense* (*very sparse for large scale networks*), therefore representation using *lists (adjacency*, *incident*, or more specially *vertex-edge list)* are much favoured instead of matrices. In network representation, a tie that indicates a relation can be *undirected* (e.g. adjacency of words in literature) or *directed* (e.g. communication network; network of hyperlinked documents). A directed tie can be again *reciprocal*, i.e. two way relation (e.g. marital relation) or *non-reciprocal*, i.e, only one way relation (e.g. preferences of smart phones; voting in an election). In Fig. 2, a random network is represented using (a) Sociogram (nodes are colored according to their degree, and scaled according to their eigenvector centrality) (b) Adjacency matrix (c) Incident matrix (d) Adjacency list (e) Incident list and (f) Vertex-Edge list.

Beside actor properties, various relational properties such as *weight* (e.g., frequency of communication, strength of relation), *ranking* (e.g., first preferance, second preferance, third preferance etc.), *type* (e.g., friend, relative, co-worker), *properties depending on the structure of the rest of the network* (e.g., betweenness) are used to assign value to links.

Recently some advance models for network representation are becoming common in practice. These informative structures are efficient to handle large complex network and are much richer than old representations (holds not only information about relation but also keep track of other features) such as, *weighted matrix* ($V_{i,j} \leftarrow weight \forall i, j \in \{1..n\}$; or *edge* $\leftarrow$ {*source actor, destination actor, weight*}) is used to represent relation strength; *fuzzy cutoff* is used to reevolute relation strength. Characteristic function using crisp and fuzzy technique is presented in equation 1 and 2 respectively.

Crisp characteristic function

$$V'_{i,j} = \begin{cases} 1, & v_i \text{ is related to } v_j \\ 0, & v_i \text{ is not related to } v_j \end{cases} \quad (1)$$

Fuzzy characteristic function

$$V'_{i,j} = \begin{cases} 1, & v_i \text{ has strong positive relationship with } v_j \\ \gamma = f(V_{i,j}) \in]0,1[, & v_i \text{ is related to a certain extent with } v_j \\ 0, & v_i \text{ is not related to } v_j \end{cases} \quad (2)$$

The *gexf* file format (http://gexf.net/format/) supports network namespace awareness, hierarchical structure, dynamicity and easy implementation; *GML (Graph Modelling Language)*, *GTL (Graph Template Library)* based on *Standard Template Library* is another representation supporting arbitary graph, node, or edge information.

Due to enhancement of *Web 2.0*, online documents are created using *semantic web frameworks* consisting of a *graph model (RDF)*, a *query language (SPARQL)*, and *definition systems (RDFS and OWL)* to represent and exchange knowledge online. These frameworks provide rich structure than raw graph and enables researchers to take all web elements into consideration instead of only connectivity.

3 Different Aspects of Social Network Analysis

Social network analysis can be viewed as the mapping and measuring of relationships and flows between people, groups, organizations, computers or other information / knowledge processing entities. The analysis can tell us about the channel of information flow; how first things are moving across the network; who are the most important persons; which community or group of people is more effective as compared to others; how diseases spread; shopping habits; degree of influence etc.

In the analysis of complete network, a distinction can be made between -

- *Descriptive methods*, also using graphical representation.
- *Analysis procedures*, often based on decomposition of several matrices.
- *Statistical models* based on probability distribution.
- *Model based learning* using machine learning techniques.

Graphical representation by displaying a sociogram as well as a summary of graph theoretical concepts provides a first description of social network data. But when network become significantly larger, the visualization approach may not be efficient enough for deep analysis.

Some properties of social networks such as *degree, density, size, diameter, distance, geodesic, reachability, group, centrality* etc. are very important for any network knowledge mining. In the succeeding section we have presented different measures and properties of social network.

3.1 Different Measures of Social Network

3.1.1 Measures Based on Connection

Degree

Degree is the simplest measure which focuses on a single actor at a time and is estimated by number of links tapped with a vertex. For an undirected graph, degree of a vertex v_i, is the number of links connected with the vertex and is denoted by $deg(v_i)$. For the directed network the measure is further extedded as *in-degree* and *out-degree*. In-degree is defined as the number of incoming links incident to the vertex and out-degree as the number of links directed from the vertex. In equation 3, and 4 we have mathematically represented in-degree ($deg_{in}(v_i)$), out-degree ($deg_{out}(v_i)$) of vertex v_i.

$$deg_{in}(v_i) = \sum_{j=1}^{|V|} V_{ji} \quad (3)$$

$$deg_{out}(v_i) = \sum_{j=1}^{|V|} V_{ij} \quad (4)$$

In some analysis, the degree is measured by *degree sequence* or *degree distribution*. Degree sequence is an ordered list of degrees (in-, out- for directed) of each actor. Degree distribution is a frequency count of the occurrence of each degree. In some network, preferably World Wide Web (WWW) and some social networks are found to have degree distribution that follows *power law* [7]: $P(k) \sim k^{-\gamma}$, where γ is a constant. Such networks are called *scale-free network* and are being attracted for their structural and dynamic properties. *Log-scales* are used to plot power graph.

For the network presented in Fig. 3a. the degree sequences are as - In-degree sequence: [2, 2, 1, 1, 1, 1, 1, 0]; Out-degree sequence: [2, 2, 2, 1, 1, 1, 0, 0];

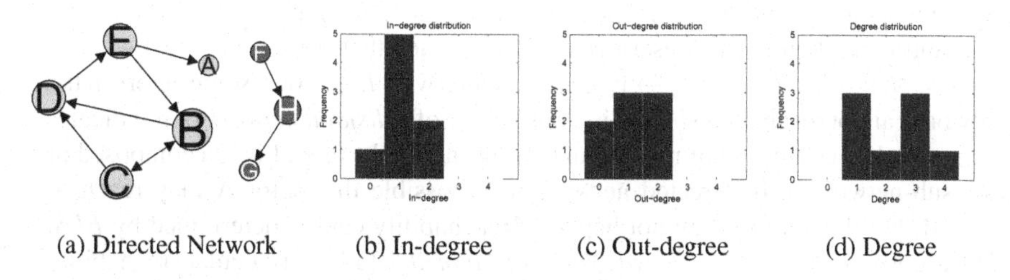

(a) Directed Network (b) In-degree (c) Out-degree (d) Degree

Fig. 3 Directed network and digree distributions

and Degree sequence (Undirected): [4, 3, 3, 3, 2, 1, 1, 1]. The degree distributions are - In-degree: [(2, 2), (1, 5), (0, 1)]; Out-degree: [(2, 3), (1, 3), (0, 2)]; degree (undirected) [(4,1), (3, 3), (2, 1), (1, 3)]. In Fig. 3b, 3c, and 3d we have shown in-degree, out-degree, and degree distribution for the network respectly.

Radiality is a measure based on degree analysis to which an individual's network reaches out to others in the network and provides degree of influence. Integration and radiality measuring the extent of an individual's connectedness proposed by Valente and Foreman [8].

Some research observed that *actors with large degree is more likely connected to the actors that also have large degrees*. This suggests that in social networks, the popular users' friends are also tend to be popular.

Density

Density (Δ) of a network of actors indicates cohesion of the network. Network density is useful to measure the health and effectiveness of the network. For a valued network, density is represented by actual number of connections of the network divided by maximum possible connection of the network. The growing density is a feature of individual's networks, not of network as a whole. [9] describs that *the network size and density do not correlate*. Inversly some other describe *Density of a network grow as network increase over time* [10]. Sometimes network density describes the portion of the potential connections in the network.

$$\begin{aligned}\Delta &= \frac{\text{Numer of actual connections}}{\text{Maximum possible connections}} \\ &= 2 * \frac{\sum_{i=1}^{|V|} deg(v_i)}{|V|(|V|-1)}, \quad \text{for undirected network} \\ &= \frac{\sum_{i=1}^{|V|} deg(v_i)}{|V|(|V|-1)}, \quad \text{for directed network}\end{aligned} \tag{5}$$

Reachability

Reachability indicates the connectedness of any two actors of the network regardless of number of hops required in between. Reachability indicates whether two actors of a network are connected or not in either a direct or indirect pathway of any length. Rechability mathemeticaly described as $R_{s,t} = 1$ such that $\exists(v_s \rightarrow v_1 \rightarrow v_2 \rightarrow \ldots \rightarrow v_k \rightarrow v_t)\ \forall i = \{1 \ldots k\}$ and $(v_i \rightarrow v_{i+1}) \in E$; Otherwise $R_{s,t} = 0$. If some actors in the network cannot reach others, there is a possibility of a *disjoint network* or it indicates the network is composed of more than one sub-networks (e.g., Fig. 3a composed of two sub-networks). In directed network it is possible that actor A may reach to actor B, but the reverse may not happen. Reachability can be determined by *BFS*, *DFS* algorithm [11, 12, 13]. *Floyd-Warshall Algorithm* [14] can calculate rechability matrix of all actors in $O(|V|^3)$ time. *Thorup's Algorithm* [15] (preprocess time of $O(|V|log|V|)$ to calculate datastructure of size $O(|V|log|V|)$), *Kameda's Algorithm* [16] (preprocess time of $O(|V|)$ to calculate datastructure of size $O(log|V|)$) can work faster to calculate rechability matrix.

Connectivity

Connectivity defines the social connection strength between two social elements and is measured by minimum number of nodes that would have to remove in order to make those elements unreachable from each other. Multiple pathway to reach two actors establish much *stronger connectivity* among them and confirms efficient delivery of information.

Bridge

Bridge in the form of a tie connects two networks and has a high importance of analysis due to capability of convey information across different network. Spread of information highly depends upon number of bridges available with an organization. Malcolm Gladwell in his book 'The Tipping Point' [17], characterizes people that habitually act as bridges as connectors. Mathematically, a bridge is an edge e, present in two sub-networks G_1 and G_2, and removing e make G_1 and G_2 disjoint. i.e.,

$$\begin{aligned} &\exists\, e \in E, \text{ such that,} \\ &G_1 \cup G_2 = G, \quad \text{when } \exists\, e \\ &G_1 \cap G_2 = \phi, \quad \text{when } \nexists\, e \end{aligned} \tag{6}$$

In social network a *local bridge* is a bridge, removing which the network will be connected but distance of end vertices will increase. A *local bridge of degree K* having the distance K between two end vertices, after removing the local vertices.

Reciprocity and Transitivity

In network analysis when a relation exists only in between two actors, is known as *reciprocal* or *dyadic relation*. In a network a very considerable portion of the relationships are reciprocal. In such a structure a good deal of unmediated pair-wise communication can occur. The reciprocal property might suggest a rather non-heretical structuring of network fields, and a field in which there may local and particular pair-wise relationship, rather than a single monolithic structure of establishment.

Triadic relations involving three actors are also important in network analysis. In triadic relation a common property known as *transitivity*, exists among actors. The transitivity principle holds that, if A is tied with B, and B is tied with C; then there is a high probability that A tied with C. The idea like balance and transitivity is that triadic relationships should tend toward transitivity as an equilibrium condition. *Balance theory* [18] is a notation of such relationship.

In Fig. 2a, {{A,B}, {A,C}, {B,C}, {C,D}, {D,E}, {E,F}, {E,G}, {E,H}, {F,G}, {F,H}, {G,H}} are in reciprocal relation and {{A,B,C}, {E,F,H}, {E,G,H}, {F,G,H}} are in triadic relation.

3.1.2 Measures Based on Size and Distance

Size, Diameter, & Distance

Size of a network is equivalent to the number of actors present in that network. The size of a network plays a crucial role in the analysis due to uses of limited resources and capacities (that is being used by actors for building and maintaining ties). Size of network is denoted by N and mathematically represented by $|V|$.

Sometimes *diameter* is also used to represent the size of a network. A diameter of a network is the length of the longest chain (chain in shortest path) between any two vertices of a network. It *measures the power of committee.*

Distance is the collective weights required to reach from one actor to another. Generally, in relational analysis distance indicates the number of actors or edges present between two actors which forms the shortest path.

In Fig. 2a, the size of the network is 8, whereas the diameter is 5 $\{A \to C \to D \to E \to H \to F\}$. Distance of $\{D,G\}$ is 2, $\{B,H\}$ is 4.

Paths and Walks

In a network, *walk* describes distance between two actors with precision. Traversal of alternate sequence of actors and links starting and ending with actors, in which each actor is incident with a tie is known as walk. Mathematically, a walk in a network is a finite sequence $W = (v_1 \to e_1 \to v_2 \to e_2 \to \ldots \to v_{k-1} \to e_k \to v_k)$ whose terms are alternatively actors and links such that, for $1 \leq i \leq k$, the links has ends v_i and v_{i+1}. The actor v_1 is called *origin* and v_k as the terminus of walk W. In a network, walk analysis is essential to know about network flow and connectivity of social elements.

A walk becomes a *path* when no actor is repeated and becomes a *trail* when no tie is repeated (vertex may repeat). Length of a path measured by the number of links associated to form that path.

For a path P in a graph $G = (V,E)$ having edge weights $\omega : E \to \mathbb{R}$, the weight of the path denoted by ω(P) is defined as the sum of the weights of the edges that makes P. A path from v_i to v_j in G is considered as *shortest path* (with respect to ω), if it's weight is least among all possible paths from v_1 to v_j. The length of a shortest path from v_i to v_j is called the *shortest-path distance between* v_i *and* v_j.

A *single source shortest path problem (SSSP)*, which finds shortest distance of all actors from a specific actor, is defined as follows: In G an actor $v_s \in V$ (the source), SSSP computes shortest paths from v_s to all other vertices in the graph. The problem is only well-defined if the graph does not contain a cycle of negative weight. When edge weights are non-negative, an SSSP such as *Dijkstra's algorithm* [19, 20] can find shortest path in $O(|E| + |V| log|V|)$ time. An efficient implementation such as *Bellman-Ford-Moore algorithm* [21] with time $O(|V||E|)$ can be use to handle cycle of negative length. BFS [11, 12] finds shortest path in linear time $O(|V| + |E|)$, when edge weights are unique.

All-pairs shortest paths problem (APSP), computes the shortest-path distances among all pairs of actors, provided that G does not contain a cycle of negative length. The *Floyd-Warshall algorithm* [22] can calculate the shortest-path matrix of size $|V| \times |V|$ in $O(|V|^3)$ time, or by $|V|$ SSSP computations in time $O(|V||E| + |V|^2 log|V|)$ using *Johnson's algorithm* [23].

These algorithms work for both directed and undirected graphs. Shortest path among two actors is also known as *geodesic*. In a values network there may present more than one geodesic. *Geodesic distance* is the length of the geodesic. *Eccentricity* of an actor v_i, is defined as the largest geodesic distance between v_i and all other actors in the network. The longest geodesic distance between all pair of vertices of a connected network is alternatively known as the *diameter* of the network.

3.1.3 Measure Based on Network Structure

Structural Similarity

Two distinct actors that may not be directly connected, but occupy a similar position in the structure are treated as *similar in structure*. As such they have similar interests in outcomes that related to positions in the structure. Similarity must be conditioned on visibility. The effect of similarity might be conditional on communication frequency. For example in Fig. 4. actor-1 and actor-34 are structurally similar as they act as center of two clubs. Structural similiraty alternatively known as *positional equivalence*.

Structural Cohesion

Structural cohesion is the measure of cohesion among social groups. Cohesion is associated with strong interpersonal ties like keenship or friendship. Members of a cohesive group are likely to be aware of each other's opinions, because information diffuses quickly within the group. Groups encourage (through balance) reciprocity and compromise. This likely increases the salience of opinions of other group members, over non-group members. Structural cohesion ($\alpha_{i,j}$) for a pair of actors can be calculated using equation 7 [24].

$$\alpha_{i,j} = 1 - (1 - \rho_{i,j}) \prod_{k=1}^{|V|} (1 - \rho_{i,k}\rho_{k,j}) \quad \forall i \neq j \neq k \tag{7}$$

where $\rho_{i,j} = {}^{V_{i,j}} / _{\Sigma V_{i,j}}$ is the relative strength of an interpersonal tie.

Structural Centrality

Centralization or *centrality indices* or *centrality* are *skew distribution*, which is meant to quantify an intuitive feeling that some actors or ties are central than others within the network. Centrality measures relative importance or power of a social element by assigning real values to it and can be interpreted as influence, prestige or control.

Some of centrality measures require to have connected network(s), otherwise computation will be an issue. As an example, disconnected network components create unreachable issue for shortest path(s) based centralities. This yields infinite distance for closeness centrality, and zero shortest path distance for betweenness centrality.

In any network, the *reachability* measure ranks network actors according to the number of neighbours, or the cost they require to reach all other vertices from it. These centralities are directly based on notion of distances within a graph, or on notion of neighbourhood. Measure of centrality can be local or global. Different centrality measures such as *Degree centrality, Betweenness centrality, Closeness centrality, Eccentricity centrality, Eigenvector centrality* are discussed below.

Degree Centrality

Degree centrality $C_D(v_i)$ is based on degree and is important to identify the structural centrality of an actor. For a directed network, degree centrality can be *in-degree centrality* or *out-degree centrality*.

$$\begin{array}{ll} \text{Degree centrality of a vertex } v_i & C_D(v_i) = deg(v_i) \\ \text{In-degree centrality} & C_{iD}(v_i) = D^-(v_i) \\ \text{Out-degree centrality} & C_{oD}(v_i) = D^+(v_i) \end{array} \tag{8}$$

Sometimes degree centrality of a directed network is the sum of in-degree and out-degree for the vertex v_i. It is considered as a local measure as it considers only neighbours. By degree centrality analysis it is possible to find out immediate importance of some actor. In-degree indicates prestige, popularity, admiration, leadership, give orders, like-dislike, trust or distrust; and out-degree indicates grievance of an actor. Ego (when a single actor gets attention for the analysis is known as ego) and alters (set of actors that are tied with ego) based analysis are based on degree centrality. A paper that is being cited by other papers has higher prestige. Calculation degree centrality takes $O(|V|^2)$ time in an average dence adjacency representation and $O(|E|)$ for sparse representation.

In equation 9, Freeman's general formula for network centralization which renges from 0 to 1 is presented.

$$C_D(G) = \frac{\sum_{i=1}^{|V|} [C_D(v^*) - C_D(v_i)]}{[(|V|-1)(|V|-2)]} \tag{9}$$

where $C_D(v^*)$ is the actor with highest degree centrality.

Betweenness Centrality

Betweenness centrality can be defined as capturing brokerage, and the number of pairs individuals would have to go through as an actor in order to reach one another in the minimum number of hops. Betweenness centrality qualifies the number of times an actor or tie acts as a bridge in all shortest path.

$$C_B(v_i) = \frac{j\langle k}{V_{jk}(v_i)/V_{jk}} \tag{10}$$

Where V_{jk} is number of geodesics connecting v_j and v_k, and $V_{jk}(v_i)$ is the number that actor v_i is on.

The normalized form of betweenness centrality for undirected network is as equation 11.

$$C'_B(v_i) = C_B(v_i)/[(|V|-1)(|V|-2)/2] \tag{11}$$

A *flow betweenness centrality* is the degree to which a node contributes the sum of maximum flow between all pairs of nodes (except itself).

Closeness Centrality

Closeness is based on the length of average shortest path between an actor and other actors in the network. For example finding a location for a shopping mall where the total distance in the region should be minimum for customers. This would make traveling more convenient for customers which would increase business. Hence the *closeness centrality defines the median of the network*. The main focus is to measure closeness of a social entity with other entities in a network. Entities having small distance are considered as more important than those having a higher total distance. Various closeness based measures have been developed, see for example [25, 26, 27, 28, 8, 29, 30].

$$C_C(v_i) = \frac{1}{\sum_{j=1}^{|V|} d(v_i, v_j)} \tag{12}$$

where $d(v_i, v_j)$ represents distance between actors v_i and v_j.

Eccentricity Centrality

Eccentricity centrality will give high centrality value to vertices that are at short maximum distances to every other reachable vertex. For example, a hospital is located at vertex $v_i \in V$. We denote the maximum distance from v_i to any random vertex v_j in the network, representing a possible incident, as the eccentricity $e(v_i)$ of v_i, where $e(v_i) = max\{d(v_i, v_j) : v_j \in V\}$. The problem of finding an optimal location can be solved by determining the minimum over all $e(v_i)$ with $v_i \in V$.

In graph theory, the set of vertices with minimal eccentricity is considered as the centre of G. The eccentricity centrality for isolated vertices is taken to be zero.

Image Hage and Harary [31] proposed a centrality measure based on the eccentricity which is presented in equation 13.

$$C_E(v_i) = \frac{1}{e(v_i)} = \frac{1}{max\{d(v_i, v_j) : v_j \in V\}} \tag{13}$$

This measure is consistent with general notion of vertex centrality, since $e(v_i) - 1$ grows if the maximal distance of v_i decreases. Thus, for all vertices $v_i \in V$ of the centre of $G : C_E(v_i) \geq C_E(v_j)$ for all $v_j \in V$.

Eigenvector Centrality

Eigenvector centrality is based on relative score and is a measure of influence of an actor in the network. The main concept of eigenvector is to find out how central an actor is, based on neighbours centrality within network. It uses concept such as, high scoring actors contribute more score for the actor. Eigenvector centrality C_V of actor v_i is calculated using equation 14.

$$C_V(v_i) = \frac{1}{\lambda} \left(\sum_{v_t \in V(v_i)} v_t \right) \tag{14}$$

where $V(v_i)$ is the set of nodes connected to v_i, and λ is a constant.

Google's Pagerank is a variant of eigenvector centrality which is calculated as equation 15.

$$PageRank(v_i) = \alpha \sum_{j}^{|V|} V_{i,j} \left(\frac{v_j}{\sum_j^{|V|} V_{i,j}} \right) + \frac{1-\alpha}{|V|} \tag{15}$$

Bonacich eigenvector centrality [32] tells that, an actor is more central when there are more connection within actor's local network otherwise the actor is more powerful.

Hubbell and Katz centrality for a directed network is another eigenvector centrality where each actor has a weight regardless of its network position [33, 34]. Mathemetically Katz centrality defined as

$$v_i = \sum_{k=1}^{\infty} \sum_{j=1}^{|V|} \alpha^k (V_{i,j})^k \quad (16)$$

where α is an attenuation factor in (0,1).
or, alternatively,

$$v_i = \alpha \sum_{j=1}^{|V|} V_{i,j}(v_i + 1) \quad (17)$$

Structural Hole

Structural holes is a conception proposed by sociologist Ronald Burt [35], which refer to the absence of ties between two parts of a network. In [36], Zhang et al. proposed a *generalized structural hole finding algorithm by bisection in social communities*. In "Social Network Analysis in Enterprise", Lin et al. [37] presented structural holes that measure the degree to which a person's network is redundant. They use equation 18 to calculate structural hole.

$$P_{i,j} = \frac{log(V'_{i,j})}{\sum_k log(V'_{i,k})}$$
$$StructuralHoles_j = 1 - \sum_j \left(P_{i,j} + \sum_q P_{i,q}P_{q,j}\right)^2, \quad q \neq i,j \quad (18)$$

where, $P_{i,j}$ is the normalized tie strength.

3.2 Essential Problems and Algorithms

Connected Components

A network is said to be *connected* if there exist at least one path between any two pairs of actors, otherwise the network is *disconnected*. When a *diode* (direct link) exists between any two pairs of actors, then the network is *strongly connected* and in sociology it is named as *clicks*. The set of actors that form a strongly connected component is known as *reciprocated*. For example in Fig. 2a. {A,B,C}, {E,F,G,H} are reciprocated. *Depth First Search (DFS)* or *Breadth First Search (BFS)* with a time complexity of O(V+E) is used to test connectivity of the network. The *connected component* of a network is the induced sub-network ($V' \leq V$) which is connected and maximal. A directed network is said to be *weakly connected* when the network is disconnected but, the undirected form is connected (i.e. every node can be reachable from every other node by following links irrespective of direction). Connectivity of the directed network can be tested with a modified DFS proposed by Robert Tarjan [13].

In the directed graph presented in Fig. 3a; the strongly connected components are

$$\{B,C,D,E\}, \{A\}, \{G,H\}, \{F\}$$

and weakly connected components are

$$\{A,B,C,D,E\},\{F,G,H\}$$

When the largest component encompasses a significant fraction of the network, it is called the *giant component*. In Fig. 2a, vertx set {E, F, G, H} makes the giant compmonent for the network.

Graph k-Connectivity

An graph $G = (V,E)$ is called *k-vertex-connected,* if $|V| > k$ and $G - X$ is connected for every $X \subset V$ with $|X| < k$. The vertex-connectivity called k of G is the largest integer such that G is k-vertex-connected. Similarly, a network is *k-edge-connected* if $|V| > 1$. G-Y is connected for every $Y \subset V$ with $|Y| < k$. The edge connectivity of G is the largest integer k such that G is k-edge-connected. A 0-connected graph is a disjoint graph and 1-connected is connected.

Network Flow

Network flow is a directed network $G = (V,E)$, where each link has a non-negative capacity and direction of flow. Flow represented by two distinct vertices $v_s, v_t \in V$ designated as the source and the target respectively. A flow from $v_s \rightarrow v_t$, or an *s-t-flow* for short, is a function f satisfying the following constraints:

Capacity constraints: $\forall e \in E : 0 \leq f(e) \leq u(e)$.
Balance conditions: $\forall v \in V : \{v_s, v_t\}$.

The problem of computing a flow having maximum value is called the *max-flow problem*. The max-flow problem can be solved in time $O(|V||E|log(|V|^2/|E|))$ using Goldberg and Tarjan algorithm [38]. A *cut* is a partition $(V_S, V_S') \in V$ into two nonempty subsets V_S and V_S'. A cut (V_S, V_S') is an *s-t-cut*, for $v_s, v_t \in V$, if $v_s \in V_S$ and $v_t \in V_S'$. The *capacity of a cut* is the sum of the capacities of the links with origin in V_S and destination in V_S'. A *minimum s-t-cut* is an s-t-cut whose capacity is minimum among all s-t-cuts. It is easy to see that the value of an s-t-flow can never be larger than the capacity of an s-t-cut. In theory of network, a classical result flow states that the maximum value and the minimum capacity are the same. It also has been observed that maximum s-t flow value is equal to the capacity of a minimum s-t-cut. The problem of computing a flow of minimum cost such as Ford-Fulkerson [39, 40] algorithm can be solved in polynomial time. The maximum flow approach focuses on the vulnerability or redundancy of connection between pair of actors - kind of a *"strength of the weakest entities"*. One notation of how totally connected two actors are, i.e., how many different actors in the neighbourhood of a source lead to pathways to a target. The flow approach suggests that the strength of my tie to you is no stronger than the weakest link in the chain of connections, where weakness means a lack of alternatives.

4 Statistical Models for Social Network Analysis: Probability and Random Walks

Since 1970's the major direction in network research was to model probabilities of the relational ties between interacting units (actors), though in the beginning only very small groups of actors were considered. Extensive introduction to earlier methods is provided by Wassman and Faust [41]. *Correlation analysis, Markov Random Fields (MRFs)* introduced by Frank and Strauss [42] and *Exponential Random Graphical Models (ERGMs)*, also known as p^* [43] are most popular in statistical modeling. Snijders et al. [44] extended the ERGM in order to achieve robustness in the estimated parameters. In statistical literature, a network is defined as consisting $|V|$ number of actors and information about binary relations between them. There are several problems with existing models such as *degeneracy analysed* by [45] and *scalability* mentioned by several sources [46, 47]. In [44] a new ERGM based specification attempts to handle unstable likelihood by different parameterization of the models. Freeman's general formulation for centralization can use gini coefficient of standard distribution.

Clustering Coefficient

A *clustering coefficient* is a measure of degree to which nodes in a network tend to cluster together i.e. it is a property of a node in network which indicates the measure of connectivity of neighbourhood nodes from a particular node. A higher clustering coefficient always indicates a greater cliquishness. Clustering coefficient of an actor is represented as the probability of two randomly selected neighbours connected to each other.

Probabilistic Network Model

In some research, probability based model is used to simulate random graph which performs exactly equivalent of real social network. *Random graph models (hypothesized model)* are also used for simple representation of complex network. Using probability model it is possible to derive properties mathematically and to predict properties and outcomes in a simulated environment.

Erdös-Renyi's Random Graphs

Beside number of nodes ($|V|$), *Erdös-Renyi's simplest random graph model* uses probability of any two nodes that share an edge (P) or total number of edges in the graph (M). In the probabilistic model, *degree distribution can be calculated using binomial distribution as networks are formed with diodes*. Therefore, probability that an actor has K degree can be calculated with a binomial distribution (Poission when limit P small; Normal limit large N) as expressed in equation 19.

$$\begin{aligned} P_K &= B(|V|-1;K;P) && \textit{Binomial} \\ &= \binom{|V|-1}{K} P^K (1-P)^{|V|-1-K} \\ P_K &= \frac{Z^K e{-Z}}{K!} && \textit{Poisson} \\ P_K &= \frac{1}{\sigma\sqrt{2\pi}} e^{-\frac{(K-Z)^2}{2\sigma^2}} && \textit{Normal} \end{aligned} \tag{19}$$

With an constant probability P of any two actors of a network are being connected, it is observed that - if the network size increases then the average degree also increases.

With P, the mean of the network or *average degree* can be calculated using equation 20 and *Variance in degree* by equation 21.

$$z = (|V|-1) \times P \tag{20}$$

$$\delta^2 = (|V|-1) \times P \times (1-P) \tag{21}$$

Power Law Networks

Some large scale network distributions such as distribution of World Wide Web uses a *Heavy-Tailed distributions* (where extreme values are still quite common). These networks are known as *power law networks*. They can be modeled using power functions. Pareto distribution is one such example. Power law networks are considers as *scale-free networks* as the slope of the qdistribution is same through the distribution.

5 Groups and Substructures in Social Networks

One of the most common interest of structural analysis is to find the *substructures* that may present in a network. Many of the approaches to understand the structure of network emphasize on how dense connections are compounded and extended to develop larger *cliques* or *sub-groupings*. This process is generally known as community detection in network.

A *community detection* is the process of detecting group/sub-group in the network. In social media there are two types of groups are available - *Explicit Group* (formed by user subscription) and *Implicit Group* (implicitly formed by social interaction). Roughly community detection methods can be divided into 4 categories (but not exclusive).

Node centric community. Each node in the group satisfies certain criterion.
criterion Group centric community. Consider the connections within a group as a whole. The group has to satisfy certain criterion without zooming to node level.
Network centric community. Partitions the whole network into several disjoint sets.
Hierarchy centric community. Constructs a hierarchical structure of communities.

In table 1, we have presented various techniques for identifying groups and sub-groups in network. Details of these techniques can be found in [48, 49].

Table 1 Techniques for detecting group and sub-groups in social network

Complete mutuality	
Clique	Group of actors who have all possible ties among themselves. Finding clique in a complete network is a NP hard problem.
Reachability of members	
K-clique	Actors are connected to every members of the group at a maximum geodesic distance of K.
K-clans	K-cliques that all paths among members occur by the way of other members of K-clique.
K-club	A sub-structure of diameter K.
Nodal degrees	
K-plex	K-plex is a minimal sub-network where each actor of the induced sub-network is connected to at least N-K other vertices, where N is size of the induced sub-graph.
K-core	Actors are connected to k of members of the group
Relative frequency of within outside ties	
LS set	Set of actors whose proper subsets has more ties to its component within the group than outside.
Lambda set	Set of actors who if disconnected, would most greatly disrupt the flow among all of the actors.
Component	Parts of sociogram that are connected within but disconnected with other components.
Cut points	Nodes which if removed, the structure will be divided into disconnected systems.
Block	The divisions into which cut points divided a group.

5.1 Bipartite Cores - Identifying Communities

If two disjoint subset of actors $V_L, V_R \in V$ of a network $G = (V, E)$ can be partitioned into two non-empty disjoint sebsets in such a way that, $V_L \cup V_R = V$ and $V_L \cap V_R = \emptyset$ and each links $e \in E$ has one end in V_L and another in V_R, then G is a *bipartite network.* Bipartite networks are much informative for identifing communities. For an instance, in a conference network (bipartite as people and conferences can partitioned into two distinct subsets) if A and B attended condference C - then there is a high chance that A knows B, or their nature of work is similar, or they share some common properties.

6 Machine Learning Based Approaches for Networked Data Mining

Traditional network researches rely on binary or weighted relation among different social entities, and always ignore underlying reason for making relation. In this approach, it is not possible to analyse different attributes (behaviours/properties of an element) in a single analysis. For a research decomposed of several social properties, each property requires individual analysis. Hence, in-depth analysis become much complicated with traditional approach. For instance, in a collaboration network, it is absolutely easy to find different communities using community detection algorithms or by visualization. But finding the real cause (e.g. nature of job, subject of work, location, gender, age, etc. or probability of importance of different attributes) behind community formation and understanding community dynamics is significantly challenging. It also needs enough direct human interaction in different phase of analysis. Hence, presence and capability of domain expert is almost important. Alongside the multi-behavioural analysis, the large data volume also works as an additive complexity for networked knowledge mining.

Machine learning consists of several mathematical, statistical, and computational techniques that has significant capability of finding patterns from historical data and are capable for *descriptive* and *predictive analysis*. In the last decards ML based techniques are heavily used in both industry and research. They are quite capable of performing even in uncertain situations. They are able to *model, classify, group* and *predict* data more accurately. ML is very effective for *large scale, multivariate* and *fuzzy analysis*. ML has become one of the widely used information processing framework.

Recently people realized that the implication of ML and related techniques for the network analysis can be rich and this realization opens up a new prospective in network research. The revolution started with multi behavioural analysis to large scale processing to parallel execution for the mining of complex social and media data. Researchers started finding different *groups or clusters, real cause of community formation*, on basis of which they started *modelling network, location-aware social network mining* and perform *business prediction*. Various ML based techniques are used for *recommendation, online advertising, market analysis and prediction, human resource and organizational collaboration*, and many more.

In the below section, we have discussed several machine learning based network mining and knowledge exploration methods including steps of network data preparation for executing machine learning techniques. We also have presented some machine learning based findings.

6.1 Social Analytical Data Preparation with Machine Learning

Data logs about user activities from heterogeneous platform such as *Online Social Networks (OSNs), Web Logs, Content Management Systems (CMSs), Tracking systems (such as GPS, Call logs), Mobile and desktop apps*, need to be integrated. Activity logs from these sources can be available in different formats, e.g., *relational data tables, flat files, HTML DOM, XML*, or *JSON*. Therefore, *data preprocessing* is essential for making collected data suitable for analysis and undergo into several steps. In broad view, preprocessing consists of *data cleaning, transformation, selection*, and *integration*. Sometimes these steps involve iterative phases to reach ultimate structural and optimized format for improving data quality.

Data Cleaning

Data cleaning (also known as *data cleansing* or *scrubbing*) consists of set of methods for removing noise or unusual tuples, and fixing missing values in the dataset to improve the quality of data. For example, in the process of preprocessing social data regarding preference of electronic gadgets and brand value, it is essential to segregate advertisements (unless and until they have no direct impact on analysis) from collected data, otherwise advertisements may lead to inadequate outcomes. Sometimes generative models such as *Clean Model (MC), Noise Model (MN), Corruption matrix (MR)* [50, 51] are used for machine learning based cleaning. Due to large number of possible corruption matrices (i.e. $O(2^{(|V|\times D)})$, where D is number of attributes), *hill-climbing method* is used to optimize search. *EM based Gaussian mixture model* [52] is used to deal with missing data. *Bayesian method* is used for cleaning data as discussed in [53].

Transformation

Dealing with multi-source problems requires restructuring of schemas to achieve a *schema integration*, including steps such as *splitting, merging, folding* and *unfolding attributes and tables. Conflicting representations* need to be resolved at instance level and overlapped data must be addressed.

Data transformations [54] are needed to support any changes in the *structure, representation* or *content of data*. Data format may vary from source to source, while they are collected from heterogeneous sources. Therefore merging them into a common format may need data transformation. For example, the data value 'India', 'Ind', or 'IN' can be representative value for India. Sometimes there are symbolic differences present in data values or there may be value differences due to *typo* or *misspelling*. For instance, the attribute qualification may contain 'PhD', 'Ph.D.', 'Ph D', 'MSC', 'M.SC', ' MSc', 'MS' as values. Although these data values show 7 different data classes, it actually belongs to two different qualifications. Clustering algorithms (e.g., *Key collision with keying functions* like *fingerprint, ngram-fingerprint, metaphone3, cologne-phonitic*; *nearest neighbour with distance algorithm* such as *levenshtein, PPM*) [55] can be applied to group data and then transform them to distinct sets. Some analysis attribute values need

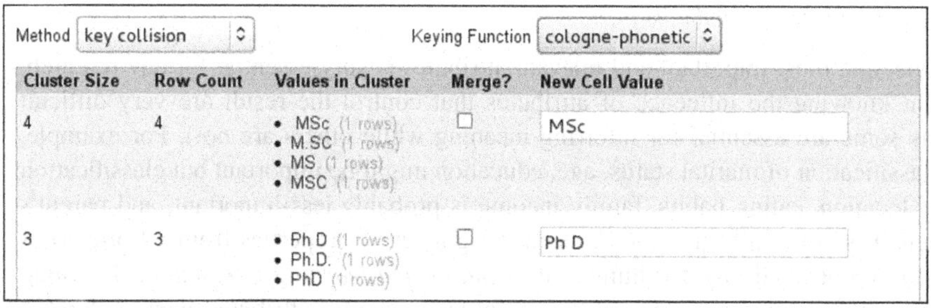

Method key collision   Keying Function cologne-phonetic

Cluster Size	Row Count	Values in Cluster	Merge?	New Cell Value
4	4	• MSc (1 rows) • M.SC (1 rows) • MS (1 rows) • MSC (1 rows)	☐	MSc
3	3	• Ph D (1 rows) • Ph.D. (1 rows) • PhD (1 rows)	☐	Ph D

Fig. 4 Using Google refine, above example data are clustered for transformation and to assign each clusters a uniform label

to be transformed to parent class. For example, 'Apple iPhone', 'LG Nexus 4', 'Samsung galaxy', 'Moto X' can be transformed to 'Smart Phone', while 'Apple iPad', 'ASUS Nexus 7' can be transformed as 'Tablet PC'. Some research groups focus on data transformation technique which is based on *schema matching* [56, 57]. To determine complete or partial matches, a *Fuzzy matching (approximate join)* that finds similar records based on a matching rule becomes necessary, e.g., specified declaratively or implemented by a user-defined function [58, 59]. A matching rule may state that persons are likely to be similar if name and portions of their address match. The measure of similarity between two instances, often calculated by a numerical value between 0 and 1, usually depends on application characteristics. For string constituent, (e.g., customer and company name) exact matching and fuzzy approaches based on *wildcards, character frequency, edit distance, keyboard distance* and *phonetic similarity (soundex)* are useful [58, 60, 61]. More complex string matching approaches consider abbreviations as are presented in [62]. Common information retrieval approaches use both string and text data matching. WHIRL is an example of such category which uses *cosine distance in the vector-space model* for determining the degree of similarity between text elements [63].

A set of transformation steps has to be specified and executed to resolve various *schema-* and *instance-level data quality* problems that are reflected in the data sources at hand. For multisource data, *structure transformation* or *standardization* is important for the equality. For example, those sites which use date format as their convenience, they may represent date in MM-DD-YYYY, YYYY/MM/DD, DD-MM-YY etc. format; Gender as Male/Female, M/F or 1/2; Relation as binary (0/1) or weighted (some numerical value). Structure transformation like *attribute split* has been adapted for preprocessing social data.

Language Transformation is equally important as structure transformation for the appropriate language for analysis. Otherwise, important tuples may be ignored due to lack of understanding.

Selection & Integration

Transformed data need to be select and integrate them in common store (wirehouse). The selection & integration can be (i) *Attribute Selection*, and (ii) *Instance Selection*.

Selecting Useful Attributes

Selecting most important and relevant attributes is very essential for any research. But knowing the influence of attributes that control the result are very difficult (as some are essential for inferring meaning while others are not). For example, classification of marital status, age, education might be important but classification of location, eating habits, family income is probably less-important, and parent's height is not at all important. Selecting right set of properties from all properties is a complicated task for human, thus *automated learning is essential.* Learning by *correlation matrix* is one of the useful technique for attribute selection. Such a matrix of size $|D| \times |D|$ represents correlation of each variable pair, where $|D|$ is the number of distinct properties of a social entity. The *correlation coefficient* between variable i and variable j is represented in cell (i, j). The correlation (ρ_{D_i,D_J}) between properties D_i and D_j is calculated as -

$$\rho_{D_i,D_J} = \frac{1}{1-n}\left[\frac{(D_i - \mu_{D_i})(D_j - \mu_{D_j})}{\sqrt{\left(\delta^2_{D_i}\right)\left(\delta^2_{D_j}\right)}}\right] \tag{22}$$

where, n is the number of samples in the dataset, μ_{D_i} and μ_{D_j} denotes the mean of property D_i and D_j respectively. From the correlation matrix, properties having at least some threshold will be selected as useful. Other methods like *principal component analysis (PCA)* [64] also used for selecting attribute.

Instance Selection

A number of authors focused on the problem of *duplicate identication* and *elimination*, e.g., [58, 65, 60, 61, 66, 62].

Filter and wrapper [67, 68] are two core aspects of instance selection. *Filter* evaluates data reduction but doesn't consider activities. On contrary, *wrapper* approaches explicitly emphasize the ML aspect and evaluate results by using the specific ML algorithm and *inlier* (a data value that lies in the interior of a statistical distribution and error) to trigger instance selection. RT [69] is a *distance based* and LOF [70] is for *density based outlier* detection and cleaning approach. Multivariate data cleaning is more difficult, but is essential in a complete analysis [71, 72]. *Wrapper based* approach by Brodley and Friedl [73] identifies and eliminates mislabelled instances to improve quality of training data. In first step, they try to learn candidate's instances using "m learning algorithms" to tag correctly or incorrectly labelled instances. Then they build a classifier to remove mislabelled instances. Filtering can be based on one or more of the m base level classifiers' tags. [74] highlights instance selection mechanism by *optimizing sample size* while maintaining quality of data. The reduction technique enables a learning algorithm to work more effectively. According to the experiments by Oates and Jensen [75], *decision tree based instance reduction* is used before knowledge mining. They also have demonstrated that, as the amount of data grows, the rate of increase in accuracy slowsdown. A *consolidative framework* developed by Reinartz [76] uses an application of statistical sampling technique to draw initial sample, followed by clustering techniques to create subsets of instances by grouping similar kind of instances. For each of these subsets, the prototyping step selects or constructs a smaller set of representative prototypes. These set of prototypes constitute the final output of instance selection.

To deal with missing data [77], domain experts need to choose techniques such as: Ignoring Instances with unknown feature values; Replace with feature's modal value; replace with most common feature value (feature value, which occurs the most within the same class); Mean substitution; Regression or classification methods (Missing value prediction based on available data by regression or classification model); Hot deck imputation (replaces value with identical instances); Treating missing values as special values (e.g., "unknown" is treated as a new value for the features those contain missing values).

Tools such as *Open Refine* (formally, Google Refine) is one of the best suited tool which implements machine learning techniques for online and offline data collection and cleaning. Other tools such as *ETL tools, D-Dup, WizRule, data-mining suite* also perform well.

6.2 Machine Learning for Networked Data Mining

Methods of networked data mining using machine learning techniques can be broadly classified as -

1. Descriptive modelling
   a. Network analysis
   b. Group/community detection
2. Predictive modelling
   a. Link prediction
   b. Attribute prediction

Clustering, Classification, and *Regression* techniques can be employed for *unsupervised, semi-supervised* or *supervised network analysis*. Based on the type of samples, specific learning technique can be applied. For example, *classification for categorical datasets, regression for numerical data*, and *clustering for unlabelled data*.

Clustering Algorithm for Community Detection

Machine Learning based Community detection using Clustering technique can be divided into 4 primary group -

1. Hierarchical clustering
2. Partitional clustering

3. Distribution-based clustering
4. Density-based clustering

Hierarchical Clustering

Hierarchical clustering technique based on connectivity is one of the highly reliable method for finding previously unknown communities within a network. The learning algorithm divides the network into a hierarchy of groups and subgroups according to a specific weight function. The data can be then represented according to the level of hierarchy (closer items are linked together) in a tree like structure known as *dendrogram*. The hierarchical clustering is most appropriate when background of the network data such as number of communities are previously unknown and groups of most similar actors need to be identified according to the convenience of researcher. Apart from the *distance functions*, different linkage criterion such as *single-linkage, complete-linkage*, or *average-linkage* (Unweighted pair group method with arithmetic mean) function can be used for hierarchical clustering. Furthermore, hierarchical clustering can either be agglomerative or divisive. The *agglomerative technique* proceeds by adding closer actors to the dendrogram, while *divisive technique* works by removing links and creating divisions. The Girvan-Newman algorithm which is based on divisive technique, detects communities in complex systems. In Fig. 5 (taken from 'Hierarchical structure and the prediction of missing links in networks' by Clauset et al. [78]), we have represented communities with dendrogram. As this technique does not provide unique partition of clusters, experts having domain knowledge are essential for choosing appropriate clusters. In general, building a dendrogram takes $O(|V|^3)$ time, that makes learning process very slow for large scale network analysis.

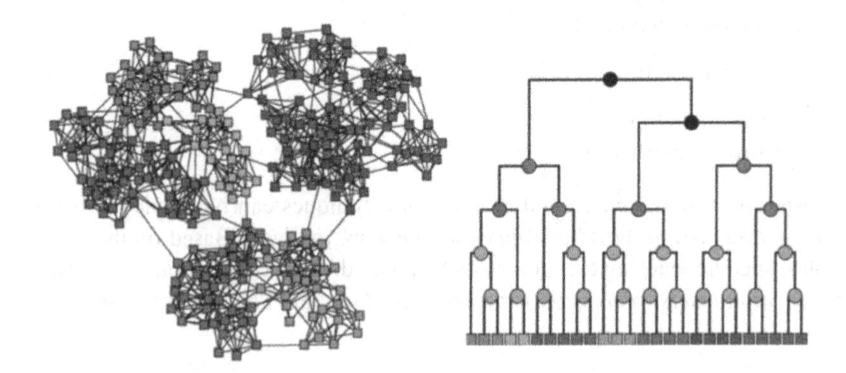

Fig. 5 A hierarchical network with structure on many scales, and the corresponding hierarchical random graph. (Image has been taken from [78])

Partitional Clustering

Partitional or *Centroid based clustering* such as *k-means, k-medians, k-medoids, k-means++, fuzzy c-means* uses a centre vector to find clusters from data sets. The big disadvantage of these learning technique are:

1. *Most of the algorithms require number of partitions - K, as an input to the algorithm*
2. *The partitions are based on initial selection of centres and tend to converge on locally optimum solutions.*

Genetic algorithms are used along with existing techniques to achieve a *global optimum solution*. Centroid based clusters partition the data space into a structure known as a *"Voronoi diagram"*. *Fuzzy c-means* technique allows an actor to be present in multiple clusters (soft assignment) with a membership value. To test the best cluster, *cluster validity indices* are used.

Distribution Based Clustering

Distribution based clustering technique is based on statistical distribution models. Clusters are created with objects belong to same distribution. Unless constraints are put on model, the technique suffers from *overfitting*. Fixed number of distributions are used to overcome this problem. For example, *Gaussian mixture model* (using the expectation-maximization algorithm) uses a fixed number of Gaussian distributions which are initialized randomly. These algorithms also suffer from local optimum solutions, hence multiple execution may require for better solution. Along with the complex clustering model these techniques capture correlation and dependence of attributes.

Density Based Clustering

In *Density based clustering*, clusters are defined with high dense areas. They are useful to separate anomalies and community boundaries. Density based algorithms such as *Density-based spatial clustering of applications with noise (DBSCAN), EnDBSCAN, Ordering Points to Identify the Clustering Structure (OPTICS), SE-QOPTICS, Partitioning Around Medoids (PAM), Clustering LARge Applications (CLARA), DeLi-Clu, density-link-clustering, EM clustering, HiSC, ERiC,* [79, 80, 81] are heavily used machine learning technique which are applied in the field of social network analysis. *Clustering Large Applications based on RANdomized Search (CLARANS)* [80], *Balanced Iterative Reducing and Clustering using Hierarchies (BIRCH)* [82, 83], *canopy clustering, subspace clustering, correlation clustering, CLIQUE* [84] and *Subspace Clustering (SUBCLU)* [85] algorithms are developed to support recent craze towards *big data*.

Classification

To deal with large graphs in the context of online social networks, a subset of actors may need to be labeled. The labels may indicate demographic values, interest, belief or other characteristics. ML based classifiers such as *Decision Tree, Nearest Neighbor, Bayesian Classification, Artificial Neural Networks, Support Vector Machine (SVM)* etc. [86, 87, 88] can propagate lable for instances. Classifiers are used to model network data from historial dataset, so that they can be able to classify new instances correctly. Models like Decision Tree generates a rule base where each arm of the tree defines a rule; whereas other classifiers generate mathemetical model. Aside the traditional classifiers, semi-supervised learning or transductive learning methods based on random walk in graph such as *Path Ranking Algorithm (PRA)* introduced by Lao and Cohen [89], *Markov chain* [90], *Factor Graph Model* [91] are much suitable for network mining and dependency modeling.

Predictive Modelling

Statistical regression models are used to estimate relationship among entities and for prediction too. Both linear and nonlinear models are useful for network modeling. *Estimation and inference methods* from the field of statistical relational learning have been successfully applied in single-network domain. *Collective learning models* improve accuracy in prediction as they are using correlation between class labels. *Collective Inference Models (CIM)* are used for relational classification as they make simultaneous judgement about same variable for a set of related data instances [92]. *Guilt-by-association model* [93], is a particularly simple CIM that uses average probability of neighbors nodes. In equation 23 the particular model function is presented.

$$P(D_i = C|N_i) = \frac{1}{z} \sum_{v_j \in N_i} W_{i,j}.P(D_j = C|N_j) \tag{23}$$

where N_i is the set of neighbours of D_i for random variable $D_1, D_2, \ldots, D_n$.

Markov Random field (MRF) [94] is popular for modelling partially labelled network with a random field. They are suitable with undirected network and are similar to Bayesian Network in representation and dependencies. Therefore, it can be calculated by equation 24.

$$P(D_i = d_i|D_j = d_j; i \neq j) = P(D_i = d_i|N_i) \tag{24}$$

where N_i is the set of neighbours of D_i for random variable $D_1, D_2, \ldots, D_n$.

Network model generation using *Probabilistic Relational Model (RBN, RMN, AMN, RDN)* and *Probabilistic Logic Model (BLP, MLN)* are useful. Other estimation techniques such as *MCMC* (e.g., *Gibbs sampling* [95]), *Iterative classification* [96], *Relaxation labelling* [97], *Loopy belief propagation* [98], *Graph cut methods* [99] are also important in network analysis. These estimation can be a maximum posteriori joint probability distribution of all free variables or marginal distributions for some or all parameters related to some likelihood scoring.

For predicting link and attribute following catagories can be used -

Link Prediction

1. Network-only Bayesian classifier (nBC), Multinomial naïve Bayes on the neighboring class labels.
2. Network-only link-based classifier, logistic regression based on a node's "distribution" of neighboring class labels, DN(v_i) (multinomial over classes)
3. Relational-neighbor classifier (weighted voting) equation
4. Relational-neighbor classifier (class distribution) equation

Attribute Prediction

1. More general attribute correlation
    a. Correlation with network structure
    b. Correlation among different attribute of related entities
2. Learning models of network data
    a. Disjoint learning
    b. Joint learning
3. Methodology issues
    a. Potential biases due to network structure and autocorrelation
    b. How to evaluate models
    c. How to understand model performance

6.3 Some Highlited Research and Findings Using Machine Learning

Domingos et al. [100] applied social network analysis based data mining techniques for identifying potential customers for the melioration of marketing. They have used network value as an influential characteristics instead of intrinsic value of the customer. In their model, customers are conceived as actors of the network and influences of one customer on all other customers is modelled as a Markov Random Field. Using MRF, they calculated the probability of the event that a user buys a product, given that some marketing actions are taken (e.g. discount is offered). They tested their method on a collaborative filtering database for marketing motion pictures named EachMovie.

With an assumption that people like their friends, Johnson et al. [101] inferred a hidden attribute of an actor by investigating their friends attribute values. With the inferred attribute they were able to predict quite accurately, whether a Facebook user is homosexual by looking at friend's gender and sexuality. The consequence of the research furnished that friends-circle convey a prominent quantity of information about users within network.

Using Bayesian networks, He et al. [102] modelled casual relations among people in social networks. They studied the impact of prior probability, influence strength, and society openness to the inference accuracy on a real OSN. The Bayesian network is constructed to infer value of a given attribute of a given actor, where each node corresponds to an actor represents its attribute(s); while friendship among those actors are represented using links. They also considered multiple hop inference.

Using modified Naïve Bayes learning technique, Lindamood et al. [103] predict privacy sensitive trait information with actor traits and link structure. To increase level of privacy, they proposed removal of K most representative attributes from the network, and L most representative links from each actor.

More tightly connected actors of a social network can be clustered into communities. It is observed that actors from the same community share same properties, based on that hypothesis actor's attribute inference methods can be built. Mislove et al. [104] extrapolated attributes based on global community detection algorithm. They used normalized mutual information metric for measuring the proximity of community structures. Their main contribution is the attribute inference based on a new local community detection algorithm, which uses the metric normalized conductance. They used two Facebook datasets for testing and succeeded to infer user attributes with high accuracy in certain cases, depending on the strength of the community.

Using Semi-Supervised Learning (SSL) technique Mo et al. [105] proposed a model that uses fewer labelled samples and relatively large amount of unlabelled samples. The model is perfectly suited for network data as they contain less quantity with much hidden information (as publicly available information are not always complete). They have provided two specific attack models: (a) a graph-based attack model, based on local and global consistency; and (b) a co-training model consisting of a graph based SSL algorithm for learning link structure, and a supervised learning algorithm for personal information found in attributes.

In [106] Mo et al. presented another attack model using graph-based SSL. In this model, specification of community consistency was added along with the specifications of local and global consistency as presented in [105]. They discovered strongly connected communities (cliques) by applying clustering algorithm. Taking all these three consistency into consideration, they also presented a closed-form and an iterative learning algorithm.

Thomas et al. [107] examined how insufficient group privacy control uncovers sensitive information. They aggregated information disclosed from heterogeneous actors in group and able to infer attributes such as gender, relationship status (single label classification), political and religious views, favourite medias, books (multi label classification). They proposed two classifiers - (a) Multinomial logistic regression for single label classifications, and linear regression for multi label classifications, which are based on actor's relationship information. (b) The other classifier is based on Facebook wall content, in which word frequencies are used as input of a multinomial logistic regression. They concluded to ensure multi-party privacy, Facebook's privacy model can be adapted.

In social networks, people subscribe to their likes such as news, blog etc.; join groups such as group of certain music band, fan club of a certain person, etc. These activities highlight one's involvement and interests. Zheleva et al. [108] used such important subscription based information and using the concept of homogeneity within a group, they inferred hidden attribute values of actors. They proposed several methods, and the method using Support Vector Machine (SVM) proven to be one of the best technique. Alike of Zheleva et al.'s classification only model, Kótyuk and Buttyan in [109] considers regression along with classification. They also used each user attribute as neural network input and a correlation-matrix-based method, which allows automatic variable selection. Thus Kótyuk and Buttyan's model using machine learning techniques supports automated inference system allowing both classification and regression.

In [110], Fang and Qian designed a platform oriented framework for popular science study based on social computing. For data acquisition and for search engine, they included features like subject search and wiki technique in the framework. They have used linearly separable model and SVM classifier with non-linear decision function as described in equation 25. Improved lucence Score Mechanism as presented in equation 26 is used to improve indexing by integrating social factors into it.

$$M(x) = sgn\left(\sum_{i=1}^{l} a_i^* y_i f(x, x_i) + b^*\right) \tag{25}$$

$$Score(t,d) = Score(t,d) \times s(t) \times sociality(d.etime, d.click_rate, d.enum) \tag{26}$$

Particularly, s(t) indicates the variance of the scores that users give to the search results; sociality(d.etime, d.click_rate, d.enum) indicates the social threshold of a document; d.etime is the last time that a term was edited; The JSPWiki engine is used to improve scientific quality of students and promote quality-oriented education.

In [111], Amershi et al. presented ReGroup as an end-user interactive machine learning system for helping people by creating custom groups in online social networks as demanded. Suggestions of ReGroup's group members are considered by a Naïve Bayes classifier, where each friend is represented by a set of feature-value pair. Then probability of each friend being a member of the group 'g' is computed via following Bayesian formulation:

$$\begin{aligned} P(g|d_1, d_2, \ldots, d_n) &= \frac{P(d_1, d_2, \ldots, d_n|g)P(g)}{\sum_{g'=g,g} P(d_1, d_2, \ldots, d_n|g)P(g)} \\ P(d_1, d_2, \ldots, d_n|g) &= \prod_{i=1}^{n} P(d_i|g) \end{aligned} \tag{27}$$

where, $P(d_1, d_2, \ldots, d_n|g)$ is the likelihood of a friend with feature values $d_1, d_2, \ldots, d_n$ being a member of g, $P(g)$ is the prior probability of any friend belonging to g and the denominator is the probability of seeing the set of feature values in the data and serves as a normalizing constant.

They have addressed unlearnable groups, typically the result of missing information.

In [112], Velardi et al. describes a new content-based model for social network analysis. In this paper instead of relation they focus on communicative content. They applied text mining and clustering technique to gather content of communication as emergent semantics of social network and to identify most popular compositions.

In [113] Pinheiro and Helfert assorted Artificial Neural Network (ANN) and Social Network Analysis to improve customer loyalty. The neural network model is suitable to establish the likelihood assigned to churn from the historical data, but insufficient to calculate influence of customer. Social network approach has been applied to enhance the knowledge related to the customer influence in internal community. Using SNA companies are able to model user's behaviour and finds influential users. Finally they have built a combined model to measure customer's loyalty for telecommunication network.

In [114], Zhou developed a prototype of trust based recommendation system, where connections and trust between users are extracted and obtained using social network analysis.

In [115], Bauer et al. investigated collaboration from user's behaviour. They worked on Wikipedia edited history data set to study whether people act more randomly in their editing behaviour while they are interacting with others in user and article talk pages.

Wang et al.'s showed that an individual's behavior is more random while they are interacting. Entropy [116] is a way to measure randomness. Systems with higher entropy are less predictable. Entropy is given by:

$$\sum p(d_i) lob_b(p(d_i)) \tag{28}$$

Conditional entropy [117] is a variation where entropy is computed separately for different situations and then combined into a single value. Conditional entropy is given by:

$$H_A^2(i|j) = -\sum_{j=1}^{M_A} P_A(j) \sum_{j=1}^{M_A} P_A(i|j) log P_A(i|j) \tag{29}$$

Recent availability of open mobile platforms (e.g., Apples iPhone, Google's android, and Microsoft's windows mobile) and location aware online social network data logs (e.g., address in user profile; location of IP for web access logs) makes GPS and location data handy to the researchers. The location-based social network data (LSN-data) and activity stream are useful for spatial and temporal analysis. For example, purchasing nature of sudden geometry from time-to-time and modelling them for prediction; copying nature of cultures among societies - e.g., middle class society try to follow higher class society.

Zhang et al. [118] presented a spatial-temporal based neighbourhood detection algorithm. They used HoodSquare for modelling and recommending neighbours in location based social network, where a similarity metric is used to assess the homogeneity of a geographic area, and a geographic navigation to detect the boundaries of a city's neighborhoods. In first step, they employed OPTICS algorithm [79] for finding high relative density based clusters with spatial feature. Then they devised a similarity metric of homogeneity across the city's geographic points. To measure similarity among cells, authors calculated homogeneity indices for each cell and named it as H_Index. For each cell $p \in Grid$, they flagged all the cells within a certain radius r of that cell by creating a set of neighboring cells $N_{p,r}$. Then, they calculated the H_Index value with respect to p and r, which is formally defined as:

$$H\_Index(p,r) = \frac{\sum_{n\ inN_{p,r}} cos(d_p,d_n) \times smooth(n)}{|N_{p,r}|} \tag{30}$$

where $cos(d_p,d_n)$ is the cosine similarity between vectors of the two grid cells p and $n \in N_{p,r}$, and smooth(n) is a 2-d isotropic Gaussian smoothing function of size r^2 centered at p. They calculated similarity with respect to weights for all grid cells $n \in N_{p,r}$. Thereafter, they used an iterative method to increase threshold value and calculated pass list as well as group list using the threshold value. They grouped the points within a certain distance.

As described in [119], Li and Chen collected data about profiles, activity updates, mobility characteristics, social graphs, and the correlation among different attributes. They studied that the degree distribution of collected data also follows power law, and high-degree users are likely to have more numbers of mobile, and post updates frequently. Using X-Means [120] clustering algorithm, they classified users into four groups (home, home-vacation, home-work and all other users) according to the user's mobility pattern. Using expectation-maximization (EM) as the clustering algorithm, which supports maximum likelihood estimates of parameters in probabilistic models, they invented 5 groups (inactive, normal, active, mobile, trial). To predict users' movement, they used Markov-based location predictor developed by Song and Kotz [121]. The median accuracy of the prediction was 49% and there were 17% users (who post most updates in a single place) whose location could be predicted with at least 99% accuracy.

With 100 mobile phones with customized softwares, the Reality Mining project at MIT studied mobile social systems. These software recorded call history to identify social links, and bluetooth scans to calculate geographical proximity [122].

Using an online opiqnion forum and a location based social networking service, Wang et al. [117], presented that people act more randomly when working as part of a group.

In [123], Zheng et al. used social network collaborative filtering (SNCF) for recommendation. In that paper they have shown purchased based SNCF (where they first select user-base using SN, and then apply CF for recommendation), proximity-based SNCF, and a hybrid model based on both techniques (where the non-zero purchase based scores are boosted in a linear combination with the proximity-based scores). They employed the concept of similar purchase pattern of other users for recommendation.

In [124] Backstrom and Leskovec presented a supervised model for predicting and recommending links in social networks. The supervised random walk model combines information from network structure with actor and link level characteristic.

In [125] Benchettara et al. used supervised ML for link prediction in bipartite graph. They were inspired from (a) Product recommendation in an e-commerce site, which used bipartite graph and (b) Academic collaboration recommendation. Li and Chen [126] proposed a graph kernel-based machine learning approach for recommendation as link prediction in bipartite graph.

By extending Kleinberg's greedy routing procedure Ratti and Sommer [127] proposed an efficient algorithm that effictively calculate shortest path in spatial social network. Experimental evaluations on social network(s) obtained from real-world mobile and landline phone communication data demonstrate that such adaptations can efciently compute accurate estimate for shortest-path distance.

In [128] Stofa and Michalik focused on conversational content in the context of safety in social networks. They analyzed interpersonal communicational content and style based approach for the identification of author.

Charabarti et al. [129] uses Markov Random Field for web-page classification; Domingos and Richardson [130] uses for viral marketing; Pandit et al. in [131] uses for eBay auction fraud. In the first part of the paper [132] by Brunelli and Fedrizzi, the natural connection between fuzzy relations and social network analysis is shortly discussed and justified. After the introductory part, attention is drawn towards the natural extension of relations to the m-dimensional case. Thus, fuzzy m-adjacency relations are introduced and their link with fuzzy binary relations is formalized. In order to estimate m-ary relations, OWA operators appear as valuable tools.

7 Conclusion

In this chapter, we have discussed important concepts of social network analysis and use of machine learning based techniques for network knowledge mining. In the initial section, we have discussed core network analysis concepts, measures, and techniques using visualization, graph-matrix based and statistical analysis. In the later section, we have discussed machine learning based network data preparation and relational knowledge mining. Although thousands of work on basic network mining has been developed using structural analysis, we have highlighted significant technological differences and importance of machine learning based network analysis over traditional network analysis. We have presented the relevance of machine learning based analysis in the current scenario towards accuracy, effectiveness in unknown situations, and large scale data processing. We have shown how to learn model of a network using clustering, classification and to use these models for attribute-link prediction. Some of the machine learning based findings are also presented in the last part of the chapter.

References

1. Moreno, J.L.: Emotions mapped by new geography. New York Times 3, 17 (1933)
2. Zachary, W.W.: An information flow model for conflict and fission in small groups. Journal of Anthropological Research, 452–473 (1977)
3. Biemann, C.: Chinese whispers: an efficient graph clustering algorithm and its application to natural language processing problems. In: Proceedings of the First Workshop on Graph Based Methods for Natural Language Processing, pp. 73–80. Association for Computational Linguistics (2006)
4. McGuffin, M.J.: Simple algorithms for network visualization: A tutorial. Tsinghua Science and Technology 17(4), 383–398 (2012), doi:10.1109/TST.2012.6297585
5. Crawford, C., Walshaw, C., Soper, A.: A multilevel force-directed graph drawing algorithm using multilevel global force approximation. In: 2012 16th International Conference on Information Visualisation (IV), pp. 454–459 (2012), doi:10.1109/IV.2012.78
6. Wikipedia: Graph drawing – wikipedia the free encyclopedia (2013) (Online: accessed September 11, 2013)
7. Faloutsos, M., Faloutsos, P., Faloutsos, C.: On power-law relationships of the internet topology. In: ACM SIGCOMM Computer Communication Review, vol. 29, pp. 251–262. ACM (1999)
8. Valente, T.W., Foreman, R.K.: Integration and radiality: measuring the extent of an individual's connectedness and reachability in a network. Social Networks 20(1), 89–105 (1998)
9. Kunegis, J.: On the spectral evolution of large networks. Ph.D. thesis, Koblenz, Landau (Pfalz), Univ., Diss. (2011)
10. Bakshy, E., Karrer, B., Adamic, L.A.: Social influence and the diffusion of user-created content. In: Proceedings of the 10th ACM Conference on Electronic Commerce, EC 2009, pp. 325–334. ACM, New York (2009), `http://doi.acm.org/10.1145/1566374.1566421`, doi:10.1145/1566374.1566421
11. Yoo, A., Chow, E., Henderson, K., McLendon, W., Hendrickson, B., Catalyurek, U.: A scalable distributed parallel breadth-first search algorithm on bluegene/l. In: Proceedings of the ACM/IEEE SC 2005 Conference on Supercomputing, p. 25. IEEE (2005)
12. Korf, R.E., Schultze, P.: Large-scale parallel breadth-first search. In: AAAI, vol. 5, pp. 1380–1385 (2005)
13. Tarjan, R.: Depth-first search and linear graph algorithms. SIAM Journal on Computing 1(2), 146–160 (1972)
14. Hougardy, S.: The floyd–warshall algorithm on graphs with negative cycles. Information Processing Letters 110(8), 279–281 (2010)
15. Thorup, M.: Compact oracles for reachability and approximate distances in planar digraphs. Journal of the ACM (JACM) 51(6), 993–1024 (2004)
16. Kameda, T.: On the vector representation of the reachability in planar directed graphs. Information Processing Letters 3(3), 75–77 (1975)
17. Gladwell, M.: The tipping point: How little things can make a big difference. Hachette Digital, Inc. (2006)
18. Feld, S.L.: The focused organization of social ties. American Journal of Sociology, 1015–1035 (1981)
19. Dijkstra, E.: A note on two problems in connexion with graphs. Numerische Mathematik 1(1), 269–271 (1959), `http://dx.doi.org/10.1007/BF01386390`, doi:10.1007/BF01386390

20. Crauser, A., Mehlhorn, K., Meyer, U., Sanders, P.: A parallelization of dijkstra's shortest path algorithm. In: Brim, L., Gruska, J., Zlatuška, J. (eds.) MFCS 1998. LNCS, vol. 1450, pp. 722–731. Springer, Heidelberg (1998)
21. Bellman, R.: On a routing problem. Quart. Appl. Math. 16, 87–90 (1958)
22. Floyd, R.W.: Algorithm 97: Shortest path. Commun. ACM 5(6), 345 (1962), `http://doi.acm.org/10.1145/367766.368168`, doi:10.1145/367766.368168
23. Johnson, D.B.: Efficient algorithms for shortest paths in sparse networks. J. ACM 24(1), 1–13 (1977), `http://doi.acm.org/10.1145/321992.321993`, doi:10.1145/321992.321993
24. Friedkin, N.E.: Structural bases of interpersonal influence in groups: A longitudinal case study. American Sociological Review, 861–872 (1993)
25. Sabidussi, G.: The centrality index of a graph. Psychometrika 31(4), 581–603 (1966)
26. Bavelas, A.: Communication patterns in task-oriented groups. The Journal of the Acoustical Society of America 22(6), 725–730 (1950)
27. Beauchamp, M.A.: An improved index of centrality. Behavioral Science 10(2), 161–163 (1965)
28. Moxley, R.L., Moxley, N.F.: Determining point-centrality in uncontrived social networks. Sociometry, 122–130 (1974)
29. Nieminen, U.: On the centrality in a directed graph. Social Science Research 2(4), 371–378 (1973)
30. Botafogo, R.A., Rivlin, E., Shneiderman, B.: Structural analysis of hypertexts: identifying hierarchies and useful metrics. ACM Transactions on Information Systems (TOIS) 10(2), 142–180 (1992)
31. Hage, P., Harary, F.: Eccentricity and centrality in networks. Social Networks 17(1), 57–63 (1995)
32. Bonacich, P.: Some unique properties of eigenvector centrality. Social Networks 29(4), 555–564 (2007)
33. Katz, L.: A new status index derived from sociometric analysis. Psychometrika 18(1), 39–43 (1953)
34. Borgatti, S.P.: Centrality and network flow. Social Networks 27(1), 55–71 (2005)
35. Burt, R.S.: The social structure of competition. Networks and Organizations: Structure, Form, and Action, 57–91 (1992)
36. Zhang, E., Wang, G., Gao, K., Zhao, X., Zhang, Y.: Generalized structural holes finding algorithm by bisection in social communities. In: 2012 Sixth International Conference on Genetic and Evolutionary Computing (ICGEC), pp. 276–279 (2012), doi:10.1109/ICGEC.2012.98
37. Lin, C.Y., Wu, L., Wen, Z., Tong, H., Griffiths-Fisher, V., Shi, L., Lubensky, D.: Social network analysis in enterprise. Proceedings of the IEEE 100(9), 2759–2776 (2012), doi:10.1109/JPROC.2012.2203090
38. Goldberg, A.V., Tarjan, R.E.: A new approach to the maximum-flow problem. Journal of the ACM (JACM) 35(4), 921–940 (1988)
39. Dantzig, G.B., Ford, L.R., Fulkerson, D.R.: A primal-dual algorithm for linear programs. Linear Inequalities and Related Systems (38), 171–181 (1956)
40. Ford, L.R., Fulkerson, D.R.: A simple algorithm for finding maximal network flows and an application to the Hitchcock problem. Rand Corporation (1955)
41. Wasserman, S.: Social network analysis: Methods and applications, vol. 8. Cambridge University Press (1994)
42. Frank, O., Strauss, D.: Markov graphs. Journal of the American Statistical Association 81(395), 832–842 (1986)

43. Wasserman, S., Pattison, P.: Logit models and logistic regressions for social networks: I. an introduction to markov graphs andp. Psychometrika 61(3), 401–425 (1996)
44. Snijders, T.A., Pattison, P.E., Robins, G.L., Handcock, M.S.: New specifications for exponential random graph models. Sociological Methodology 36(1), 99–153 (2006)
45. Handcock, M.S., Robins, G., Snijders, T.A., Moody, J., Besag, J.: Assessing degeneracy in statistical models of social networks. Tech. rep., Working paper (2003)
46. Hoff, P.D., Raftery, A.E., Handcock, M.S.: Latent space approaches to social network analysis. Journal of the American Statistical Association 97(460), 1090–1098 (2002)
47. Smyth, P.: Statistical modeling of graph and network data. In: IJCAI Workshop on Learning Statistical Models from Relational Data, Citeseer (2003)
48. Tang, L., Liu, H.: Community detection and mining in social media. Synthesis Lectures on Data Mining and Knowledge Discovery 2(1), 1–137 (2010)
49. Yang, Q.: Community detection and graph-based clustering. Powerpoint Presentation (2010)
50. Kubica, J., Moore, A.: Probabilistic noise identification and data cleaning. In: Third IEEE International Conference on Data Mining, ICDM 2003, pp. 131–138 (2003), doi:10.1109/ICDM.2003.1250912
51. Rahm, E., Do, H.H.: Data cleaning: Problems and current approaches. IEEE Data Engineering Bulletin 23 (2000)
52. Ghahramani, Z., Jordan, M.I.: Supervised learning from incomplete data via an em approach. In: Advances in Neural Information Processing Systems, Citeseer, vol. 6 (1994)
53. Schwarm, S., Wolfman, S.: Cleaning data with bayesian methods. Final project report for CSE574, University of Washington (2000)
54. Rundensteiner, E.: Special issue on data transformation. IEEE Techn. Bull. Data Engineering 22(1) (1999)
55. Morris, T., Verlic, M.: Clustering in depth: Methods and theory behind the clustering functionality in openrefine (2013), `https://github.com/OpenRefine/OpenRefine/wiki/Clustering-In-Depth`
56. Abiteboul, S., Cluet, S., Milo, T., Mogilevsky, P., Siméon, J., Zohar, S.: Tools for data translation and integration. IEEE Data Eng. Bull. 22(1), 3–8 (1999)
57. Milo, T., Zohar, S.: Using schema matching to simplify heterogeneous data translation. In: VLDB, Citeseer, vol. 98, pp. 24–27 (1998)
58. Galhardas, H., Florescu, D., Shasha, D., Simon, E.: Declaratively cleaning your data using ajax. In: Journees Bases de Donnees, Citeseer (2000)
59. Hellerstein, J.M., Stonebraker, M., Caccia, R.: Independent, open enterprise data integration. IEEE Data Eng. Bull. 22(1), 43–49 (1999)
60. Hernández, M.A., Stolfo, S.J.: Real-world data is dirty: Data cleansing and the merge/purge problem. Data Mining and Knowledge Discovery 2(1), 9–37 (1998)
61. Li Lee, M., Lu, H., Ling, T.-W., Ko, Y.T.: Cleansing data for mining and warehousing. In: Bench-Capon, T.J.M., Soda, G., Tjoa, A.M. (eds.) DEXA 1999. LNCS, vol. 1677, pp. 751–760. Springer, Heidelberg (1999)
62. Monge, A.E., Elkan, C., et al.: The field matching problem: Algorithms and applications. In: KDD, pp. 267–270 (1996)
63. Cohen, W.W.: Integration of heterogeneous databases without common domains using queries based on textual similarity. In: ACM SIGMOD Record, vol. 27, pp. 201–212. ACM (1998)
64. Jolliffe, I.: Principal component analysis. Wiley Online Library (2005)
65. Galhardas, H., Florescu, D., Shasha, D., Simon, E.: Ajax: an extensible data cleaning tool. ACM SIGMOD Record 29(2), 590 (2000)

66. Monge, A.E.: Matching algorithms within a duplicate detection system. IEEE Data Eng. Bull. 23(4), 14–20 (2000)
67. Jankowski, N., Grochowski, M.: Comparison of instances seletion algorithms II. Algorithms survey. In: Rutkowski, L., Siekmann, J.H., Tadeusiewicz, R., Zadeh, L.A. (eds.) ICAISC 2004. LNCS (LNAI), vol. 3070, pp. 598–603. Springer, Heidelberg (2004)
68. Jankowski, N., Grochowski, M.: Comparison of instances seletion algorithms I. Algorithms survey. In: Rutkowski, L., Siekmann, J.H., Tadeusiewicz, R., Zadeh, L.A. (eds.) ICAISC 2004. LNCS (LNAI), vol. 3070, pp. 598–603. Springer, Heidelberg (2004)
69. Knorr, E.M., Ng, R.T.: A unified notion of outliers: Properties and computation. In: KDD, pp. 219–222 (1997)
70. Breunig, M.M., Kriegel, H.P., Ng, R.T., Sander, J.: Lof: identifying density-based local outliers. In: ACM Sigmod Record, pp. 93–104. ACM (2000)
71. Hadi, A.S.: Identifying multiple outliers in multivariate data. Journal of the Royal Statistical Society. Series B (Methodological), 761–771 (1992)
72. Rocke, D.M., Woodruff, D.L.: Identification of outliers in multivariate data. Journal of the American Statistical Association 91(435), 1047–1061 (1996)
73. Brodley, C.E., Friedl, M.A.: Identifying mislabeled training data. arXiv preprint arXiv:1106.0219 (2011)
74. Liu, H., Motoda, H.: Feature selection for knowledge discovery and data mining. Springer (1998)
75. Oates, T., Jensen, D.: The e ects of training set size on decision tree complexity. In: Proceedings of the Fourteenth International Conference on Machine Learning, Citeseer (1997)
76. Reinartz, T.: A unifying view on instance selection. Data Mining and Knowledge Discovery 6(2), 191–210 (2002)
77. Lakshminarayan, K., Harp, S.A., Samad, T.: Imputation of missing data in industrial databases. Applied Intelligence 11(3), 259–275 (1999)
78. Clauset, A., Moore, C., Newman, M.E.: Hierarchical structure and the prediction of missing links in networks. Nature 453(7191), 98–101 (2008)
79. Ankerst, M., Breunig, M.M., Kriegel, H.P., Sander, J.: Optics: ordering points to identify the clustering structure. ACM SIGMOD Record 28(2), 49–60 (1999)
80. Ng, R.T., Han, J.: Clarans: A method for clustering objects for spatial data mining. IEEE Transactions on Knowledge and Data Engineering 14(5), 1003–1016 (2002)
81. Chen, Y., Reilly, K., Sprague, A., Guan, Z.: Seqoptics: a protein sequence clustering system. BMC Bioinformatics 7(suppl. 4), S10 (2006)
82. Zhang, T., Ramakrishnan, R., Livny, M.: Birch: an efficient data clustering method for very large databases. In: ACM SIGMOD Record, vol. 25, pp. 103–114. ACM (1996)
83. Zhang, T., Ramakrishnan, R., Livny, M.: Birch: A new data clustering algorithm and its applications. Data Mining and Knowledge Discovery 1(2), 141–182 (1997)
84. Duan, D., Li, Y., Li, R., Lu, Z.: Incremental k-clique clustering in dynamic social networks. Artificial Intelligence Review 38(2), 129–147 (2012)
85. Kailing, K., Kriegel, H.P., Kröger, P.: Density-connected subspace clustering for high-dimensional data. In: Proc. SDM, vol. 4 (2004)
86. Lotte, F., Congedo, M., Lécuyer, A., Lamarche, F., Arnaldi, B., et al.: A review of classification algorithms for eeg-based brain–computer interfaces. Journal of Neural Engineering 4 (2007)
87. Kotsiantis, S.B., Zaharakis, I., Pintelas, P.: Supervised machine learning: A review of classification techniques (2007)

88. Han, J., Kamber, M., Pei, J.: Data mining: concepts and techniques. Morgan Kaufmann (2006)
89. Lao, N., Mitchell, T., Cohen, W.W.: Random walk inference and learning in a large scale knowledge base. In: Proceedings of the Conference on Empirical Methods in Natural Language Processing, pp. 529–539. Association for Computational Linguistics (2011)
90. Jaakkola, M.S.T., Szummer, M.: Partially labeled classification with markov random walks. Advances in Neural Information Processing Systems (NIPS) 14, 945–952 (2002)
91. Xu, H., Yang, Y., Wang, L., Liu, W.: Node classification in social network via a factor graph model. In: Pei, J., Tseng, V.S., Cao, L., Motoda, H., Xu, G. (eds.) PAKDD 2013, Part I. LNCS, vol. 7818, pp. 213–224. Springer, Heidelberg (2013)
92. Jensen, D., Neville, J., Gallagher, B.: Why collective inference improves relational classification. In: Proceedings of the Tenth ACM SIGKDD International Conference on Knowledge Discovery and Data Mining, pp. 593–598. ACM (2004)
93. Benjamin, R., Parham, P.: Guilt by association: Hla-b27 and ankylosing spondylitis. Immunology Today 11, 137–142 (1990)
94. Rozanov, Y.A.: Markov random fields. Springer (1982)
95. Carter, C.K., Kohn, R.: On gibbs sampling for state space models. Biometrika 81(3), 541–553 (1994)
96. Besag, J., York, J., Mollié, A.: Bayesian image restoration, with two applications in spatial statistics. Annals of the Institute of Statistical Mathematics 43(1), 1–20 (1991)
97. Rosenfeld, A., Hummel, R.A., Zucker, S.W.: Scene labeling by relaxation operations. IEEE Transactions on Systems, Man and Cybernetics (6), 420–433 (1976)
98. Murphy, K.P., Weiss, Y., Jordan, M.I.: Loopy belief propagation for approximate inference: An empirical study. In: Proceedings of the Fifteenth Conference on Uncertainty in Artificial Intelligence, pp. 467–475. Morgan Kaufmann Publishers Inc. (1999)
99. Delong, A., Boykov, Y.: A scalable graph-cut algorithm for nd grids. In: IEEE Conference on Computer Vision and Pattern Recognition, CVPR 2008, pp. 1–8. IEEE (2008)
100. Domingos, P., Richardson, M.: Mining the network value of customers. In: Proceedings of the Seventh ACM SIGKDD International Conference on Knowledge Discovery and Data Mining, pp. 57–66. ACM (2001)
101. Johnson, C.Y.: Project 'gaydar': An mit experiment raises new questions about online privacy. Boston Globe (2009)
102. He, J., Chu, W.W., Liu, Z.V.: Inferring privacy information from social networks. In: Mehrotra, S., Zeng, D.D., Chen, H., Thuraisingham, B., Wang, F.-Y. (eds.) ISI 2006. LNCS, vol. 3975, pp. 154–165. Springer, Heidelberg (2006)
103. Lindamood, J., Heatherly, R., Kantarcioglu, M., Thuraisingham, B.: Inferring private information using social network data. In: Proceedings of the 18th International Conference on World Wide Web, pp. 1145–1146. ACM (2009)
104. Mislove, A., Viswanath, B., Gummadi, K.P., Druschel, P.: You are who you know: inferring user profiles in online social networks. In: Proceedings of the Third ACM International Conference on Web Search and Data Mining, pp. 251–260. ACM (2010)
105. Mo, M., Wang, D., Li, B., Hong, D., King, I.: Exploit of online social networks with semi-supervised learning. In: The 2010 International Joint Conference on Neural Networks (IJCNN), pp. 1–8. IEEE (2010)
106. Mo, M., King, I.: Exploit of online social networks with community-based graph semi-supervised learning. In: Wong, K.W., Mendis, B.S.U., Bouzerdoum, A. (eds.) ICONIP 2010, Part I. LNCS, vol. 6443, pp. 669–678. Springer, Heidelberg (2010), http://dx.doi.org/10.1007/978-3-642-17537-4_81

107. Thomas, K., Grier, C., Nicol, D.M.: unfriendly: Multi-party privacy risks in social networks. In: Atallah, M.J., Hopper, N.J. (eds.) PETS 2010. LNCS, vol. 6205, pp. 236–252. Springer, Heidelberg (2010)
108. Zheleva, E., Getoor, L.: To join or not to join: the illusion of privacy in social networks with mixed public and private user profiles. In: Proceedings of the 18th International Conference on World Wide Web, pp. 531–540. ACM (2009)
109. Kotyuk, G., Buttyan, L.: A machine learning based approach for predicting undisclosed attributes in social networks. In: 2012 IEEE International Conference on Pervasive Computing and Communications Workshops (PERCOM Workshops), pp. 361–366 (2012), doi:10.1109/PerComW.2012.6197511
110. Fang, W., Qian, M.: Design of a platform of popular science education based on social computing. In: International Conference on Computational Science and Engineering, CSE 2009, vol. 4, pp. 897–902. IEEE (2009)
111. Amershi, S., Fogarty, J., Weld, D.: Regroup: Interactive machine learning for on-demand group creation in social networks. In: Proceedings of the SIGCHI Conference on Human Factors in Computing Systems, pp. 21–30. ACM (2012)
112. Velardi, P., Navigli, R., Cucchiarelli, A., D'Antonio, F.: A new content-based model for social network analysis. In: 2008 IEEE International Conference on Semantic Computing, pp. 18–25. IEEE (2008)
113. Pinheiro, C.A.R., Helfert, M.: Mixing scores from artificial neural network and social network analysis to improve the customer loyalty. In: International Conference on Advanced Information Networking and Applications Workshops, WAINA 2009, pp. 954–959. IEEE (2009)
114. Zhou, L.: Trust based recommendation system with social network analysis. In: International Conference on Information Engineering and Computer Science, ICIECS 2009, pp. 1–4. IEEE (2009)
115. Bauer, T., Garcia, D., Colbaugh, R., Glass, K.: Detecting collaboration from behavior. In: 2013 IEEE International Conference on Intelligence and Security Informatics (ISI), pp. 13–15. IEEE (2013)
116. Shannon, C.E.: A mathematical theory of communication. ACM SIGMOBILE Mobile Computing and Communications Review 5(1), 3–55 (2001)
117. Wang, C., Huberman, B.A.: How random are online social interactions? Scientific Reports 2 (2012)
118. Zhang, A.X., Noulas, A., Scellato, S., Mascolo, C.: Hoodsquare: Modeling and recommending neighborhoods in location-based social networks. arXiv preprint arXiv:1308.3657 (2013)
119. Li, N., Chen, G.: Analysis of a location-based social network. In: International Conference on Computational Science and Engineering, CSE 2009, vol. 4, pp. 263–270. IEEE (2009)
120. Pelleg, D., Moore, A.W., et al.: X-means: Extending k-means with efficient estimation of the number of clusters. In: ICML, pp. 727–734 (2000)
121. Song, L., Kotz, D., Jain, R., He, X.: Evaluating location predictors with extensive wi-fi mobility data. In: Twenty-Third Annual Joint Conference of the IEEE Computer and Communications Societies, INFOCOM 2004, vol. 2, pp. 1414–1424. IEEE (2004)
122. Eagle, N., Pentland, A.S., Lazer, D.: Inferring friendship network structure by using mobile phone data. Proceedings of the National Academy of Sciences 106(36), 15,274–15,278 (2009)
123. Zheng, R., Wilkinson, D., Provost, F.: Social network collaborative filtering. Stern, IOMS Department, CeDER (2008)

124. Backstrom, L., Leskovec, J.: Supervised random walks: predicting and recommending links in social networks. In: Proceedings of the Fourth ACM International Conference on Web Search and Data Mining, pp. 635–644. ACM (2011)
125. Benchettara, N., Kanawati, R., Rouveirol, C.: Supervised machine learning applied to link prediction in bipartite social networks. In: 2010 International Conference on Advances in Social Networks Analysis and Mining (ASONAM), pp. 326–330. IEEE (2010)
126. Li, X., Chen, H.: Recommendation as link prediction in bipartite graphs: A graph kernel-based machine learning approach. Decision Support Systems 54(2), 880–890 (2013), http://www.sciencedirect.com/science/article/pii/S0167923612002540, doi:http://dx.doi.org/10.1016/j.dss.2012.09.019
127. Ratti, C., Sommer, C.: Approximating shortest paths in spatial social networks. In: 2012 International Conference on Privacy, Security, Risk and Trust (PASSAT), and 2012 International Confernece on Social Computing (SocialCom), pp. 585–586 (2012), doi:10.1109/SocialCom-PASSAT.2012.132
128. Stefa, J., Michalik, P.: Conversational content in the context of safety of social networks. In: 2013 IEEE 9th International Conference on Computational Cybernetics (ICCC), pp. 137–140 (2013), doi:10.1109/ICCCyb.2013.6617576
129. Chakrabarti, S., Dom, B., Indyk, P.: Enhanced hypertext categorization using hyperlinks. In: ACM SIGMOD Record, vol. 27, pp. 307–318. ACM (1998)
130. Domingos, P.: Mining social networks for viral marketing. IEEE Intelligent Systems 20(1), 80–82 (2005)
131. Pandit, S., Chau, D.H., Wang, S., Faloutsos, C.: Netprobe: a fast and scalable system for fraud detection in online auction networks. In: Proceedings of the 16th International Conference on World Wide Web, pp. 201–210. ACM (2007)
132. Brunelli, M., Fedrizzi, M.: A fuzzy approach to social network analysis. In: International Conference on Advances in Social Network Analysis and Mining, ASONAM 2009, pp. 225–230 (2009), doi:10.1109/ASONAM.2009.72

Testing Community Detection Algorithms: A Closer Look at Datasets

Ahmed Ibrahem Hafez, Aboul Ella Hassanien, and Aly A. Fahmy

Abstract. Social networks of various kinds demonstrate a strong community effect. Actors in a network tend to form closely-knit groups; those groups are also called communities or clusters. Detecting such groups in a social network (i.e., community detection) remains a core problem in social network analysis. Among the challenges that face the researchers to come up with advanced community detection methods, there is a key challenge, which is the validation and evaluation of their methods. The limited benchmark data available, the lack of ground truth for many of the available network datasets, and the nature of the social behavior factor in the problem, turned the evaluation process to be very hard. Accordingly, understanding such challenges may help in designing good community detection methods. This chapter presents testing strategies for community detection approaches and explores a number of datasets that could be used in the testing process as well as stating some characteristics of those datasets.

Keywords: Social network analysis, Community detection, Method evaluation, Social network datasets.

Ahmed Ibrahem Hafez
Faculty of Computer and Information, Minia University,
Minia - Egypt,
Scientific Research Group in Egypt (SRGE), `http://www.egyptscience.net`
e-mail: `ah.hafez@gmail.com`

Aboul Ella Hassanien
Faculty of Computers and Information, Cairo University,
Cairo - Egypt,
Scientific Research Group in Egypt (SRGE), `http://www.egyptscience.net`
e-mail: `aboitcairo@gmail.com`

Aly A. Fahmy
Faculty of Computers and Information, Cairo University,
Cairo - Egypt
e-mail: `aly.fahmy@gmail.com`

M. Panda, S. Dehuri, and G.-N. Wang (eds.), *Social Networking*
Intelligent Systems Reference Library 65,
DOI: 10.1007/978-3-319-05164-2_4, © Springer International Publishing Switzerland 2014

1 Introduction

Networks are used to model and represent many real world systems. This fact has made complex network analysis a popular research area. Collaboration networks, the Internet, biological networks, communication networks, and social networks are just some examples of such complex networks. A common feature of complex networks is community structure [1], i.e., groups of nodes in the network that are more densely connected internally than with the rest of the network. Communities play special roles in the structure-function relationship, and detecting communities (or modules) can help identify substructures that may correspond to important functions in the network. Therefore, detecting community structure is one of the most important problems in the field of complex network analysis.

Many methods have been developed for the community detection problem. These methods use tools and techniques from disciplines like physics, biology, applied mathematics, and computer and social sciences. Results of a recent survey can be seen in [2]. One of the most known algorithms proposed to date is the Girvan-Newman algorithm, which introduces a divisive method that iteratively removes the edge with the greatest betweenness value [3]. Some improved algorithms have also been proposed [4, 5]. These algorithms are based on a foundational measure criterion of community; namely *modularity*, which is a popular quality function that was proposed by Newman [3]. The larger the modularity value, the more accurate the community partition. Other methods use *spectral clustering* [6] derived from the problem of graph partition in which the objective is to minimize the cut function in the network. The community detection problem can be formalized as an optimization problem so well-known optimization techniques such as *Genetic Algorithm* [7, 8] can be applied to the problem. Community detection can be also formulated as an inference problem so statistical and inference techniques can be applied such as Hastings [9] applied belief propagation to the problem and Newman [10] use *mixture model* and *expectation maximization (EM)* [11] to infer communities assignments.

A social network can be modeled as a graph $G = (V,E)$, where V is a set of nodes or vertices, and E is a set of edges that connect two elements of V. In the social network context nodes represent persons or actors, and an edge represents a relationship or a tie between two persons. Moreover, $n = |V|$ is the number of nodes and $m = |E|$ is the number of edges. A community c in a network is a group of nodes having a high density of edges among the nodes and a low density of edges between different groups. A community structure of a network is a set $C = c1, c2, \cdots ck$ of such groups where k is the number of the groups. Community detection involves identifying the number of k communities or groups in a network and assigning communities membership for each node. Typically, the number of community's k is unknown. Community assignment can be hard; each node is assigned to one group. The previous type of assignment is *exclusive* i.e. nodes are allowed to belong to only one community. Alternatively, assignment can be *soft*, where each node is assigned a degree of membership per community indicating the extent to which that node belongs to that particular community. Soft assignment could mean just an observer belief of how much a particular node belong to each community, in this case it also consider an example of exclusive assignments but with uncertainty. Soft assignment could also mean an overlapping situation in which nodes could belong to more than one community. Overlapping is more practical in social life because a single person can be associated with more than one community with different degrees of association.

Despite the challenges [12] that researchers face to come up with a good community detection algorithm, evaluating the algorithm is another major challenge they face. The lack of known social network datasets, the nature of such datasets, and the lack of a ground truth of those datasets are some of the challenges that could affect evaluating a community detection methodology. There exist some well-known datasets and their ground truth i.e. the original community structure is known, however the nature of the dataset make it possible to have a more than one good community structure, which makes evaluation even harder. It's expected that a community detection algorithm would find the true community structure of the network, however it could return another structure, which raises the question "Is that mean that the algorithm fail to recover a correct solution?". This could be the case, but in some situations the algorithm could find even better community structure than the original community structure of the network under study. Hence, understating the nature of the network datasets could help in this judgment. Another challenge in evaluation is that some of the networks datasets do not have a ground truth i.e. there is no information of their community structure, thus there is nothing to compare the result of an algorithm with, and in this case evaluation could be established by measuring the quality of the community structure. Quality measures like *modularity* [3] and *conductance* [13] and many other that capture the intuition of communities are used to measure how good a community structure is. By using such measures, different algorithms could be compared by measuring the quality of its output. Many quality measures have been proposed over the years, and there is no straightforward way to compare these quality measures based on their definitions, so which one is better is still open question for a recent comparative study of most popular measures see [13, 7].

This chapter introduces testing strategies for community detection approaches as well as exploring a number of social network datasets that could be used in the testing process along with stating some characteristics of those datasets.The remainder of this chapter is organized as follows. In Section 2, some of the most popular quality measures used in the field were stated. Also, similarity measures used to compare community structures were illustrated. Section 3, describes some social network datasets used in testing and evaluating community detection methods and shed some light on their network natures. Finally, section 4, discusses conclusions.

2 Evaluating Community Detection Methods

Evaluating an algorithm in general requires two parts. The first part is to make sure it outputs a correct result. The second part is to evaluate its performance in term of running time and memory usage. In community detection problem case the first part of the evaluation is unclear due to the nature of problem and the networks dataset under study. One evaluation strategy is to measure the goodness of its output using a quality measure of communities. Another strategy is to compare the method output to a ground truth of the input network using a similarity measure. Testing strategies of community detection methods could be categorized into one of the following cases:

- There is no knowledge of the output, so a quality measure is used to evaluate the method.
- There is knowledge of a possible solution, so a similarity measure is used to measure how similar is the method's output to that possible solution.
- There could be more than one possible solution, in this case a similarity measure and a quality measure can be used in the evaluation process.

The remaining of the section is devoted to discuss key testing strategies of community detection algorithms.

2.1 Quality Measures

A good quality measure of a community should be able to distinct a community alike a group of nodes in a network from just randomly connected a group of nodes. Usually a measure will consider the two most important characteristics of a community; namely *internal edges* representing the number of connections/edges inside the community and *external edges* representing the number of connections/edges that members of a community make with the rest of the network. The first characteristic, internal edges, should be high and the second one, external edges, should be minimized in order the group to be a good community structure.

Let G be the network graph with n nodes and m edges. A community structure of the network is a set $C = c_1, c_2, \cdots c_k$ with k groups. Let c_i be a selected community from C , n_c is the number of nodes inside community c_i , m_{c_i} is the number of edges inside the community c_i and s_{c_i} is the number of edges leaving community c_i i.e. edges that connect members of c_i to other communities in the network. Let $f(c)$ be the quality measure function that estimate the quality of a given community c.

Some of the quality measure that is widely used or have been frequently used in the context of community detection are defined as follows:

- **Conductance:** measures the fraction of total edge volume that points outside the group [13]. Better communities should have a lower conductance value. Conductance is measured as shown in equation (1).

$$f(c) = \frac{s_c}{(2m_c + s_c)} \tag{1}$$

.

- **Normalized Cut:** is the normalized fraction of edges leaving the cluster [14]. It is measured as shown in equation (2).

$$f(c) = \frac{s_c}{(2m_c + s_c)} + \frac{s_c}{(2(m_c - m) + s_c)} \tag{2}$$

.

- **Average-ODF:** is the average out-degree fraction (ODF) of nodes' edges pointing outside the cluster [15] and it is measured as shown in equation (3).

$$f(c) = \frac{1}{n_c} \sum_{u \in c} \frac{\{(u,v) : v \notin c\}}{d(u)} \tag{3}$$

- **Modularity Q:** [3] measures the number of within-community edges relative to a null model of a random graph with the same degree distribution. Modularity is measured as shown in equation (4), where A_{ij}=1 if there is an edge between v_i and v_j and 0 otherwise. $\partial(v_i, v_j)$=1 if v_i and v_j are in the same community and $d(v_i)$ is the degree of node v_i.

$$Q = \frac{1}{2m} \sum_{ij} (A_{ij} - \frac{d(v_i)d(v_j)}{2m}) \partial(v_i, v_j) \tag{4}$$

The previous Modularity definition works for unweighted networks, Newman [16] suggested an extension to the definition for weighted network in which A_{ij} represents the weight of the edge between node i and node j. Also, the weighted degree of node v_i is

$$d(v_i) = \sum_j A_{ij}$$

and

$$m = \frac{1}{2} \sum_{ij} A_{ij}$$

Regarding the first three quality measures, conductance, normalized cut, and average out-degree fraction, the lower the value the better the community structure. As for modularity quality measure, the higher the value the better the community structure.

2.2 *Similarity Measure*

To compare the accuracy of the resulting community structures, *Normalized Mutual Information (NMI)* [17] has been used in order to measure the similarity between the true and the detected community structures. NMI similarity measure proposed by Danon et al. [17] is inspired from information theory. It is based on defining a confusion matrix N, where the rows correspond to the real communities, and the columns correspond to the found communities. The member of N, N_{ij} is simply the number of nodes in the real community i that appear in the found community j. The number of real communities is denoted c_A and the number of found communities is denoted c_B, the sum over row i of matrix N_{ij} is denoted $N_{.i}$, and the sum over column j is denoted $N_{j.}$. A measure of similarity between the partitions, based on information theory is calculated as shown in equation (5):

$$NMI(A,B) = \frac{-2\sum_{i=1}^{CA}\sum_{j=1}^{CB} N_{ij} log(\frac{N_{ij}N}{N_{i.}N_{.j}})}{\sum_{i=1}^{CA} N_{i.} log(\frac{N_{i.}}{N}) + \sum_{j=1}^{CB} N_{.j} log(\frac{N_{.j}}{N})} \tag{5}$$

3 Social Network Datasets

In traditional data mining and Machine Learning (ML) tasks a training-testing model of evaluation is used in which some of the data are used in the training process, then the rest of the available data are used in testing; similar to the classification problem. However, in social network analysis things are different, datasets are limited and lack a ground truth solution to the problem to compare against, and it might be there is more one solution to the problem as well.

A synthetic benchmark can be used to test community detection method such as Girvan and Newman (GN) benchmark [1], Fan et al. [18] weighted version of GN benchmark or LFR benchmark [19], the problem with such benchmarks is that they lack the random and complex social behavior of people. So a real life social network is much preferable to use in testing.

This section highlights a number of social network datasets that can be used in evaluating community detection methods. For each dataset, we visualized the network along with its possible community structures. Visualization has be done using Gephi [20] via utilizing ForceAtlas2 layout algorithm [21]. Datasets are available to use in [21, 22].

3.1 The Zachary Karate Club

One of the most used datasets in the field of social network analysis. The Zachary Karate Club data was first analyzed by Wayne Zachary in 1977 [23]. The network is one of the smallest and extensively used dataset in social network analysis. Zachary used these data and an information flow model of network conflict resolution to explain the split-up of this group following disputes among the members. The karate club was observed for a period of three years, from 1970 to 1972. In addition to direct observation, the history of the club prior to the period of the study was reconstructed through informants and club records in the university archives. At the beginning of the study, there was a conflict between the club president and the karate instructor over the price of karate lessons. As time passed, the entire club became divided over this issue, and the conflict became translated into ideological terms by most club members and that led to the division of the club into two groups.

The network is undirected network and consists of 34 nodes and 78 edges. In the original study Zachary constructed two networks from the observations. One version is unweighted binary in which an edge indicates the existence of a relationship between two people in the club. The other is a weighted version in which a weight w is associated with each edge; w represents the strength of a relationship between two people i.e. number of situations in and outside the club in which interactions occurred. The network describe a real life scenario this make the connectivity pattern in the graph acutely reflecting the community structure of the club after the division making the network a good candidate for validating a community detection method, however its small size make a poor judgmental call of a community detection method.

Figure 1a is a visualization of the binary network describing the relations between club members. An observer of the figure can easily classify the network into two communities, except for a few nodes that lays in-betweens. For example, node 10 only have two connections one with node 34 and one with node 3, without any farther information; it is a fifty/fifty chance that node 10 could belong to either community. However, we could increase the odds by saying that node 10 will end up with the nodes that has more influence - people like to connect to a person with high authority, a famous person or a person with good connections - so in this case it is more likely that node 10 will chose to be with node 34 after the division. However if node 10 has a strong connection with node 3, it is more likely that node 10 will chose to be with node 3.

Figure 1b is a visualization of the weighted version of network depicting relations strength between club members. With this information in hand, an observer of the figure can easily detect communities in the network and be sure about his classification. Now we know that node 10 has a stronger connection with node 34 than its connection with node 3 and node 34 is more important than node 3 - node 34 has more connections - so it is clear now that node 10 will end up with node 34 after the division. So, let's look at some of the quality measures of the community structure shown in figure 1. Since most community detection methods find a problem to correctly classify node 10 in the network, so we compare the original community structure with two other community structure. The first community structure (C1) has node 10 with node 1 in the same community. The second community structure (C2) has node 9 with node 1 in the same community.

Table 1 summarizes the quality measures of those different community structures. The modularity quality measure has been calculated for the binary network and the weighted network. Regarding the normalized mutual information similarity measure; moving one node from one community to another result in the same NMI values for **C1** and **C2** compared to the original community structure of the network. The original community structure of the network as shown in figure 1 is the best solution in term of all quality measures calculated in table 1. **C1** community structure has approximately the same modularity, conductance and average-ODF values as the original one. **C2** community structure has lower modularity value than **C1** and the original community structure since in **C2**; more edges are crossing communities' boundaries. Also, **C2** has a higher normalized cut, conductance and average-ODF values than **C1** and the original community structure.

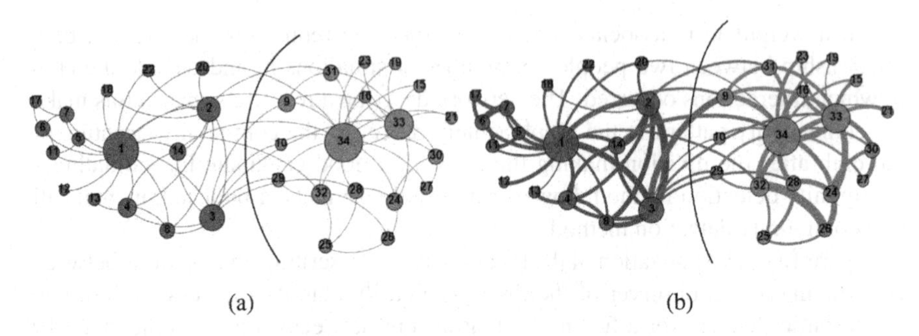

Fig. 1 Zachary Karate Club network datasets: (a) the binary version of the network, (b) the weighted version of the network

Table 1 Quality measure of different community structures of Zachary Karate Club network

Community structure	NMI	Modularity		Normalized cut	Conductance	Average-ODF
		binary	weighted			
Original	1	0.421	0.455	0.306	0.137	0.206
C1	0.837	0.422	0.451	0.343	0.137	0.206
C2	0.837	0.408	0.443	0.411	0.152	0.234

3.2 The Bottlenose Dolphin Network

This dataset was compiled by Lusseau [24] and is based on observations over a period of seven years of the behavior of 62 bottlenose dolphins living in Doubtful Sound, New Zealand. A relationship between two dolphins was established by statistically significant frequent association. The network split naturally into two groups.

Figure 2 visualizes the Bottlenose Dolphin network. Figure 2(a) shows the original division of the network from the original study by Lusseau [24]. Figure 2(b), (c) and (d) show three different divisions of the network where in figure 2(b) the network is divided into 2 communities by moving 2 nodes -node 31 and node 40- from the left community to the right one. In figure 2(c) the network is divided into 4 communities and in figure 2(d) the network is divided into 5 communities. Table 2 summarizes some quality measures of Bottlenose Dolphin network's community structures shown in figure 2.

Regarding the modularity values of the different community structures shown in figure 2, we can see that the structure shown in figure 2(d) has the best modularity value; however its NMI value compared to the original community structure is low. As for the rest of the quality measures we found that the original community structure shown in figure 2(a) has the best Normalized cut, Conductance and Average-ODF values since it has fewer edges crossing community's boundaries.

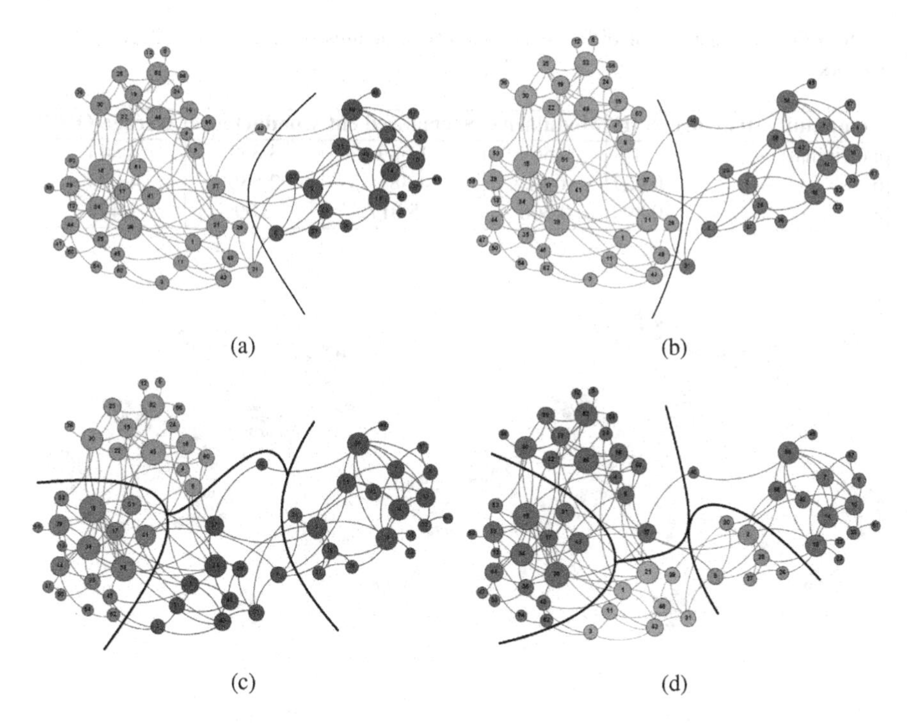

Fig. 2 Different community structures of the Bottlenose Dolphin network dataset

Table 2 Quality measure of different community structures of Bottlenose Dolphin network

Community structure	NMI	Modularity	Normalized cut	Conductance	Average-ODF
a (original)	1	0.395	-0.049	0.047	0.093
b	0.814	0.406	0.073	0.053	0.109
c	0.576	0.533	-0.384	0.59	0.884
d	0.526	0.535	0.406	0.863	1.239

3.3 American College Football Network

This network [1] represents football games between American colleges during a regular season in Fall 2000. Nodes in the graph represent teams and edges represent regular-season games between the two teams they connect. What makes this network interesting is that it incorporates a known community structure. The teams are divided into conferences containing around 8-12 teams each. Games are more frequent between members of the same conference than between members of different conferences, the network is divided into 12 conferences.

In the original division of the network shown in figure 3(a) there are few independent teams that do not belong to any conference so they are grouped to the same conference but do not play any match with each other. Most community detection

Table 3 Quality measure of different community structures of American College football network

Community structure	NMI	Modularity	Normalized cut	Conductance	Average-ODF
a (original)	1	0.563	1.589	3.358	4.843
b	0.927	0.61	4.381	2.536	4.073
c	0.899	0.608	-15.017	2.09	3.497

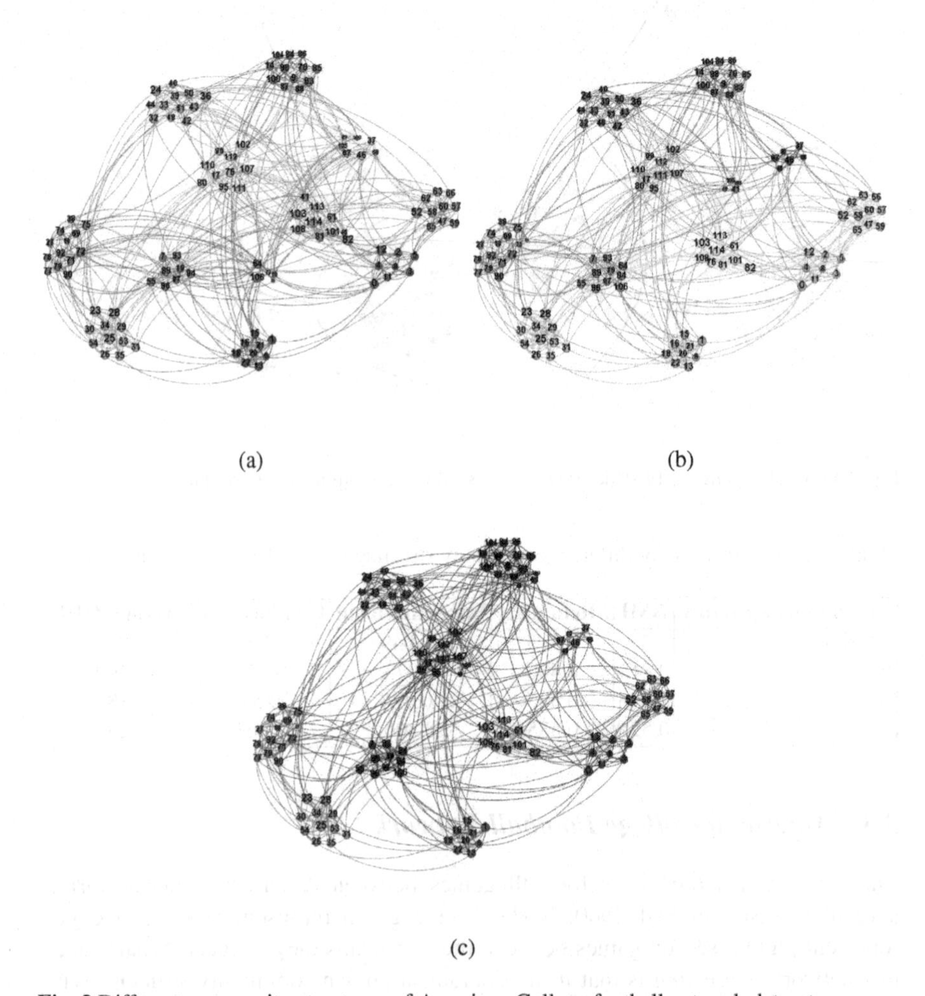

Fig. 3 Different community structures of American College football network dataset

method will fail to correctly classify those few independent teams; however there is an obvious connection pattern in the network, so those teams should be grouped with the conference with which they are most closely associated i.e. has more connection with the team.

Table 3 summarizes some quality measures of American College football network's community structures shown in figure 3. The original conference division figure 3(a) of the network has a low Modularity value since the independent teams are groups together and there are no edges between them. The second division of the network in figure 3(b) is somehow has a better modularity value in which the independent teams now regroups with the conference which they have more connections with for example node 106 will be in the conference which contains node 79 as node 106 has more connections with this conference's nodes than other conferences. The network division shown in figure 3(c) has a better Normalized cut, conductance and Average-ODF than the other two divisions.

3.4 Online Social Networks

Online social networks, such as Facebook or Google+, is a platform to build social networks or social relations among people who, for example, share interests, activities, backgrounds, or real-life connections. A social network service consists of a representation of each user (often a profile), his/her social links, and a variety of additional services. Most social network services are web-based and provide means for users to interact over the Internet, such as e-mail and instant messaging. Social networking sites allow users to share ideas, pictures, posts, activities, events, and interests with people in their network due such online social behavior online communities are formed where online users tend to form communities that group users who share some comment interest. Users usually form groups or circles in online social network.

In Facebook a user can create lists of friends that are from the same context like high school friends, collage friends, co-workers, family and etc. this lists form a community of people with the same interest or people that know each other form the same context. A community detection method that can maintain and update such lists and circles automatically will ease the process for the user. Leskovec [25] propose a method that use the graph connectivity patterns and user's features - a user's likes, interests, places, etc. - to automatically detect circles and updates user's friends lists. In their study they collect data for the Facebook website, the data was collected from survey participants using a Facebook application [26]. The dataset includes node features (profiles), circles, and ego networks. The ago network consist of a user's -the ego node- friends and their connections to each other. Each ego network has a ground truth that can be used to validate the result of a method. The problem with the ego network ground truth is that it was identified by the ego user himself, making it valid form the user point of view. Another problem with this data is that has been collect for an online social network, online social behavior of people is not similar to real social. People online may connect to people they do not know thus an ego network may be almost fully connect due this online behavior. This problem may cause a community detection method to fail finding a community structure if it depend only on the connectivity of the network.

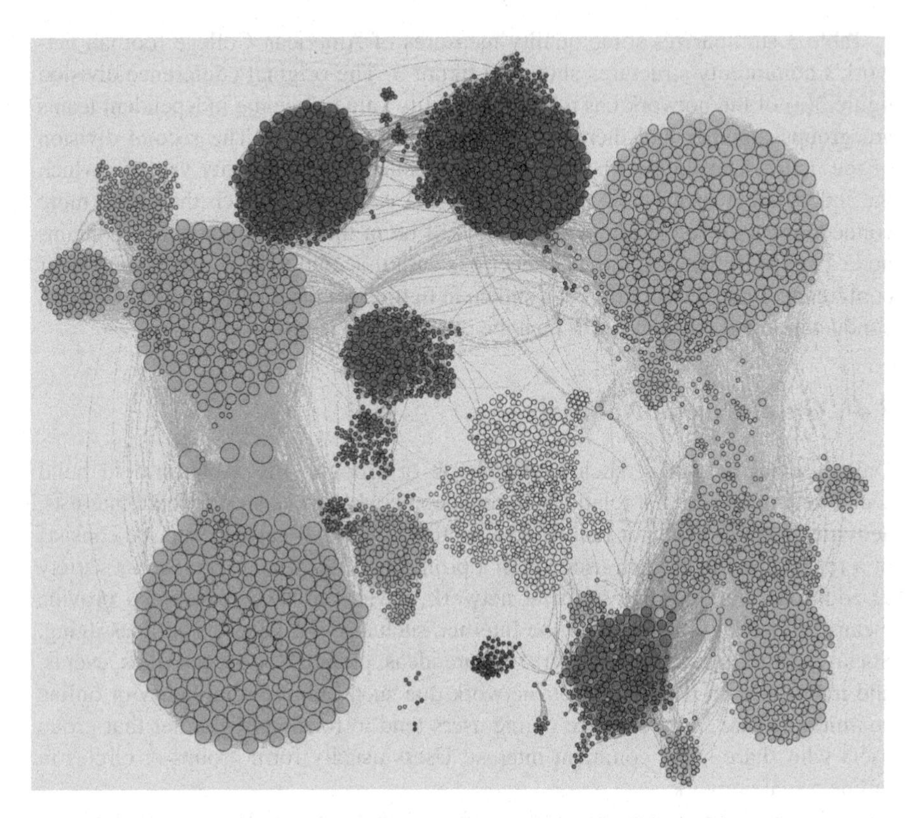

Fig. 4 Online social network: 10 ego networks from Facebook site divided into 10 communities according to the 10 ego's friends

Table 4 Quality measures values of the community structures shown in figure 4

Community structure	Modularity	Normalized cut	Conductance	Average-ODF
10-communities	0.723	0.065	0.332	0.452

The dataset can be used to test a community detection method; one could use only the ego network without the features of the user to investigate the community structure of the ego network. We combine the 10 Facebook ego networks from [25] into one big network. We remove the ego nodes and their edges, to make it clear for visualization. The result network is undirected network which contain 3959 nodes and 84243 edges. In figure 4 we visualized the combined network and highlight the top level community structure of network; which is the ego network itself. Simply we assume that each ego network is a community. If a node appears in more than one ego network, we count the node's neighbors in each ego network and the ego network with the maximum number of neighbor will be the node's community. This community structure can be break into smaller communities, as we can observe the group of nodes that on the left can be farther divided into 4 communities. This dataset represents a good challenge for a community detection method. As we can observe that the distribution of the friends number in each ego networks is not uniform, one ego network contain only 50 node and anther ego network contains over 1000 nodes. Also we can observe the formation of the closely-knit groups in which users from the same groups almost connect to each other user from the same group.

Now let's look at the quality measures of suggested community structure of combined network. Table 4 summarizes the quality measures values calculated for the network. We can observe that the community structure shown in figure 4 has a high modularity value which indicate that it is a good community structure also it have a low values for normalized cut , conductance and average-ODF, which also indicate that it is a good community structure.

In the study conducted in [25], they also collected data for Google+ and Twitter sites. Google+ and Twitter social networks differ from Facebook networks as they are directed networks. The data collected in the study lacks a good ground truth, since the ground truth is identified by the ego user who volunteer to share his/her social network in the study and some users may fail to assign some of his friends to any community. However, the data can be used to test a community method leaving the evaluation process to any of the quality measures that best suit the problem.

4 Conclusion

In social network analysis, detecting communities, with cohesive groups of actors who interact with each other more frequently than with those actors outside the groups, is one of the fundamental tasks. Hechter say in [27] "the founders of sociology claimed that the causes of social phenomena were to be found by studying groups rather than individuals". Community detection also is important to many real world applications such as the grouping of customers with similar interests in social media marketing or huge network compression for a better visualization and understanding of the network. Researchers from multiple disciplines are working to come up with the state of art community detection methods. One major challenge faces researchers in this filed of community detection is the methods of evaluation. That is due to the limited datasets and lack of ground truth for such datasets. Understating the nature of the problem and the social behavior underling in the network datasets may result in a better community detection method also the evaluation process of different community detection methods require a comparative and extensive study for a better judgments of those methods.

References

1. Girvan, M., Newman, M.E.J.: Community structure in social and biological networks. Proceedings of the National Academy of Sciences 99(12), 7821–7826 (2002)
2. Fortunato, S.: Community detection in graphs. Physics Reports 486(3-5), 75–174 (2010)

3. Newman, M.E.J., Girvan, M.: Finding and evaluating community structure in networks. Physics Rev. E 69(2), 026113 (2004)
4. Radicchi, F., Castellano, C., Cecconi, F., Loreto, V., Parisi, D.: Defining and identifying communities in networks. Proceedings of the National Academy of Sciences of USA 101(9), 2658–2663 (2004)
5. Clauset, A., Newman, M.E.J., Moore, C.: Finding community structure in very large networks. Phys. Rev. E 70(6), 066111 (2004)
6. Luxburg, U.: A Tutorial on Spectral Clustering. Statistics and Computing 17(4), 395–416 (2007)
7. Hafez, A.I., Ghali, N.I., Hassanien, A.E., Fahmy, A.A.: Genetic Algorithms for community detection in social networks. In: 2012 12th International Conference on Intelligent Systems Design and Applications (ISDA), pp. 460–465 (2012)
8. Pizzuti, C.: A multi-objective genetic algorithm for community detection in networks. In: 21st International Conference on Tools with Artificial Intelligence, pp. 379–386 (2009)
9. Hastings, M.B.: Community detection as an inference problem. Phys. Rev. E 74(3), 035102 (2006)
10. Newman, M.E.J., Leicht, E.A.: Mixture models and exploratory analysis in networks. Proceedings of the National Academy of Sciences 104(23), 9564–9569 (2007)
11. Dempster, A.P., Laird, N.M., Rubin, D.B.: Maximum likelihood from incomplete data via the EM algorithm. Journal of the Royal Statistical Society, Series B 39(1), 1–38 (1977)
12. Tang, L., Liu, H.: Community Detection and Mining in Social Media. Morgan & Claypool Publishers (2010)
13. Leskovec, J., Lang, K., Mahoney, M.: Empirical Comparison of Algorithms for Network Community Detection. In: ACM WWW International Conference on World Wide Web (2010)
14. Shi, J., Malik, J.: Normalized Cuts and Image Segmentation. IEEE Transactions on Pattern Analysis and Machine Intelligence 22, 888–905 (1997)
15. Flake, G.W., Lawrence, S., Lee Giles, C.: Efficient Identification of Web Communities. In: Sixth ACM SIGKDD International Conference on Knowledge Discovery and Data Mining, pp. 150–160 (2000)
16. Newman, M.E.J.: Analysis of weighted networks. Phys. Rev. E 70(5), 056131 (2004)
17. Danon, L., Diaz-Guilera, A., Duch, J., Arenas, A.: Comparing community structure identification. Journal of Statistical Mechanics: Theory and Experiment 9, 09008 (2005)
18. Fan, Y., Li, M., Zhang, P., Wu, J., Di, Z.: Accuracy and precision of methods for community identification in weighted networks. Physica A: Statistical Mechanics and its Applications 377(1), 363–372 (2007)
19. Lancichinetti, A., Fortunato, S.: Benchmarks for testing community detection algorithms on directed and weighted graphs with overlapping communities. Phys. Rev. E 80(1), 016118 (2009)
20. Bastian, M., Heymann, S., Jacomy, M.: Gephi: An Open Source Software for Exploring and Manipulating Networks. In: International AAAI Conference on Weblogs and Social Media (2009)
21. Jacomy, M., Heymann, S., Venturini, T., Bastian, M.: ForceAtlas2, A Continuous Graph Layout Algorithm for Handy Network Visualization. Medialab Center of Research (2012)
22. Network DataSets (2013), `http://www.personal.umich.edu/mejn/netdata`
23. Stanford Large Network Dataset Collection (2013), `http://snap.stanford.edu/data/index.html`

24. Zachary, W.W.: An information flow model for conflict and fission in small groups. Journal of Anthropological Research 33(4), 452–473 (1977)
25. Lusseau, D.: The emergent properties of dolphin social network. Proceedings of the Royal Society of London. Series B: Biological Sciences 270(suppl. 2), S186–S188 (2003)
26. McAuley, J.J., Leskovec, J.: Learning to Discover Social Circles in Ego Networks. In: NIPS, pp. 548–556 (2012)
27. Leskovec, J.: Social Circles in Ego Networks (2013), http://snap.stanford.edu/socialcircles/
28. Hechter, M.: Principles of Group Solidarity, ch. 2. University of California Press (1988)

Societal Networks: The Networks of Dynamics of Interpersonal Associations

B.K. Tripathy, M.S. Sishodia, and Sumeet Jain

Abstract. Social networks have become popular and useful tools now a day. In the beginning researchers in social networks concentrated in studying static versions. However, the inclusion of time variations made the social networks dynamic. In the present day scenario the social networks are more dynamic than static. The introduction of societal networks by J. Fiksel is an evolutionary step in the study of social networks which originated the concept of dynamic social networks. The term societal network was used by Fiksel in order to distinguish it from the diversified ways in which the term 'social network' was used prior to the 1980s. There seems to be little work done following his approach. In this chapter we introduce the social structures which form the basis of Fiksel's concept and also present his results. Some further study was made by Acharya using the graph theoretic concepts and also he proposed some directions of research. Dynamic social networks are studied recently from different angles. We also present some of these works and propose possible direction of research in this interesting field of research in social networks.

1 Introduction

An individual may be seen as an open system, existing and capable of existing only through processes of exchange with the environment. Thus, theories of human behaviour and of human relationships are in many ways analogous to those of system theory as applied to institutions. The individual is not just a single activity system with an easily defined primary task, but a multi-task system capable of multiple activities [29].

"Individual" has little meaning as a concept except in relationships with others [18]. He or she uses them and vice versa to express views, take action and play

B.K. Tripathy · M.S. Sishodia · Sumeet Jain
SCSE, VIT University, Vellore, TN, India
e-mail: tripathybk@vit.ac.in,
{mayanksshishodia, sumeet.jain44}@gmail.com

M. Panda, S. Dehuri, and G.-N. Wang (eds.), *Social Networking*
Intelligent Systems Reference Library 65,
DOI: 10.1007/978-3-319-05164-2_5,

roles. The individual is a creature of the group, the group of the individuals. The terms 'individual' and 'group' in somewhat equivalent sense, 'equivalent' limited to their behaviours and functions, for extending it beyond is fraught with the danger of conveying similarity in their physical characteristics such as, say, both having the same weight. We will see the advantage of this approach in studying some very important socio-psychological phenomena by developing hypothetical models to enrich our conceptual resources to understand them better. Towards this end, first we need the following definition of a 'group' due to Alderfer:

A human group is a collection of individuals ([3,4,31]).

1. Who have significantly interdependent relations with each other.
2. Who perceive themselves as a group by reliably distinguishing members from nonmembers, whose group identity is recognized by non-members.
3. Who, as group members acting alone or in concert, have significantly interdependent relations with other groups
4. Whose roles in the group are therefore a function of expectations from themselves, from other group members and from non-group members.

This idea of a group begins with individuals who are interdependent, moves to the sense of the group as a significant socio-psychological object ([7],[8]) whence boundaries are confirmed from inside and outside, recognizes that the group-as-a-whole is an interacting unit through representatives or by collective action and returns to the individual members whose thoughts, feelings and actions are determined by forces within the individual and from both group and non-group members ([24]). This conceptualization of a group makes every individual member into a group representative whenever he or she deals with members of other groups and treats every transaction among individuals as at least in part, an intergroup event.

In the study of intergroup relations ([12],[34]), which in a generic sense can cover interpersonal relations too invoking the above mentioned approach, psychology has fallen into the problem of taking characteristics of interaction patterns and talking about them as though they were attributes of the entities engaged in the interaction makes distinction between characteristics of the actors in interaction and the characteristics of interaction patterns, what we observe in a society is the latter; Smith prefers to analyse the latter since we can only 'see' behaviour of the entities engaged in the interaction whereas we cannot actually 'see' interactions. To observe interactions demands, then, a derivative, analytical approach that enables us to focus on what transpires, invisibly, in the space between the interactions.

Secondly, the problem about the patterns that are dynamic ([5]) is that the moment attempt to represent fluid, changing phenomena in words, there is a tendency for the pattern to become reified. Then we find the rich, dynamic, movie-like patterns becoming reduced to impoverished static snapshots. This tendency has been most evident in sociology, which treats dynamic interactions as the focal point of analysis, yet conceptualizes them as crystallized, institutionalized and stabilized properties of social structure. As a result, structured characteristics come to be viewed by sociologists not as crystallized interactions but as entities possessing empirical and conceptual reality themselves.

Perhaps, it is because of the complex nature of the socio-psychological situations mentioned above:

i. Focus of social network theorists attention has been mostly on 'ego-centred" networks, that is, the set of associations radiating from a particular individual
ii. They have tended to employ a static type of analysis which fails to account for the continuous fluctuations in socio-psychological attitudes, and
iii. They have avoided forming general conclusions about the patterns that they have observed.

The organisational structure of this chapter is as follows:
In section 2 we provide the background of dynamic social networks. In section 3 we present the sigraphs and sidigraphs, which are popular and accurate models for representation of social systems. Also, in this section we discuss on several concepts like dynamism, cognitive balance and clustering of nodes have been introduced. A brief formal introduction to social networks is presented in section 4. The cream of this chapter, societal networks and their development with analysis is presented in section 5. In section 6, we deal with two recent works in dynamic social networks. Section 7 is interesting for persons who would like to do some research work in the field of societal networks and other dynamic social networks. Some concluding remarks are provided in section 8 and it follows by references to several source materials consulted during the compilation of the chapter.

2 Background

A notable exception is one of the earliest network models to represent behavioural dynamics due to French [17], in his theory of social power. The basic aim of this approach is to explore the extent to which the influence process in groups can be explained in terms of patterns of interpersonal relations. The process of influence in a group takes place gradually over a period of time. In his model, the power structure and the communication channels of the group are translated into a process of influence over time. The influence has been characterised into two categories; the direct influence, which is exerted on another person by direct communication and the indirect influence, which is exerted on another through the medium of one or more other persons. Basically the definition of power used here is defined as, the power of A over B is equal to the maximum force which A can induce on B minus the maximum resisting force which B can mobilize in the opposite direction. The power structure of a group is represented conceptually in terms of the mathematical theory of directed graphs. Using directed graphs to represent power structure in a group, he examined the opinion patterns that evolved from various influence configurations to represent power structure in a group. Rapport [29, 30] laid foundation for probabilistic social network models through his investigation of information transfer in random nets. Lorrain and White [27] gave a rigorous exposition, using the theory of categories of structural classifications in networks of binary oppositions with their avowed aim to achieve a "global" understanding of social networks by studying the underlying structure. Subsequently, the idea of cognitive structural balance first introduced by Cartwright and Harary [10] was placed in a sociometric setting and generalised by Holland and Leinhardt [23], who invoked the concept of partial order.

Fiksel [14] put a step forward by development of a network model explicitly incorporating the dynamics of interpersonal associations thereby expanding the scope of the model to encompass large groups of interrelated individuals as well as permitting a description of patterns of evolution within such groups. It has powerful simplicity.

3 Graphs and Digraphs

In this section we introduce some definitions and notations to be used in this chapter.

Definition 0.1. **(i)** A graph is a discrete structure formed by a set V whose elements are called vertices and a collection E is essentially a set of elements called edges, each of which is a pair of elements of V.
(ii) A graph G is then taken to be the ordered pair (V, E) where V:= V(G) and E:= E(G) are respectively the vertex set and the edge set of the graph G.
(iii) G is called finite if V as well as the collection E(G) are finite and infinite otherwise.

Definition 0.2. **(i)** A digraph (directed graph) is a discrete structure formed by a set V whose elements are called vertices and a collection E, is essentially a set of elements called arcs, each of which is an ordered pair of elements of V
(ii) A diagraph D is then taken to be the ordered pair (V, E) where V:= V(D) and E:= A(D) are respectively the vertex set and the arc set of the digraph D.
(iii) D is called finite if V as well as the collection A(D) are finite and infinite otherwise. Here A(D) is taken to denote the set of arcs of a digraph D.
(iv) If e = (u, v) A(D) then u is called the tail and v is called the head of the arc. We say that u is adjacent (or, joined) to v, written symbolically as $u \to v$.
(v) If $u \to v$ and $v \to u$ then (u, v), (v, u) is called a symmetric arc.
(vi) we may regard any graph as a symmetric diagraph in which one has $u \to v \Leftrightarrow v \to u$.
(vii) A digraph in which there is at most one arc joining a vertex u to a vertex v, for all u, $v \in V(D)$, is called a 1-digraph or a simple digraph.

A social network that represents a group of persons or smaller groups of them endowed with interpersonal or intergroup relationships existing amongst them [6].

Suppose A and B are two persons in a social system. The relationship between them can be represented by a two-vertex signed digraph $S = (V,E;\sigma)$ where V = {A, B}, E = {(A,B),(B,A)}, $\sigma(A,B)$ = +1, and $\sigma(B,A)$ = -1. Such a social system is called an antisymmetric dyad.

Definition 0.3. A sigraph ([20],[32]) is an ordered pair $S = (S^u, \sigma)$, where $S^u :=$ $G = (V,E)$ is a graph called the underlying graph of S and $\sigma : E \rightarrow \{+,-\}$, called a signing, is a function from the edge set E of G into the set {+, -} of signs, the set {+, -} may either be treated simply as a set of symbols called 'colours', when σ is regarded as an edge colouring of G or as the involutory group {-1, +1} defined by a multiplication table in which case σ is treated as a valuation of the edges of G. Further, $E^+(S)$ will denote the set of edges S^u of that are mapped by σ to the element +1 and $E^-(S) = E - E^+(S)$. The elements of $E^+(S)$ are called positive edges and those of $E^-(S)$ are called negative edges of S ; in generic terms, we say that the edges are signed (by σ).

Definition 0.4. A sidigraph ([20],[32]) is an ordered pair $S = (S^u, \sigma)$, where $S^u :=$ $G = (V,E)$ is a digraph called the underlying digraph of S and $\sigma : E \rightarrow \{+,-\}$, called a signing, is a function from the arc set E of G into the set {+, -} of signs, the set {+, -} may either be treated simply as a set of symbols called 'colours', when σ is regarded as a arc colouring of G or as the involutory group {-1, +1} defined by a multiplication table in which case σ is treated as a valuation of the arcs of G. Further, $E^+(S)$ will denote the set of arcs of S^u that are mapped by σ to the element +1 and $E^-(S) = E - E^+(S)$. The elements of $E^+(S)$ are called positive arcs and those of $E^-(S)$ are called negative arcs of S; in generic terms, we say that the arcs are signed (by σ).

3.1 Representations of Sigraphs and Sidigraphs

In most of the applications, one needs to assign weights to the vertices or edges of a graph ([16]), whence one calls such a 'weighted' configuration an undirected network; weights as such may be taken from any arbitrary set of labels which might, in particular, be real numbers or subsets of a set or elements of an algebraic structure such as a group or a semigroup, etc., depending on real-life applications for which appropriate network models have to be constructed.

Thus, in most of the applications, one needs to assign weights to the nodes or arcs of a digraph, whence one calls such a 'weighted' configuration directed network; weights as such may be taken from any arbitrary set of labels which might, in particular, be real numbers or subsets of a set or elements of an algebraic structure such as a group or a semigroup, etc., depending on real-life applications for which appropriate network models have to be constructed.

A sidigraph is all-positive if all its arcs are positive and is all negative if all its arcs are negative. It is said to be homogeneous if it is either all-positive or all-negative and heterogeneous otherwise. Unless otherwise mentioned any digraph is regarded as an all-positive sidigraph.

For a sidigraph S, the sets of its positive arcs and negative arcs are denoted by $E^+(S)$ and $E^-(S)$ respectively. Thus $E(S) = E^+(S) \bigcup E^-(S)$. A sidigraph without orientations of its arcs (where it may be regarded as a sigraph) may also be regarded as a symmetric sidigraph in the sense that any two vertices in S are either not linked at all or are linked (that is joined by an arc) by two oppositely oriented arcs of the same sign.

3.2 Dynamics of a Social System

In the dynamics of a social system, it is natural that the nature of interpersonal interactions keeps changing due to various sociopsychological aspects such as variations in the perceptions, sentiments, actions, attitudes and circumstantial behaviours in the system. How do such fluctuations in interpersonal interactions affect the social system as a whole?

This question is of paramount importance toward developing deeper insights into the dynamics of social systems so as to be able to understand not only their complex behavioural responses but also the nature of intergroup interactions.

One fundamental aspect of dynamics of a social system [13]is the change in its very structure as it evolves over a period of time. This may happen in a number of complex ways as, for instance, departure of a member from the social group or arrival of a new person as a member of the social group or ever fluctuating dyadic relations.

As dyadic relations form a fundamental component of social change the effect of fluctuations in dyadic relationships on the structure of a social system is of profound interest for study due to its generally intricate dynamics in real-life situations.

Hence, considering the seminal role of dyadic interactions in the process of development of more enduring relationships between the members of the dyads in a social group [26], a sociograms was defined by Acharya and Joshi ([1],[2]) as the 'structure' of a social group specified by certain number of dyads existing in various states of interpersonal interaction, each interaction characterised qualitatively as being positive or negative in the evaluation of the individuals about the nature of their interaction with others. To be closer to real-life situations, it was assumed there that each evaluative relationship of an individual A with an individual B, represented by the ordered pair (A, B) and called an arc, in the social group is basically ambivalent in the sense that it might simultaneously contain positive as well as negative evaluations of A about B which is denoted by writing $(A,B)^{\pm}$ or showing them independently as $(A,B)^{+}$ and $(A,B)^{-}$ as per convenience in a given context so that one has $(A,B)^{\pm} = \{(A,B)^{+},(A,B)^{-}\}$. Over a period of time, by the process of reciprocity even the pair $(B,A)^{\pm}$ may develop eventually, whence the quadruple $\{(A,B)^{+},(A,B)^{-},(B,A)^{+},(B,A)^{-}\}$ is called the fully developed dyad (or fdd in short). However, during any epoch of time, any one of the dyadic combinations displayed in Table 1 ([1],[2]) below may exist between the individuals A and B, where $<A,B>$ represents the state in which there is no interaction between A and B whence A and B are regarded to be either in a state of indifference or nonacquintance with each other. This particular state is considered to be the natural opposite of the state of fdd mentioned above.

In general, the sixteen possible dyadic combinations between A and B may be paired off in such a way that each pair consists of the mutual opposites of each other as listed.

Table 1 Evaluative states of dyadic interactions and their natural duals

State of Dyadic interaction	Dual state of corresponding dyadic interaction
$<A,B>$	$\{(A,B)^+, (A,B)^-, (B,A)^+, (B,A)^-\}$
$(A,B)^+$	$\{(A,B)^-, (B,A)^+, (B,A)^-\}$
$(A,B)^-$	$\{(A,B)^+, (B,A)^+, (B,A)^-\}$
$(B,A)^+$	$\{(A,B)^+, (A,B)^-, (B,A)^-\}$
$(B,A)^-$	$\{(A,B)^+, (A,B)^-, (B,A)^+\}$
$\{(A,B)^+, (B,A)^-\}$	$\{(A,B)^-, (B,A)^+\}$
$\{(A,B)^+, (B,A)^+\}$	$\{(A,B)^-, (B,A)^-\}$
$\{(A,B)^+, (A,B)^-\}$	$\{(B,A)^+, (B,A)^-\}$

Thus, the socio-psychological states of 'ambivalence' and 'indifference' between individuals are represented in each dyad of a sociograms. A random instance of a sociograms that might thus exist during a given epoch of time can be depicted as a signed multi-digraph, called ambisidigraph; it must be noted however that ceaselessly ongoing social interaction processes continually change the structure of social groups over longer periods of time. These structural fluctuations in social systems are generally highly complex and occur not only due to innumerably many cognitive and behavioural responses of individuals and subgroups that constitute the social systems, but also due to effects of the 'external environment'; these considerations force one to look for mathematical modeling of socio-psychological phenomena that are responsible for the dynamical behaviour of the social systems.

Study of sidigraphs rose to prominence due to the discovery by Harary [20] that they serve as apt prototype models for the study of the notion of structural balance in a social group endowed with dyadic interactions characterised by a qualitatively dichotomic nature (represented by a positive or a negative sign) of interaction between pairs of members (dyads) in the social group, generalizing such considerations earlier on triads; in fact taking off from this classic paper. Harary generalised Heider's notion of cognitive balance in social systems as basically a structural feature of their underlying networks configured in such a way that every semicycle of such a network contains an even number of negative arcs and derived the following structural criterion for balance in sigraphs (sidigraphs).

3.3 Cognitive Balance and Sidigraphs

In this section we define the notion of cognitive balance [21,34] with respect to a sidigraph. The property of balance for sigraph and sidigraphs are provided in the following result.

Theorem 0.1. *([19]) A sigraph (sidigraph) S is balanced if and only if its vertex set V(S) can be partitioned into two subsets V_1 and V_2 such that every negative edge (arc) of S joins a vertex of V_1 with one of V_2 while no positive edge (arc) does so.*

Theorem 1 can be proved by using the following result ([19]):

Proposition 1: A sigraph S is balanced if and only if for every two vertices of S, all paths joining them have the same sign.
The matrix equivalent of Theorem 2 is:

Theorem 0.2. *([19]) A signed digraph S is balanced if and only if there exists a permutation matrix P of the same order as that of its adjacency matrix A(S) such that $PA(S)P^{-1}$ appears in the block-partitioned form $[M_1 : M_2; B : B']$ where the square submatrices M_1 and M_2 are principal minors whose nonzero entries are all +1s, B is a submatrix whose nonzero entries are all -1s and B' denotes the transpose of B.*

If, instead of requiring every semicycle in a sidigraph S to contain an even number of negative arcs, it is required that every cycle in S satisfy this property then S is said to be in a state of cycle-balance. Clearly, if S is a symmetric sidigraph (i.e. a sigraph) or a strongly connected sidigraph then the notions of balance and cycle-balance in S are equivalent. Is there a structural characterisation of a cycle-balanced sidigraph in terms of partition of its vertex set? The answer to this is provided in the form of the following proposition:

Proposition 2: A sidigraph $S = (D, \sigma)$ is cycle-balanced if and only if there exists a function $\mu : V(S) \to \{-1, +1\}$ such that for every cycle $Z = (u_1, u_2, ..., u_{n-1}, u_n, u_l)$, $n \geq 2$ one has,
$\Pi_{i=1}^{n} \mu(u_i)\sigma((u_i, ui+1)) = \mu(u_1)$, where $u_{n+1} = u_1$.

3.4 The Phenomenon of Clustering

The social groups corresponding to V_1 and V_2 in Theorem 2 may be regarded as subgraphs of similar ideologies in the social system, when its structure is represented by the si(di)graph S; a digest of the theorem is contained in a more general sociological principle that such subgroups of similar ideologies in a society coexist harmoniously thriving in an "agree-to-disagree" state between any two subgroups maintaining at the same time their internal cohesion [11]; in fact, this general principle was formulated as a generalisation of Harary's notion of balance in sidigraphs called clustering by Davis. Formally, a sidigraph S is said to be clustered if its vertex set can be partitioned into 'cohesive sets' $V_1, V_2, ...$ at most one of which could be empty, such that no negative arc of S has both its head and tail in the same subset for any i whereas every positive arc has both of its ends in the same subset for some index j. The following is a well known characterisation of clustered sidigraphs.

Theorem 0.3. *([19]) A finite sidigraph S is clustered if and only if no semicycle of S has exactly one negative arc.*

As such, with our definition of the notion of clustering given above, it may be observed that Theorem holds even for infinite sidigraphs.

Notwithstanding often admonished limitations of mathematical modeling, 'social network analysts' view their enterprise as having a structural focus that is concerned with the 'topology' of the social networks, their effects on the environment how they are generated and how they might evolve through time. On the other hand, the traditional social science language of variables and relationships between variables does not fit comfortably into this paradigm. Yet this is a natural language in which system causality can be expressed.

4 Social Networks

In this section we shall discuss briefly on social networks. A social network consists of a finite set of social entities called the actors and the relation or a set of relations defined on them. The relations defined on the nodes characterise the social networks in many ways. The network analyst would seek to model these relationships to depict the structure of a group of nodes forming a social network. From the social network perspective their analysis critically depends upon its structure. Since the structures may be behavioural, social, political or economic they play a major role in their impact and their evolution. Several real life popular social networks have come up over the past few years and it has increased the importance of social networks. Online social interaction has become very popular around the globe and it expected not to fade away. The development in social network and their utility has become possible due to the advancements in computer science in the form of computing power and the spread of World Wide Web.

In order to facilitate analysis, social networks are represented in two ways. These are graphs and matrices. In the first method, the nodes of the corresponding graph represent actors in the network and the edges represent the relations between the actors. A special term called "sociograms" is used for such type of representations. It has many advantages but it becomes difficult to represent and visualise when the size of the networks go beyond a certain limit. In this the number of rows and columns is equal to the number of actors and the relationship among the nodes is represented by the elements at the corresponding places in the matrix. One major advantage of this representation of this approach is that it allows mathematical and computer tools can be applied for summarize and to find patterns there in.

The representation of a social network through graphs depends upon the information and the type of relations existing among the nodes. For example, the graph may be a labeled one if the relations are to be quantified. It may be a directed graph if the directions of the relations are to be represented.

5 Societal Networks

In this section we discuss and analyse the concept of societal networks introduced by J.Fiksel [15] in 1980 as a model for dynamic social networks. The term "societal network" to denote a mathematical system having certain well defined properties.

Basing upon the concepts introduced in Fiksel [14], a societal network was defined by Fiksel [15] as follows:

Definition 0.5. A societal network is a labeled directed network (N, A, T), where N is a finite set of nodes, T is a finite set of relational types and $A \subseteq N^2 \times T$. The set A represents all the arcs in the network which link pairs of nodes. Thus, if $(x,y,u) \in A$, then there is an arc from node x to node y bearing the relational type u.

The other specifications for the network T are as follows:
(i) In a social context, the nodes would represent individual people, while the arcs would represent directed relations. It is assumed that the network is connected.
(ii) T has the property that if $u \in T$, then there exists a reciprocal relation $u^{-1} \in T$.
Thus we are enabled to consider that the arc (x, y, u) is being equivalent to an oppositely-directed arc (y,x,u^{-1}). Some examples of this concept are, the reciprocal of "superior to" is "inferior to", so that x is superior to y if and only if y is inferior to x. When $u = u_{-1}$, we are dealing with a symmetric relation such as "friend of".
(iii) We define G(x) to be the set of nodes adjacent to x, $G(x) = \{y \in N : (x,y,u) \in A, u \in T\}$ and let R(x, y) be the relation u on the arc (x, y, u). We shall not permit self-loops, that is $x \notin G(x)$.
The network is endowed with dynamic properties as follows:
Let us suppose that each node is always in one of a finite number of possible states, denoted by the set S. At every instant of discrete time (t = 0, 1, 2) a node may undergo a change of state; furthermore, these state transitions are influenced only by the states of adjacent nodes in the network. More precisely, let $\sigma_x(t)$ denote the state of node x at time t. Then we postulate $\sigma_x(t+1) = \phi_x(\sigma_x(t); \{\sigma_y(t), R(x,y) : y \in G(x)\})$. That is, a state transition for node x is determined according to the transition rule ϕ_x, whose arguments are the previous state of node x and the previous states of all nodes adjacent to x and the relations of these nodes to x.

The proposed model simply implies that the major factors determining a change of state are the states of related individuals. Of course, it is possible that external influences are present as well, but for a wide range of social contexts it is meaningful to focus on the network.

Now, given any initial state at time t = 0, the transition rule determines the entire future evolution of all nodes in the network. In the model the interest is not in the detailed state changes undergone by any one individual, but in the long-term patterns and trends that emerge when the societal network is viewed as a whole.

5.1 The Key Principle

Local properties of a societal network affect the global behaviour of the network. Equivalently, the minute interactions of related individuals collectively influence the macroscopic characteristics of the structure to which they belong.

5.2 The Origin

This model of dynamic societal networks can be viewed as a generalisation of an earlier model developed by French [17]. In fact, French postulated a group power structure with a single type of directed relation, denoting influence of one individual over another. To represent opinion, he used a numerical scale, analogous to a continuous interval state space. Transitions in this network constituted of changes in opinion, governed by a simple arithmetic transition rule. French was able to show that, given certain initial states, patterns of opinion convergence were determined by the structure of the network.

5.3 Extension to a Formal Model

Harary [21] later formalized and extended these results and suggested as an open area the consideration of multiple relational types.

The definitions of (static) graphs and networks involve the entities, V (a set of nodes), E (a set of edges), f (a function which maps vertices to numbers) and g (a function which maps edges to numbers). In a dynamic graph is a graph which is obtained when these four entities are allowed to change over time. Harary introduced four basic kinds of dynamic graphs as follows:

Type-1 Dynamic graph: It is a node-dynamic graph or digraph, the set of vertices V varies over time. Thus some nodes may be added or removed. As edges are defined through their incident vertices, the edges incident at these nodes are removed along with the nodes which are removed.
Type-2 Dynamic graph: It is an edge-dynamic (or arc-dynamic) graph or digraph where the set of edges E varies with time. Here the edges(or arcs) may be added or removed over time.
Type-3 Dynamic graph: It is a node weighted dynamic graph where the function f varies over time leading to changes in the weights of nodes.
Type-4 Dynamic graph: It is an edge or arc weighted graph or digraph where the function g varies over time leading to changes in the weights of edges.

A derived type of dynamic graph is the fully weighted dynamic graph where both f and g vary over time. So that in this case both the weights on nodes and the weights on edges change over time.

In fact combinations of all the types mentioned above can occur. An example for this can be a computer network with changing bandwidth (Type-4), changing topology (Type-2), changing computing power (Type-1) and the participating computers in the network may be crashing and recovering or new computers getting added/ existing computers being removed to/from the network represents a dynamic graph where all four types of dynamisms being incorporated.

5.4 Structural Equivalence

A kind of the notion of structural equivalence of objects was introduced by Lorrain and White [27]. In order to introduce it we require the following additional definitions:

Definition 0.6. A category C is constituted by a class Cobj- the elements of which may be called nodes but are usually called objects- together with a class CMor- the elements of which are called morphisms-, and provided with a structure of the following type:

They defined it as follows ([23],[24]):
(i) To each ordered pair (a, b) of objects is assigned a subset (a,b)Mor of CMor. The elements of (a,b)Mor are referred to as the morphism linking a to b. If M is a morphism linking a to b, this may be indicated by writing aMb. It is understood that every morphism appears as a link for at least one pair of objects.
(ii) If objects a, b, c and morphisms M, N are such that aMb and bNc then there is a morphism MN linking a to c, called the compound of M and N. The compound MN depends only on M and N; it is independent of any particular objects a, b, c that might be involved in their concatenation. If on the other hand there are no objects a, b, c through which M and N concatenate as above, then the compound of the two morphisms is undefined, the expression 'MN' has no meaning. This operation of compounding of morphisms is called the composite operation of C.
(iii) If objects a,b,c,d and morphisms M,N,P are such that aMb, bNc and cPd then (MN)P = M(NP) so that there is no ambiguity in speaking of the compound of M, N and P through a, b, c , d; it is a unique morphism-which may as well be denoted simply by MNP, deleting the parentheses- linking a to d: a(MNP)d. This is the property of associativity of the composition operation. Of course, by 2, MNP is independent of the particular a, b, c, d involved in the concatenation of the three morphisms.

Definition 0.7. Objects a, b of a category C are structurally equivalent if for any morphism M and any object x of C, a M x if and only if b M x and x M a if and only if x M b.

In other words, a is structurally equivalent to b if a relates to every object x of C in exactly the same ways as b does. From the point of view of the logic of structure, then a and b are absolutely equivalent, i.e. they are substitutable.

Clearly the relation of structural equivalence among objects of C is an equivalence relation so that objects in Cobj can be partitioned into classes of structurally equivalent objects.

However, the structural equivalence in a societal network was modified and defined by using the notion of equivalence relation and equivalence classes generated therein by Fiksel [15] as follows:

In a societal network, if each node has a different transition rule ϕ_x then very little can be predicted about the behaviour of the network. State changes will propagate throughout the network in a chaotic fashion, with no clear pattern emerging. However, we shall concern ourselves with networks in which the nodes may be partitioned into equivalence classes, so that equivalent nodes share certain structural properties. Moreover, we shall postulate that equivalent nodes have identical transition rules. The following definitions will be useful where $'-'$ denotes set differences.

Definition 0.8. The circle of two nodes $x, y \in N$ is given by V(x, y) = $G(x) \bigcup G(y) - \{x, y\}$.

Definition 0.9. The core of x and y is given by $\bigwedge(x,y) = G(x) \bigcap G(y) - \{x,y\}$.

The above definitions have a natural extension to include three or more nodes. We shall write

$\bigvee(x_1,, x_2, ..., x_n) = \bigcup_{i=1}^{n} G(x_i) - \{x_1,, x_2, ..., x_n\}$

$\bigwedge(x_1,, x_2, ..., x_n) = \bigcap_{i=1}^{n} G(x_i) - \{x_1,, x_2, ..., x_n\}$.

Definition 0.10. In a social network, a pair of nodes x and y are said to be structurally equivalent when

(10.1) $\bigvee(x,y) = \bigwedge(x,y)$

(10.2) for any $\upsilon \in \bigwedge(x,y), R(x,\upsilon) = R(y,\upsilon)$.

(10.3) if $y \in G(x), R(x,y) = R(y,x)$.

The structural equivalence of x and y will be indicated by $x \approx y$; this equivalence essentially means that x and y share a common set of relationships with a particular group of other nodes. Note that x and y need not be directly related.

It is easy to show that $\approx$ is an equivalence relation and induces a partition of N into disjoint equivalence classes. The key observation here is that the multitude of relations in the network can be summarized by simply observing the relations that exist among equivalence classes.

5.5 Reduced Networks

The equivalence-reduction of a network is defined as follows:

Definition 0.11. The equivalence-reduction of a network G = (N, A, T) is a network, G* = (N*, A*, T) in which

(11.1) The nodes of N* correspond to the structural equivalence classes of N.

(11.2) For $E_1, E_2 \in N^\star$, if $x \in E_1, y \in E_2$, and $y \in G(x)$ in G, then $(E_1, E_2; R(x,y)) \in A^\star)$.

Let us consider the societal networks represented in Fig. 1, where the nodes $1, 2, ..., 10$ represents employees in a department in a company of which $'1'$ is the manager; 2, 3 and 4 are the heads of the three different sections and $\{5,6,7\}$, $\{8\}$ and

{9,10} represent the set of employees working in the three sections. The technical persons working in each sections are taken as a colleague of each other. In Fig. 2, we provide the corresponding equivalence reduction of the societal network in Fig. 1.

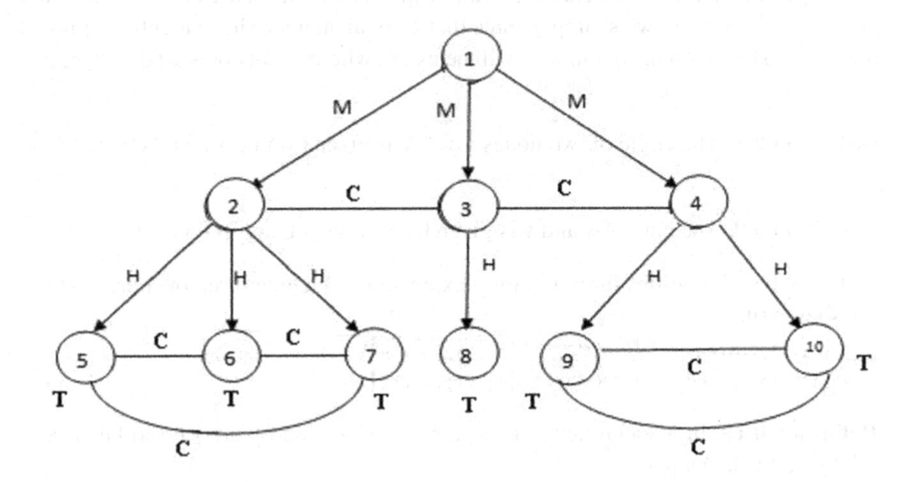

Fig. 1 Representation of nodes before reduction

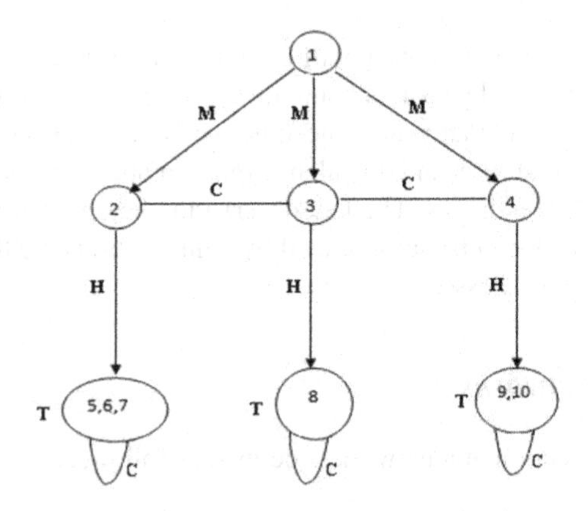

Fig. 2 Representation after reduction

The following assumption was made by Fiksel, which is crucial in the search for patterns of evolution:

Axiom 1.: Nodes which are structurally equivalent have the same transition rule. That is, $x \approx y \rightarrow \phi_x(\cdot) \equiv \phi_y(\cdot)$.

The transition rule $\phi_E(\cdot)$ of an equivalence class E, governs the state changes of any of the nodes in E , that is the relationships display similar behaviour. Formally, the identification of ϕ_x to ϕ_E can be defined as follows:

Definition 0.12. If node x belongs to equivalence class E, then $\phi_x(\cdot)$ is said to reduce to $\phi_E(\cdot)$. Given the set of arguments $\{c_F(t), R(E,F) : F \in G(E)\}$ the value of $\sigma_E(t+1)$ assigned ϕ_E by is the same as the value of $\sigma_x(t+1)$ assigned by ϕ_x when $\sigma_y(t) = \sigma_F(t)$ for $y \in F$.

The second assumption below ensures the consistency of $\phi_x(\cdot)$ and that definition of $\phi_x(\cdot)$ is mathematically sound.

Axiom 2.: If R(x, y) = R(x, z) then $\phi_x(\cdot)$ is unaltered by the interchange of y and z.

Each node E of N* has the transition rule $\phi_x(\cdot)$ and the symbol E will represents both a node of N* and the equivalence class of N, where the context resolves any ambiguity.

In Theorem 4 it is established that the reduction G* can be useful in analyzing the state dynamics of G. In fact, if the initial states of equivalent nodes are the same, then the entire future evolution of G may be determined from G*. It suggests an efficient method for determining the time-varying behaviour of a societal network G.

Theorem 0.4. *([15], Theorem 2) Let G = (N, A, T) be a societal network satisfying the initial condition:*
(4.1) If $x \approx y$ *then* $\sigma_x(0) = \sigma_E(0) = \sigma_y(0)$ *for* $x \in E, E \in N^\star$.
Then for any $t > 0$ *and for all x and E, we have*
(4.2) $x \in E \rightarrow \sigma_x(t) = \sigma_E(t)$.

From the reduction G*, the state of any node x in N at a future time t from the state of node E in N*, where can be determined. With the restriction (4.5.3), the computation required to trace the evolution of G* is significantly less than for G. The restriction (4.5.3) may not always be justified. But, it is removed in section 4.6 and the emerging patterns in the long-run evolution of the network are examined.

5.6 Class Structures

Structural equivalence is a strong property of societal networks, but it enables us to reduce the network only to a limited extent. In fact, additional reductions are possible if we take advantage of other structural symmetries in a network. Here we shall develop a weaker property of networks which is still sufficient to ensure the evolutionary pattern characterised by Theorem 4, while permitting much more efficient types of reduction.

Definition 0.13. A class structure of a societal network G = (N, A, T) is a partition of the set of nodes N into disjoint sets $C_1, C_2, ..., C_k$ called classes, satisfying the following:
For any two nodes x, y of a class C_i there exists a mapping M: $G(x) \to G(y)$ such that for $z \in G(x)$.
(13.1) R(x, z) = R(y, M(z)) and
(13.2) z is in the same class as M(z).

The mapping M is unrestricted. In a class structure, nodes of the same class have isomorphic sets of relations incident to them, but the nodes to which they are linked may be different. The notion of class structure generalises the notion of structural equivalence.

Definition 0.14. A class-reduction of a societal network G= (N, A, T) is a network $G^C = (N^C, A^C, T)$ where each node of N^C corresponds to a class in a class structure of G, $A^C \subseteq ((N^c)^2 \times T$ and $R(C_1, C_2) = R(x, y)$ for $x \in C_1, y \in C_2; C_1, C_2 \in N^C$.

It is clear that every class-structure of G defines a unique class-reduction of G. We assume here a property analogous to Axiom 1, that the transition rule $\phi_x(\cdot)$ is given by $\phi_C(\cdot)$ for $x \in C$, C being any class. The following theorem generalises Theorem 4.

Theorem 0.5. *([15], Theorem 3) Let G be a social network satisfying the initial condition:*
(5.1) $\sigma_x(0) = \sigma_y(0)$ *if* $x, y \in C$, *for each class C in a given class structure of G.*
Let G^C *be the corresponding class reduction of G, with transition rule* $\phi_C(\cdot)$ *for node C. Then for any* $t > 0$ *we have* $\sigma_x(t) = \sigma_E(t)$ *if* $x \in E$.

A comparison of the two concepts, structural equivalence and class structure shows that the former induces a partition of the set of nodes where as the later was defined in terms of a partition of the set of nodes, thus inducing an equivalence relation on the set of nodes. These concepts were applied to the modeling of societal dynamics.

5.7 *Self-balancing Networks*

The concept of a balanced network has received considerable attention in the literature of mathematical sociology. From the point of view of group dynamics, a balanced network was interpreted as having a certain structural stability due to the relationships among the members of a group. More specifically, suppose that these relationships are of two types, positive or negative. Thus the set T would consist of the two symmetric relations P and N, denoting positive and negative respectively: T = P, N where $P^{-1} = P, N^{-1} = N$.

Such a network is called a signed network. Cartwright and Harary [10] postulated that a signed network G is balanced if and only if all its cycles are positive, that is they contain an even number of N relations. This definition implies that the group members can align themselves into two opposing camps, with only P relations within each camp. This notion was extended by Davis [33] to the case where group members divide themselves into two or more "clusters". He proposed the following:

Definition 0.15. ([33]) G has a clustering if there exists a partition of N into disjoint sets (clusters), such that
(i) Nodes in the same cluster are joined by P relations
(ii) Nodes in different clusters are joined by N relations

Thus, a balanced network always has two clusters. Clustering is merely a general type of balancing. Davis also proved the following:

Theorem 0.6. *([11]) G has a clustering if and only if G contains no cycles with exactly one negative arc.*

It can be shown that every clustering is, in fact, a class structure. However, we are interested in exploring the ability of an unbalanced network to balance itself. In particular, we shall assume that all nodes have an identical transition function, and we will show that they are capable of modifying their relations until a clustering in achieved. This is an excellent example of a global objective being attained through local interpretations within the network.

Let the set of possible states for each node be $S = \{0, 1, 2\}$ and let us postulate that apart from performing state transitions, each node has the power to flip the relation on any arc from positive to negative or vice-versa. All nodes are initially in state 0. The balancing process is initiated when one node places itself in a state other than zero. The following two possible models of behaviour expressed in the form of transition functions were proposed in [15].

MODEL 1: Majority pressure/no waiting:

This model assumes a simple alignment process on the part of each node. The node x simply places itself in state 1 or 2, depending upon the states of its immediate neighbours, namely the set of nodes G(x). The "majority pressure" determines the selected state; for example if the majority of nodes y in G(x) are in state 1 with R(x, y) = p or state 2 with R(x, y) = N, then x will select state 1. The following notation will be helpful. Let us define, for I = 1 or I = 2 and u = P or u = N:

$S_{iu}(x,t) = \{y \in G(x) : R(x,y) = u, \sigma_y(t) = i\}$.

Thus, for example the set of nodes adjacent to x which are in state 2 at time t and have a positive relation to x is given by $S_{2P}(x,t)$. The transition rule ϕ_1 corresponding to this behaviour model is then summarized by the five rules ([15], Table 1). These rules indicate only those conditions which will cause a change of state or a change of relation between two nodes. Under any conditions not covered by these rules, x will remain in the same state, that is, $\sigma_x(t+1) = \sigma_x(t)$.

The first three transition rules assert that a node will select the state most compatible with its neighbours, taking into account both positive and negative relationships. In case of a tie, the choice is arbitrary. The last two transition rules assert that once a node selects states 1 or 2, it reconciles its relations with all neighbouring nodes. Fiksel provides an example ([15], Figure 4) in which a single node ia placed in state 1 at time 0. The state change propagate rapidly through network until a balanced network is reached. In this case, the relations on two arcs were changed.

The following theorem expresses in formal terms the ability of a network to become clustered:

Theorem 0.7. *([15], Theorem 5) Given a signed network G = (N, A, T), with all nodes following transition function ϕ_1 , let a single node x_0 be placed in state 1 at time t = 0. Then G will reach a stationary state in which the sets of nodes joined by P relations define a class structure. Furthermore,*
(i) *If G is not clustered, certain relations will be changed so that a clustered network results.*
(ii) *Nodes other than x_0 which are structurally equivalent will always be in the same cluster.*
(iii) *He stationary state will be reached within time $\tau = d(G) + 1$, where d(G) denotes the diameter of G.*

This elementary model of societal behaviour shows that simple rules of individual choice can lead to a balanced societal structure. Although the dynamics are somewhat mechanical, they provide a good starting point for analyzing group processes of this type. A model which might be considered more realistic is presented below; here each node may go through a period of indecision before selecting its final state. Although it may take longer for the network to become balanced or clustered, the number of changed relations will generally be fewer than in the previous model. It seems plausible that individuals will delay their decision rather than arbitrarily choosing a state which will violate half of their existing relations.

MODEL 2: Majority Pressure/Indecision

This model differs from the earlier one in a single feature: the procedure for breaking a tie. Let us expand the state space S to include a waiting state W. The transition rules for ϕ_2 are the same as for ϕ_1 except that the third rule is replaced by the following whenever

$|S_{2P}(x,t) \bigcup S_{1N}(x,t)| = |S_{2N}(x,t) \bigcup S_{1P}(x,t)|$:

If for some $z \in G(x), \sigma_z(t) = 0$ or W and $\sigma_x(t) = 0$ then $\sigma_x(t+1) = W$.
If for all $z \in G(x), \sigma_z(t) \neq 0$ or W and $\sigma_x(t) = 0$ or W then $\sigma_x(t+1) = 1$ or 2 (arbitrary choice).

Also, the other rules for change of state apply whether x is in state 0 or W.

Essentially, node x will not break a tie unless all of its neighbouring nodes are committed one way or the other. It would appear that Theorem 5 holds when ϕ_1 is replaced by ϕ_2, except that the time until G becomes balanced is bounded by the length of the longest cycle in G. However, under this model it is possible that a stalled situation is reached, in which several nodes are waiting to see what the others will do. This is illustrated in figure 5. Although a stationary state is reached, the network is not clustered and two nodes are stuck in the indecisive state.

These elementary examples show how a societal network can provide a formal model for interpersonal dynamics. Assumptions about network structure and individual behaviour are crucial in determining the global patterns of network evolution. Thus, the creation of realistic models will require a careful testing and validation of these assumptions. As Bernard and Killworth [9] have pointed out, 'models can only be useful aids to research if they proceed simultaneously with the development of better research tools." Experimental methods for observing and classifying changes in human behaviour will be a necessary prerequisite to the successful application of dynamic network models.

5.8 Long-Term Evaluation

In many types of mathematical models it is found that the effect of an initial state on the model is independent of the initial state. Keeping this in view, restriction on initial state in a social network was dropped and the behaviour of the network as t tends to infinity was investigated by Fiksel [15]. Because of transition rules the choice of initial state governs the long term evolution. However, when probabilistic transition rules are introduced, this dependence vanishes.

Theorem 0.8. *([15], Theorem 6) Let G = (N, A , T) be a societal network with arbitrary initial states. Then one of the following two outcomes holds as t $\to \infty$;*
***(a)** G enters a periodic repeating pattern of state transitions.*
***(b)** G enters a stable state in which $\sigma_x(t)$ is constant for $t > T$.*

Corollary 1.: ([15], Corollary 1) If axiom 1 holds then G has a unique stable state in which $\sigma_x(t) = \sigma_E$ for x in equivalence class E and $t \geq T$.

Corollary 2.: ([15], Corollary 2) The stable state of a societal network G is the same as the unique stable state of its equivalence reduction G*.

We remark that the uniqueness assumption is very strong; it is certainly possible to have a stable state of G in which $x \approx y$ but $\sigma_x \neq \sigma_y$.

To determine the existence of stable states, the following procedure can be used for any transition rule ϕ_x:

1. Find all state configurations for which $\{\sigma_y(t) : y \in G(x)\}$ for which $\sigma_x(t+1) = \sigma_x(t) = S_x$.

2. For each such configuration, check for the existence of $\sigma_z(t) : z \in G(y), z \neq x$ for which $\sigma_y(t+1) = \sigma_y(t)$.

3. Continue searching in a tree-like manner until either all nodes are exhausted or a node is found with no compatible stable state.

The above procedure can be accelerated by performing step 1 simultaneously for all nodes.

The question of whether a societal network will reach a stable state depends upon the initial state. In fact, the initial states separate into two groups, leading to either outcome (a) or outcome (b) of Theorem 8. Thus, in the deterministic case the choice of initial state can influence the entire evolution of the network.

The concept of a transition rule was further extended by Fiksel to allow a probability distribution over the state space S. The state transitions are Markovian in the network and not on its previous history. Instead of selecting a specific state based on its neighbouring nodes, a node enters several different states with varying probabilities. The probability distributions are defined as follows:

Definition 0.16. $P_x(A/\alpha) = Pr\{\sigma_x(t+1) = A/\sigma_{G(x)}(t) = \alpha\}$ = Probability that x selects state A given the states of all nodes in G(x). $\sigma_{G(x)}(t)$ and α denote vectors of states with dimension $|G(x)|$.

It is implicitly assumed that the distribution $P_x(\cdot|\cdot)$ is dependent upon the relations on all arcs incident to x.

Let N denote the number of nodes in N. the entire network can be viewed as a discrete Markov chain with state space S^N and with a transition function given by:
(16.1) $Q(\alpha,\beta) = Pr\{\sigma_N(t+1) = \beta|\sigma_N(t) = \sigma\}$.
Let β_x be the component of β corresponding to node x.

Assuming the state transitions of individual nodes are independent random variables, we have,
(16.2) $Q(\alpha,\beta) = \Pi_{x\in N}P_x(\beta_x|\sigma)$.

Let S* be the subset of the finite state space S which forms an irreducible Markov chain, namely the set of recurrent states. Then it is well known that,
(16.3) $lim_{t\to\infty}Pr[\sigma_N(t) = \alpha] = \pi_\alpha$.

In other words, the probability distribution of the state of the network (which is an N-vector) approaches a unique stationary distribution as time increases. Furthermore, this stationary distribution π is given by the solution of the system of equations
(16.4) $\pi_\alpha = \Sigma_{(\beta)}\Pi_{(\beta)}Q(\beta,\alpha), \forall\alpha \in S^\star$.
(16.5) $\Sigma_{(\alpha\in S^\star)}\pi_\alpha = 1$.

The following theorem was proved in [15].

Theorem 0.9. *([15], Theorem 7) Let G = (N, A, T) be a societal network with a probabilistic transition rule defined by $P_x(\cdot|\cdot)$ for each $x \in n$. Then as $t \to \infty, Pr\{\sigma_x(t) = A\}$ approaches initial state $\sigma_x(0)$ of the network G.*

The key point here is that, like in any Markov chain, the initial state of the network has only a transient effect and that the ultimate state distribution of individual nodes is determined purely by the transition mechanisms and the relations among nodes. Now, returning to the notion of structural equivalence, we find that by symmetry, if $x \approx y$ and Axiom 1 holds then $\pi_A(x) = \pi_A(y)$ for any $A \in S$. This is true since the positions of x and y in the state vector will not alter the steady-state equations from which π is derived. Thus, structurally equivalent nodes will have the same long-run state distribution, no matter what their initial state.

Previous research into probabilistic models of networks has been sparse, although Holland and Leinhardt [23] have applied Markov process models to the study of structural tendencies in sociometric data. Our probabilistic societal networks represent a class of stochastic systems ([27],[28],[29]) which have only begun to receive the attention that they deserve. One open area of particular interest is the relationship between equilibrium states of the original network and those of the reduced networks which can be derived from it. Despite their greater complexity, probabilistic networks offer a more realistic and flexible model of societal behaviour than do the deterministic networks discussed earlier.

6 Some Recent Developments in Dynamic Social Networks

In this section we discuss two aspects of dynamic social networks([35]) tackled in the recent past. These approaches are not in the same line as the societal networks proposed and discussed by Fiksel, which is presented by us above. But, these works shed some light on the importance of dynamic social networks in the modern society and some of the important properties of such networks have been obtained.

6.1 Mining Periodic Behaviour in Dynamic Social Networks

Conventional social networks are now superseded by continuous streams of dynamic interaction data or dynamic networks, which have opened the way to new techniques for analyzing the underlying populations. In order to identify the regular behaviour of social interaction which are infrequent and where the interaction pattern is hard to detect a mining algorithm for finding periodic or near periodic subgraphs in dynamic social networks was proposed by Lahiri and Wolf [25]. They also analysed the computational complexity of the problem, showing that, unlike any of the related subgraph mining problems, it is polynomial. A practical, efficient and scalable algorithm to find such subgraphs that that takes imperfect periodicity into account was also proposed.

The definition of a dynamic network used in this context is given differently in this work. It is,

Definition 0.17. A dynamic network G = $\langle G_1, G_2, ..., G_T \rangle$ is a time-series of graphs, where $G_t = (V_t, E_t)$ is the graph of interactions E_t observed at time step t, among the set of uniquely labeled entities $V_t \subseteq V$.

Definition 0.18. For an arbitrary graph F = $(V_f \subseteq V, E_f \subseteq V_f \times V_f$, its support set in G is the set of time steps where F is a subgraph of G_t, denoted $F \prec G_t$: S(F) = $t : F \prec G_t$, F is a frequent subgraph of G if $|S(F)| \geq \sigma$ where $1 \leq \sigma \leq T$ is a user-defined minimum support threshold.

Let $F^{(\sigma)}$ be the set of all frequent subgraphs of G at minimum support σ.

Definition 0.19. Definition 6.3: A frequent subgraph $F \in F^{(\sigma)}$ is maximal if there is no other frequent subgraph $F' \in F^{(\sigma)}$, where $F \prec F'$. $F_{max}^{(\sigma)} \subseteq F^{(\sigma)}$ is the set of maximal frequent subgraphs.

Definition 0.20. A frequent subgraph $F \in F^{(\sigma)}$ is closed if it is maximal at some support $\sigma \geq \sigma$. $F^{(\sigma)}_{closed} \subseteq F^{(\sigma)}$ is the set of closed frequent subgraphs.

Definition 0.21. A periodic support set or periodic subgraph embedding (PSE) of an arbitrary graph $F = (V_f, E_f)$ in G, where $V_f \subseteq V$, is a maximal, ordered set of time steps where F is a subgraph of G_t, such that the difference between consecutive time steps in the set is constant. Here, $S_P(F) = <t : F \prec G_t>$ and $\forall i : t_{i+1} - t_i = p$. The constant p is the period of F and F is a periodic subgraph if it has at least one such embedding with $|S_P(F)| \geq \sigma$, where σ is the minimum support threshold.

6.2 Periodic Subgraph Mining Problem

Given a dynamic network G and a minimum support threshold $\sigma \geq 2$, the PERIODIC SUBGRAPH MINING problem is to identify all closed periodic subgraphs in G.

It was proved that [25] the PERIODIC SUBGRAPH MINING problem in dynamic networks is in P. Also, an efficient, single-pass, polynomial time and space algorithm for mining all closed periodic subgraphs in a dynamic network was proposed.

It has been concluded in this paper that dynamic social networks have inherent periodicity and that these periodic interaction patterns can be extracted in an efficient and tractable manner.

6.3 Framework for Analysis of Dynamic Social Networks

The most common way to capture information on social interactions is a network. A major drawback of this model is that it is essentially static in that all information about the time that social interactions take place is discarded. The static nature of the model can give inaccurate and inexact information about patterns in the data. Thus decisions made based solely on the static data may be flawed [36]. The static graph representation prevents us from even asking certain fundamental questions about either the causes or consequences of social patterns.

In [36], Tanya and Jared have proposed a mathematical and computational framework that enables analysis of dynamic social networks and that explicitly makes use of information about the time that social interactions occur. Several algorithms have been presented for obtaining information about the structure of dynamic social networks in this framework. Also, the utility of these algorithms on real data has been discussed.

In this approach also, like that of Fiksel, a partition of the individuals into groups at every time step as input has been assumed. Also, it is assumed that a population of individuals is monitored in some way over a period of time. Interactions between individuals are recorded at every time step.

For completeness, we shall outline the important features of this work. This article contains a lot of open problems and is worth pursuing. For this, we need some additional notations and definitions taken from [36] as follows:

Given a population X ={ $x_1, x_2, ..., x_n$ }, we define a group to be a subset $g \subseteq X$. A set of partitions $P_1, P_2, ..., P_T$ with one partition per a time step is defined over X. Every partition P_i is a disjoint set of groups. $P(g)$ denotes the index of the partition to which g belongs. A similarity measure between two groups g and h is defined through a function $sim(g,h)$ and a turnover threshold β . The similarity measure $sim(,)$ satisfies the following properties:

(i) sim (g, g) has maximum similarity value.

(ii) sim(h, g) monotonically increases with the increase of $|h \cap g|$ when $|h| + |g|$ is fixed.

(iii) sim (h, g) monotonically increases with the decrease of $|h| + |g|$ when is fixed $|h \cap g|$.

Definition 0.22. Given partitions $P_1, P_2, ..., P_T$ of a set of individuals X, a set similarity measure sim(,), turnover threshold β and a function $\alpha(T)$, a metagroup MG is a sequence of groups MG = $\langle g_1, ..., g_l \rangle, \alpha(T) \leq l \leq T$ such that:

(i) No two groups in MG are in the same partition and the groups are ordered by the partition time steps:

$\forall i, j, 1 \leq i < j \leq l, P(g_i) < P(g_j)$,

(ii) The consecutive groups in MG are "similar" in that:

$\forall i, 1 \leq i < l, sim(g_i, g_{i+1}) \geq \beta$.

Definition 0.23. An individual $x \in X$ is a member of a metagroup MG = $\langle g_1, ..., g_l \rangle$ if the number of groups $g_1, ..., g_l$ to which x belongs is at least an priori chosen membership threshold function γ (which may be a function of T, the total number of individuals associated with MG and other parameters).

The interpretations of the three parameters α, β, γ are as persistence, turnover and membership respectively and they give meaning to a group. A weighted multipartite directed graph is used for the conceptual representation. G = $(V_1, ..., V_T, E)$, where V_i is the set of groups in partition P_i and $(g_i, g_j) \in E$ if $P(g_i) < P(g_j)$ and $sim(g_i, g_j) \geq \beta$. A metagraph in this graph is a path of length at least $\alpha(T)$ and is called a metagroup $\beta-$graph.

Algorithms have been provided to compute metagroup statistics (the number of metagroups and the average metagroup lifespan) and extremal metagroups (more persistent metagroup, most stable metagroup and largest metagroup).

Several critical network properties have been studied in this paper, though a lot more open problems are there. The problems handled include group connectivity and individual connectivity.

The group connectivity problem can be stated as:
Let $g_1,...,g_l$ be a set of groups in separate partitions ordered by their partition indices (i.e. $P(g_i) < P(g_{i+1})$). Then one can be interested in finding answers to the following questions:
(i) Is there a metagroup that contains all the groups?
(ii) Find the most persistent/stable/largest metagroup that contains all groups.
(iii) If no metagroup contains all the groups find the one that contains most.

Questions on individual connectivity similar to group connectivity above have been raised. Similarly questions on critical group set and critical individual set has been raised. These are:
(iv) CRITICAL GROUPS SET: Find the smallest set of groups whose removal leaves no metagroups (with respect to given β and α).
(v) CRITICAL INDIVIDUAL SET: Find the smallest set of individuals whose removal leaves no metagroups.
Discussions on (v): There are several greedy heuristics for this problem. In particular, one may iteratively remove the individual that:
1. Appears in the intersections of the largest number of groups that are still connected by an edge.
2. Removes the largest number or the heaviest of edges.
3. Reduces the weight of the edges by most.
4. Removes edges from the largest number of metagroups.

7 Problems for Further Study

In this section we present some open problems for further study.

We take this open area of research basing upon Fiksel's work as proposed by B. D. Acharjya (Cybernetic approach to study socio-psychological systems : Fiksel's societal network model, 2004, Lecture - 2, IASST Conference, Guwahati, India.)

7.1 Fiksel postulates that equivalent nodes have identical transition rules. It can be thought of to extend this to a more realistic model by postulating that equivalent nodes to have asymptotically identical transition rules to accommodate the long term behavior of the social system represented by the model.

The genesis of this postulation lies in the fact that in reality people exposed to the same set of social relationships are more likely to exhibit varying behaviors initially since behavior, being the response of the individual to the situation depends basically on the personality of the individual which is essentially an evolutionary characteristic; however their continuous exposure over a long period may be expected to develop in them similar behavior. This extended postulate leads to an open area for research.
7.2 In the development of the societal network and analysis through reduction the notion of equivalence relation is used. But these equivalence relations are very strict in nature. Can the requirement of an equivalence relation be relaxed to similarity relations so that the study will be wider and more applicable?

7.3 The relationships between nodes in a social network in general and a societal network in particular are imprecise. These imprecise relations can be better depicted through the several imprecise models. Also, the notion of fuzzy graph has been introduced long back and the theory has become rich by now. Can we define a model which is represented by a fuzzy graph and corresponding imprecise relations to make the study more reliable and more realistic? A list of open problems have been suggested in [36] which are based upon the model discussed in 6.2. We present below these problems.
7.4 (Individual membership) Given an individual x, find the metagroup MG which maximizes the cardinality of the set of groups in MG in which x occurs.
7.5 (Extroverts and Introverts) Find the individual who is a member of the largest (smallest) number of metagroups.
7.6 (Loyal Individuals) Given an individual, what fraction of its time is it a member of the same metagroup? Find the individuals that appear most frequently in one metagroup.
7.7 (Metagroup representative) Find out whether there is an individual who occurs more than any other individual in a metagroup MG and occurs in MG more than in any other metagroup.
7.8 (Demographic Distinction) Given a colouring of individuals (a partition), is there a property that distinguishes one colour from the others, i.e. some colour is in more metagroups, fewer metagroups, longer metagroups, more time in any metagroup?
7.9 (Critical Parameter Values) Find an estimation of the largest values of α, β for which there exists at least one metagroup. Identify the largest value of γ for which each metagroup has at least one member.
7.10 (Critical Time Moments) Determine the critical time moments.

8 Conclusion

The study of dynamism in social networks ([36]) was initiated and studied by French and later formalized through their graphical representation by Harary. But the realization of this study through a mathematical model in the form of a societal network was due to Fiksel. In this chapter our primary focus was to present the theoretical development of societal network along with the different results established in support of the theory. However, we have provided the basics of dynamism in social networks and their mathematical presentation over the years. The notion of sigraph and sidigraph, which represent the different forms of social networks are also presented. Societal networks have dynamic properties which cause fluctuations in the states of all nodes, due to the influence of nodes adjacent to them. The structure of a network is a cause of worry in their representation and analysis. One of the solutions to this is the reduction of their size such that the pattern of state changes of the original network is related directly to those of the reduced network is achieved through the concepts of "structural equivalence" and "class structure". As an illustration of how these ideas can be applied to social science research, the notion of a self-balancing network was explored. The long-term evolution of societal networks was investigated for both deterministic and probabilistic transition rules and the existence of equilibrium state distributions was demonstrated.

Recently dynamic social networks are studied extensively by researchers in the field, although the model of Fiksel has not received much attention. We have presented two such works; Mining Periodic Behaviour in Dynamic Social Networks and an efficient framework which facilitates the study of such networks have been presented. We have presented several open problems, which can be pursued for further study.

References

1. Acharya, B.D., Joshi, S.: On sociograms to treat social systems endowed with dyadic ambivalence and indifference. In: (Invited presentation at the) 17th Annual Session of the Ramanujan Mathematical Society, Banaras Hindu University, Varanasi, India (June 10-12 2002) (preprint, June 2001)
2. Acharya, B.D., Joshi, S.: On the complement of an ambisidigraph. In: R. C. Bose centenary International Symposium on Discrete Mathematics and Applications, December 20-23. Indian Statistical Institute, Kolkata (2002); (Also see Electronic Notes on discrete Mathematics (2003)
3. Alderfer, C.P.: Group and intergroup relations. In: Hackman, J.R., Suttle, J.L. (eds.) Improving life at Work. Goodyear, Santa Monica (1977)
4. Alderfer, C.P., Smith, K.K.: Studying intergroup relations embedded in organisations. Administrative Science Quarterly 27, 35–65 (1982)
5. Bacharach, S.B., Lawler, E.J.: Power and politics in organisations. Jossey-Bass, San Fransisco (1980)
6. Barnes, J.A.: Social Networks. Addison Wesley Module (1972)
7. Bateson, G.: Mind and nature. Bantam, New York (1979)
8. Bavelas, A.: A Mathematical model for group structures. Applied Anthropology 7, 16–30 (1948)
9. Bernard, H.R., Killworth, P.D.: Deterministic Models of social Networks. NITS report AD-A014, 968 (1975)
10. Cartwright, D., Harary, F.: Structural Balance: A generalisation of Heider's theory. Psych. Review 63, 77–293 (1956)
11. Davis, J.A.: Clustering and structural balance in Graphs. Human Relations 20, 181–187 (1967)
12. Doreian, P.: Mathematics and study of Social relations. Schoken Books, New York (1970)
13. Fararo, T.: Mathematical Sociology: An introduction to Fundamentals. Wiley, New York (1973)
14. Fiksel, J.: A network of Automata Model for Question Answering in Semantic Memory. Technical Report No.218. Institute for Mathematical Studies in the Social Sciences, Stanford, California (1973)
15. Fiksel, J.: Dynamic evolution in societal Networks. Jour. Mathematical Sociology (7), 17–46 (1980)
16. Flament, C.: Applications of Graph Theory to Group Structure. Prentice Hall, Englewood Cliffs (1963)
17. French, J.R.P.: A formal theory of social power. Psychology Review 63, 181–194 (1956)

18. Goffman, E.: The presentation of self in Everyday Life. Doublday, Garden City (1959)
19. Harary, F.: On the notion of balance of a signed graph. Michigan Mathematical Journal 2, 143–146 (1953)
20. Harary, F.: Structural Models: An Introduction to the theory of Directed Graphs. Wiley, New York (1963)
21. Harary, F.: Graph Theory. Addison Wesley, New York (1964)
22. Harary, F.: Graph Theoretic Methods in the management sciences. Management Science 5, 387–403 (1959)
23. Holland, P.W., Leinhardt, S.: A method for detecting structure in sociometric data. In: Leinhardt, P.W. (ed.) Social Networks. Academic Press, New York (1977)
24. Kahn, R.L., Wolfe, D.M., Quinn, R.P., Snock, J.D., Rosenthal, R.A.: Organisational Stress: Studies in Role Conflict and Ambiguity. Wiley, New York (1964)
25. Lahiri, M., Berger-Wolf, T.Y.: Berger-Wolf: Mining Periodic Behaviour in Dynamic Social Networks. In: Proceedings of 2008 Eighth IEEE International Conference on Data Mining, pp. 373–382 (2008), doi:10.1109/ICDM.2008.104
26. Lorrain, F.: Social Networks and Social Classifications. Editions Hermann, Parris (1975)
27. Lorrain, F., White, H.C.: Structural equivalence of individuals in social networks. Journal of Mathematical Sociology 1, 49–80 (1971)
28. Parzen, E.: Stochastic Processes. Holden-Day, San Francisco (1962)
29. Rapoport, A.: Contributions to the theory of random and biased nets. Bulletin of Mathematical Biophysics 19, 257–277 (1957)
30. Rapoport, A., Horvath, W.: A study of a large sociograms. Behavioral Science 6, 279–291 (1961)
31. Rice, A.K.: Individual, group and intergroup processes. Human Relations 22, 565–586 (1969)
32. Roberts, F.S.: Graph Theory and its application to problems of society. SIAM, Philadelphia (1978)
33. Shepard, R.N., Arabie, P.: Additive clustering: representation of similarities as combinations of discrete overlapping properties. Psychol. Rev. 86(2), 87–123 (1979)
34. Smith, K.K.: An intergroup perspective on individual behaviour. In: Hackman, J.R., Lawler, F.E., Porter, L.W. (eds.) Perspective on Behaviour in Organisations. McGraw-Hill, New York (1977)
35. Smith, K.K.: Social comparison process and dynamics conservation in intergroup relations. In: Research in Organisational Behaviour, vol. 5, pp. 199–233. JAI Press Inc. (1983)
36. Berger-Wolf, T.Y., Jared, S.: A Framework for Analysis of Dynamic Social Networks. In: KDD 2006, August 20-23. ACM, Philadelphia (2006), doi:1-59593-339-5/06/0008

Methods of Tracking Online Community in Social Network

Sanjiv Sharma and G.N. Purohit

Abstract. Social relationships and networking are key components of human life. Social network analysis provides both a visual and a mathematical analysis of human relationships. Recently, online social networks have gained significant popularity. This popularity provides an opportunity to study the characteristics of online social network graphs at large scale. An online social network graph consists of people as nodes who interact in some way such as members of online communities sharing information using relationships among them. In this paper a state of the art survey of the works done on community tracking in social network. The main goal is to provide a road map for researchers working on different measures for tracking communities in Social Network.

1 Introduction

A social network is a set of people (or organizations or other social entities) connected by a set of social relationships, such as friendship, co-working or information exchange. Social network analysis [1] focuses on the analysis of patterns of relationships among people, organizations, states and such social entities. Online social network provide social networking services over internet. The members of online social network can interact to each other in form of online communities. Communities are considered as groups of densely connected members that are only loosely connected to the rest of the network. Online community [2] is a subgroup where all members of online social network can share information about common interest. The evolution of communities over time is typically analyzed by observing changes in the interaction behavior of their members. The structure of community over time shows stable communities with a considerable amount

Sanjiv sharma · G.N. Purohit
Banasthali Vidyapith, Bansthali, Rajasthan, India
e-mail: er.sanjiv@gmail.com, gn_purohitjaipur@yahoo.co.in

M. Panda, S. Dehuri, and G.-N. Wang (eds.), *Social Networking*
Intelligent Systems Reference Library 65,
DOI: 10.1007/978-3-319-05164-2_6, © Springer International Publishing Switzerland 2014

of members who participate over a long time and a small amount of fluctuating members. Many researchers have been proposed various methods and algorithm for tracking community structure over time in online social network. This chapter illustrates review of all researches for tracking online community in social network.

2 Related Work

Social network analysis (SNA) [1] is an interdisciplinary methodology developed mainly by sociologists and researchers in social psychology in the 1960s and 1970s, further developed in collaboration with mathematics, statistics, and computing that led to a rapid development of formal analysing techniques which made it an attractive tool for other disciplines like economics, marketing or industrial engineering. SNA is based on an assumption of the importance of relationships among interacting units or nodes. Existing methods for tracking community is based on following review of research relevant to following problems that need to be addressed in subgroup analysis of social networks.

2.1 Subgroup Identification

The problem of identifying subgroups has been a challenging issue in sociology. Wasserman and Faust [3] defined a cohesive subgroup as a set of actors (nodes) that are relatively dense and directly connected through reciprocated (bi-directional) relationships (links). They analyzed subgroups over time with in small size of network.

Kumar [4] *et al.* addressed the problem of finding emerging subgroups on the Web using heuristic style & created larger community from union of smaller community. The method proposed by him related to degree centrality as a way of selecting potential subgroups and suitable for scalable large network. While an early example of subgroup identification and of a heuristic style of research that was designed to be scalable to very large networks, bur proposed method was not fully automated and involved some filtering and interpretation by humans.

Research on finding Web communities has tended to utilize content analysis of text and tags associated with Web pages and allow the subgroup identification task to be linked to powerful search engine algorithms. Flake [5] presented a heuristic community identification algorithm which used Web pages as topic seeds. Searches and link analysis were then used to identify a community of pages relating to the topic seeds. In similar vein, Chau [6] *et al.* developed a method identifying the communities associated with business Web sites by tracking back through the incoming links and carrying out data mining on the resulting network. Gruzd and Haythornthwaite [7] examined the links between people in threaded conversations, but also analyzed the content of the posts to infer the order in which people were talking to each other in the discussion. According to them, this created a social network representative of actual conversation rather than the less informative chain networks that show the structure of how messages connect to each other. Text in the threaded conversations can also be analyzed using natural language processing

(NLP) algorithms to identify nouns and phrases, which may be used to automatically find and characterize communities.

There has also been considerable graph-theoretic research on finding densely-connected subgraphs within larger graphs. For instance, Gibson, Kumar, and Tomkins [8] presented a method for detecting densely connected groups of servers on the Web. However, these analyses (particularly when applied on a large scale) have typically focused on static social networks.

Recently, researchers such as Tantipathananandh, Berger-Wolf and Kempe [9] have addressed the issue of finding subgroups in dynamic social networks using an optimization approach. However, their evaluations have tended to involve either synthetic datasets or networks with relatively few members (considerably fewer than 100).

Recently, Microsoft has described a supervised learning approach to identifying communities in a patent application [10]. In this type of approach (a form of machine learning), classification of objects and their relationships is conducted through initial training to determine if an object is a member of the community. The classificatory activity may utilize techniques such as Support Vector Machines [11] and feature-based object classification algorithms such as PopRank [12].

In summary, while there has been considerable work on methods of automated subgroup identification, much of it has been done on small or static networks, or has used known topics or Web pages as seeds. In addition, much of the focus has been on finding communities based on links between text and web content, rather than on links between people as reflected in their online interactions. Thus, the development of a scalable and valid method for discovering cohesive subgroups of people based on large scale online interactions within social networks, using an unsupervised learning approach, is still an open research problem.

2.1.1 Centrality

Network centrality (or centrality) is a widely used measure for identifying possible members of subgroups in networks derived from online interactions (e.g., with respect to blogs), since centrality measures how important or central an individual node is to a network. A centrality measure can be used to identify the most important people at the center of a network or those that are well connected. Numerous centrality measures such as degree [13,14] ,closeness[15,16,17], betweenness[18,19,20], information [21,22], eigenvector [23,24], and dependence centrality[25,26] have been used for characterizing the social behaviour and connectedness of nodes within networks. The logic of using centrality measures is that people who are actively involved in one or more subgroups will generally score higher with respect to centrality scores for the corresponding network. Thus, high scores of network centrality (relative to others in the social network) should be predictive of subgroup membership, although the relationship is by no means deterministic. Researchers have compared and contrasted centrality measures in various social networks. There are four measures of centrality that are widely used in network analysis: degree centrality, betweenness, closeness, and eigenvector centrality.

2.1.2 Degree Centrality

Degree centrality [27] is defined as the number of links incident upon a node (i.e., the number of ties that a node has). Degree is often interpreted in terms of the immediate risk of node for catching whatever is flowing through the network (such as a virus, or some information). If the network is directed (meaning that ties have direction), then we usually define two separate measures of degree centrality, namely indegree and outdegree. Indegree is a count of the number of ties directed to the node, and outdegree is the number of ties that the node directs to others.

For positive relations such as friendship or advice, we normally interpret indegree as a form of popularity, and outdegree as gregariousness. For a graph G:=(V,E) with n vertices, the degree centrality $C_D(v)$ for vertex v is:

$$C_D(v) = deg(v)/(n-1) \tag{1}$$

The definition of centrality can be extended to graphs. Let v * be the node with highest degree centrality in G. Let X: = (Y,Z) be the n node connected graph that maximizes the following quantity (with y * being the node with highest degree centrality in X):

$$H = \sum_{j=1}^{|Y|} C_D(y*) - C_D(y_j) \tag{2}$$

Then the degree centrality of the graph G is defined as follows:

$$C_D(G) = \sum_{i=1}^{|Y|} (C_D(v*) - C_D(v_i))/H \tag{3}$$

H is maximized when the graph X contains one node that is connected to all other nodes and all other nodes are connected only to this one central node (a star graph). In this case

$$H = (n-1)(n-2) \tag{4}$$

so the degree centrality of G reduces to:

$$C_D(G) = \sum_{i=1}^{|Y|} (C_D(v*) - C_D(v_i))/(n-1)(n-2) \tag{5}$$

2.1.3 Betweenness Centrality

Betweenness[27,28] is a centrality measure of a vertex within a graph (there is also edge betweenness, which is not discussed here). Vertices that occur on many shortest paths between other vertices have higher betweenness than those that do not.

For a graph $G := (V,E)$ with n vertices, the betweenness $C_B(v)$ for vertex v is computed as follows:

1. For each pair of vertices (s,t), compute all shortest paths between them.
2. For each pair of vertices (s,t), determine the fraction of shortest paths that pass through the vertex in question (here, vertex v).
3. Sum this fraction over all pairs of vertices (s,t). Or, more succinctly:[2]

$$C_B(v) = \sum_{s \neq v \neq t \in V}^{|Y|} \sigma_{st}(v)/\sigma_{st} \tag{6}$$

where σ_{st} is the number of shortest paths from s to t, and $\sigma_{st}(v)$ is the number of shortest paths from s to t that pass through a vertex v. Calculating the betweenness and closeness centralities of all the vertices in a graph involves calculating the shortest paths between all pairs of vertices on a graph. In calculating betweenness and closeness centralities of all vertices in a graph, it is assumed that graphs are undirected and connected with the allowance of loops and multiple edges. When specifically dealing with network graphs, oftentimes graphs are without loops or multiple edges to maintain simple relationships (where edges represent connections between two people or vertices). In this case, using Brande's algorithm [32] will divide final centrality scores by 2 to account for each shortest path being counted twice.

2.1.4 Closeness Centrality

In topology and related areas in mathematics, closeness [27] is one of the basic concepts in a topological space. Intuitively we say two sets are close if they are arbitrarily near to each other. The concept can be defined naturally in a metric space where a notion of distance between elements of the space is defined, but it can be generalized to topological spaces where we have no concrete way to measure distances. In graph theory closeness is a centrality measure of a vertex within a graph. Vertices that are 'shallow' to other vertices (that is, those that tend to have short geodesic distances to other vertices within the graph) have higher closeness. Closeness is preferred in network analysis to mean shortest-path length, as it gives higher values to more central vertices, and so is usually positively associated with other measures such as degree.

In the network theory, closeness is a sophisticated measure of centrality. It is defined as the mean geodesic distance (i.e., the shortest path) between a vertex v and all other vertices reachable from it:

$$\sum_{t \in V \; v}^{n} (d_G(v,t))/(n-1) \tag{7}$$

where n¿=2 is the size of the network's 'connectivity component' V reachable from v. Closeness can be regarded as a measure of how long it will take information to spread from a given vertex to other reachable vertices in the network.

Some define closeness to be the reciprocal of this quantity, but either way the information communicated is the same (this time estimating the speed instead of the timespan). The closeness $C_C(v)$ for a vertex v is the reciprocal of the sum of geodesic distances to all other vertices of V:

$$C_c(v) = 1/(\sum_{t \in V\ v} (d_G(v,t))) \quad (8)$$

Different methods and algorithms can be introduced to measure closeness, like the random-walk centrality [14] introduced by Noh and Rieger (2003) that is a measure of the speed with which randomly walking messages reach a vertex from elsewhere in the network a sort of random-walk version of closeness centrality. The information centrality [27,31] is another closeness measure, which bears some similarity to that of Noh and Rieger. In essence it measures the harmonic mean length of paths ending at a vertex i, which is smaller if i has many short paths connecting it to other vertices. Dangalchev (2006), in order to measure the network vulnerability, modifies the definition for closeness so it can be used for disconnected graphs and the total closeness is easier to calculate:

$$C_c(v) = \sum_{t \in V\ v} 2^{-d_G(v,t)} \quad (9)$$

2.1.5 Eigenvector Centrality

Eigenvector centrality [30] is a measure of the importance of a node in a network. It assigns relative scores to all nodes in the network based on the principle that connections to high-scoring nodes contribute more to the score of the node in question than equal connections to low-scoring nodes.

Using the adjacency matrix to find eigenvector centrality : Let x_i denote the score of the ith node. Let $A_{i,j}$ be the adjacency matrix of the network. Hence $A_{i,j} = 1$ if the ith node is adjacent to the jth node, and $A_{i,j} = 0$ otherwise. More generally, the entries in A can be real numbers representing connection strengths, as in a stochastic matrix. For the ith node, let the centrality score be proportional to the sum of the scores of all nodes which are connected to it. Hence

$$X_i = (1/\lambda) \sum_{j \in M(i)\, j=1}^{N} (X_j) = (1/\lambda) \sum A_{i,j}(X_j) \quad (10)$$

where M(i) is the set of nodes that are connected to the ith node, N is the total number of nodes and λ is a constant. In vector notation this can be rewritten as $X = (1/\lambda)AX$ or as the eigenvector equation $AX = \lambda.X$ In general, there will be many different eigenvalues ? for which an eigenvector solution exists. However, the additional requirement that all the entries in the eigenvector be positive implies (by the Perron-Frobenius theorem) that only the greatest eigenvalue results in the desired centrality measure. The ith component of the related eigenvector then gives the centrality score of the ith node in the network. Power iteration is one of many eigenvalue algorithms that may be used to find this dominant eigenvector. Principal eigenvector of the (possibly valued) adjacency matrix of a network.

Betweenness centrality is mostly used to find and measure subgroup and community membership whereas degree and closeness centrality are used for characterizing influential members. Although network centrality measures are easy to calculate using computer programs such as Pajek and UCINET [33], there has been no consensus among researchers as to the most meaningful centrality measure to use for finding subgroup members. In extremely large social networks, computational efficiency may become an issue in selecting which centrality measure to use. With respect to three commonly used centrality measures, degree centrality is the easiest to calculate, closeness centrality is more complex and betweeness centrality has the highest calculation complexity.

2.2 Clustering and Partitioning of Subgroups

Finding groups within networks is an example of a pattern recognition task. In general, pattern recognition methods may be divided into two broad types reflecting the amount of information about a potential group provided to the analysis process. In the case of supervised learning (classification) the nature of groups is known and the problem is to assign new members to those groups. Methods of classification include discriminant analysis, nearest-neighbour assignment, and a number of Bayesian approaches [35].

However, if the basis for grouping is not already known, then unsupervised learning must be used. Since the goal in this dissertation is to construct a general and scalable technique for identifying cohesive subgroups, the focus will be on unsupervised learning.

Finding cohesive subgroups within social networks in the absence of prior knowledge about grouping (i.e., in an unsupervised fashion) is a problem that has attracted considerable interest. Clique analysis and related methods look directly at the links that occur in a network and identify specific patterns of connectivity (e.g., subgroups where everyone in the subgroup has a direct connection to everyone else). Clustering and partitioning methods are less direct (but more computationally efficient) in that they base their groupings (clusters) on proximity measures (similarities or distances) derived from the connection patterns between network nodes. Both clique analysis and clustering approaches will be considered in the following subsections.

2.2.1 Clique and k-plex Analysis

Cliques and k-plexes have been used to characterize groupings in social networks [36,37,38]. Cliques are fully connected subgroups [3] where each member has a direct connection to every other member in the subgroup, thus forming a completely connected graph within the subgroup. In n-cliques, the requirement that subgroup members are fully connected is relaxed, so that each member now has to be connected to every other member in the subgroup within some maximum allowed distance which is typically small, since small-world network phenomena tend to lead to an explosion in the size of n-cliques as the size of n increases. Pure cliques tend to be rare in social networks because the criterion of full connectedness is too strict to apply to most social networks. Additionally, cliques tend to fare poorly in dynamic settings where individual edges may be added or deleted relatively quickly, leading to radical changes in the structure of inferred subgroups. However, the most important argument against cliques is that they are too restrictive and will miss many meaningful subgroupings.

The criteria for subgroup formation can be further relaxed by allowing subgroups to form in which members are not completely connected and where each node in the subgroup has direct ties to at least n-k members. The resulting structure is referred to as a k-plex. K-plex analysis has also been used for finding subgroup members in a network. However, similar to cliques, finding k-plexes in large networks is a computationally expensive and exhaustive process because it scales exponentially with the number of nodes in the network. A further problem is that the value of k needs to be selected and the most appropriate value of k for subgroup analysis in a particular social network may not be obvious.

2.2.2 Cluster Analysis

Clustering and methods initially developed for problems such as numerical taxonomy may also be applied to analysis of subgroups and communities. While clustering is a conceptually promising approach, in practice there are a huge number of clustering and other techniques that may be potentially relevant to the detection of subgroups within social hypertext networks. For instance, link analysis can be used to identify topics within clusters of web pages. Co-citation analysis has been used to rank search engine results and to find groupings in Web pages, blogs, and tags associated with web pages.

Hierarchical clustering can automate the process of finding subgroups. It groups nodes into a cluster if the nodes are similar and then successively merges clusters until all nodes have been merged into a single remaining cluster. Techniques based on hierarchical clustering have been used to quantify the structure of community in documents [39], web pages, blogs, [40] and discussion groups [34]. Hierarchical clustering using such algorithms as in, results in a hierarchy (tree) being formed where the leaves of the tree are the nodes that are clustered. The resulting trees can be visualized as dendrograms. There are different methods to perform hierarchical clustering depending on how distances between clusters are defined, as explained by Johnson [42]. In contrast to hierarchical cluster analysis, the groups formed in partitioning methods are not nested. The k-means algorithm [43] is a popular method for partitioning that is available in widely used statistical packages such as SPSS. K-means analysis has been used to detect clusters in blogs. One example of a hiearachical clustering is a correspondence tree, or a dendrogram (shown in Fig. 1), which shows how samples are grouped together. The first level shows all samples x_i as singleton clusters. As we increase levels, more and more samples are clustered together in a hierarchical manner.

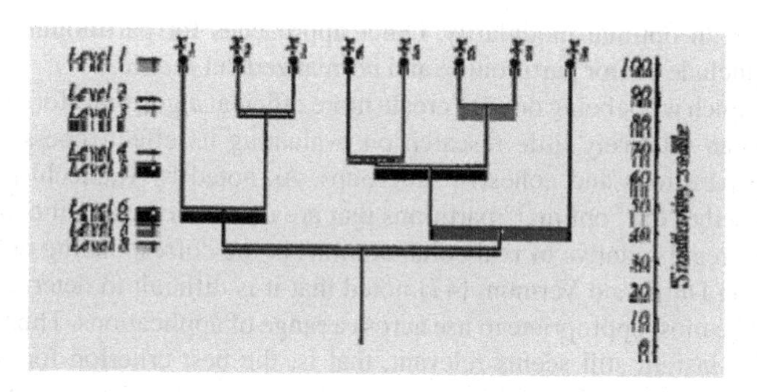

Fig. 1 Dendrogram

Partitioning methods are relatively efficient, but they require that the number of subgroups in the partition be defined prior to the analysis. On the other hand, hierarchical cluster analysis does not yield a partition and the hierarchy (dendrogram) that is output needs to be cut in order to identify a particular set of subgroups. In practice, for both partitioning analysis and hierarchical cluster analysis, the method needs to be supplemented with an additional selection criterion. For partition analysis, the method is run using a number of different values of k (i.e., number of groups in the partition) and the selection criterion is used to define which of the possible partitions should be chosen as the best subgrouping. For hierarchical clustering, the selection criterion is used to decide at which point the dendrogram should be cut in order to obtain a non-nested set of subgroups.

2.2.3 Partitioning Criteria

Orford [44] described a range of criteria for determining where to partition a dendrogram. Orford made the important point that the best criterion to use will generally vary with the problem context. In contrast to Orford's eclectic approach, recent research has tended to assess specific measures for obtaining an optimal partition. The modularity (designated as Q) discussed by Newman and Girvan [19], has been proposed as a definitive measure of the quality of clustering. Newman [24] claimed that maximizing modularity results in a set of clusters that best represents optimum subgroup structure.

Modularity has been used for finding community structure and subgroups in networks. The computational performance of different algorithms based on modularity was evaluated in and algorithms have been revised to improve over Newman's original method. Most recently, Noack [45] showed that the set of clusters found from optimum modularity, corresponded to the clusters formed by visualizing the social network using an energy layout model of pairwise attraction and repulsion between the nodes in the network. This is an important result since many researchers often use energy layout algorithms for visualizing social network structure in social network analytic software such as Pajek and NetDraw, in addition to finding clusterings with optimal modularity. Other approaches for partitioning based on optimality include vector partitioning and normalized cut metrics.

Despite much work being done to create more efficient algorithms for modularity, there has been relatively little research on evaluating its effectiveness in finding meaningful partitions and cohesive subgroups. As noted by Radicchi [46] , it is not clear whether the "optimal" partitions that are discovered using the modularity criterion are representative of real collaborations in the corresponding online communities. Van Duijn and Vermunt [47] noted that it is difficult to determine which measure is the most appropriate to use across a range of applications. Thus, Orford's [44] original insight still seems relevant, that is, the best criterion for splitting a dendrogram may depend on factors such as the type of data being collected and compared.

2.3 Similarity

Most social networks are dynamic and connections between people change naturally over time for a number of reasons including recruitment to the network, attrition, and changing relationships between members of the network. Thus even if it were possible to have a definitive measure of optimality at a single point in time, the optimal subgrouping at one point in time would be unlikely to remain optimal at later times. In general, groupings that are transitory or ephemeral will be of less interest than groupings that remain cohesive over time. How can cohesion over time be measured?

2.3.1 Existing Models of Similarity

Cohesive subgroups should have a core group of people that remain the same over different time periods. The situation is complicated by the fact that subgroups may split or merge, so that cohesiveness is not necessarily a property of a single subgroup, but may sometimes relate to a family of one or more related subgroups. However, in general, cohesive families of subgroups at one time period should be similar to corresponding subgroups at a different time period. Similarity is a topic that has received attention in a wide variety of scientific fields and a number of approaches are available for the measurement of similarity. Mathematically, similarity may be viewed as a geometric property involving the scaling or transformation necessary to make objects equivalent to each other. Similarity can be defined as the inverse of distance, with a well-known distance measure being Euclidean distance [48], which itself is a special case of a family of distance measures known as Minkowski metrics [49]. However, distance measures typically require a vector (spatial) model of the entities being compared, which is often not appropriate for comparing aggregations of nodes in a network. In developing methods to assess the similarity between different species, numerical taxonomists have developed and utilized a number of similarity measures. Many of these measures involve some sort of correlation, a construct that is conceptually related to similarity. One correlation measure is the cosine distance or dot product that measures the angle between two objects represented as vectors of numerical features. However, since features cannot always be expressed on a well-defined numerical scale, researchers (e.g., psychologists) have developed feature models of similarity that assess similarity based on a comparison of matching andmismatching features, using a set-theoretic approach. Tversky's [50] feature contrast model expressed the degree of similarity of two stimuli to a linear combination of their common and distinctive features. Gregson recommended a content similarity model where similarity was expressed as the ratio of the intersection of the features for the objects being compared to the union of their features. A simplified version of the content similarity model is the Jaccard coefficient (first proposed in 1901 by Paul Jaccard [51]), which is defined as the size of the intersection divided by the size of the union of the objects being compared.

Johnson [42] proposed the ultrametric distance as a way of measuring distance within a hierarchy. For comparing two different clustering hierarchies,one heuristic method for estimating similarity consists of converting each hierarchy to a matrix of ones and zeros where the ones represent the parent-child links in each hierarchy. The similarity between two hierarchies is then estimated as the correlation between the two corresponding matrices of ones and zeroes. A more formal approach is to use quadratic assignment to assess the similarity between two partitions. Quadratic assignment is a combinatorial approach, where simulation is used to create a sampling distribution of possible shuffles of a partition in terms of a correlation or regression statistic between the original partition and each shuffled version. The similarity observed between two partitions is then compared with that sampling distribution to see how extreme or notable the observed statistic actually is. Other related work by Falkowski [52] focused on finding community instances using similarity.

2.4 Behavioural Measures of Community

In contrast to the fully automated methods discussed thus far, human judgment can also be used in inferring subgroupings. In some cases human judgment, as reflected in the structuring of online communities and websites, can be inferred without additional data collection effort. For instance, researchers have used corroborating events, groups and categories inherent in the structure of online communities such as LiveJournal, DBLP and IMDB [48,53] to validate inferred community structure. User behaviour has also been studied in virtual communities. For instance, content analysis of interactions such as newsgroups, e-mail, or blogs may be used to find evidence of group or community membership. While these methods of validating subgroupings or inferred community are relatively indirect, they have the benefit of being easy to apply. Alternatively, human judgments may be collected in validating the obtained groupings, but with greater effort. In the following subsections, two methods for collecting behavioural judgments relevant to subgroup validation are highlighted. While a broad range of ethnographic and behavioural assessments may be generally relevant to the problem at hand, the following subsections will focus on behavioural assessment methods that have been developing specifically to address the problem of validating social networks and cohesion within those networks.

2.4.1 Ties and Connectivity

Social networks can also be constructed by asking people directly who they are connected to and what groups or communities they belong to. Responses to such questions can then be used to build a social network based on actual ties. In this way, subgroup members may be inferred using egocentric networks in which networks are centered on an individual. Two instruments used in sociology for building social networks based on actual ties are the Social Network List (SNL) developed by Hirsch [54] and the Social Support Questionnaire (SSQ) developed by Sarason [55], which are based on name generator techniques in which participants are asked to name all the people that they know or communicate with (considered as ties). In the SNL, respondents are asked to delineate up to 20 network members they feel are "significant" and with whom they have contact at least once during a 4-6 week period. In the SSQ, respondents are asked to list members who provide specific kinds of supportive exchanges such as"who would give you useful suggestions to help you to avoid making mistakes". The SNL,and/or SSQ, as well as asking people directly who they are connected to or what groups or communities they belong to, could then be used to qualitatively evaluate the strength of the membership and cohesiveness of subgroups found from the previous sections. A variety of approaches have been suggested to address this problem and the corresponding research literature on centrality, clustering, and optimization methods for finding subgroupings are reviewed. This review will include a critical analysis of the limitations of past approaches.. First, the method proposed by Chin and Chignell called SCAN [52,56] (Social Cohesion Analysis of Networks), where a combination of heuristic methods is used to identify subgroups in a manner that can potentially scale up to very large social networks. then, The DISSECT [58] (Data-Intensive Socially Similar Evolving Community Tracker) method proposed by Chin and Chignell where multiple known subgroups within a social network are tracked in terms of similarity-based cohesiveness over time. The DISSECT method relies on cluster analysis of snapshots of network activity at different points in time followed by similarity analysis of subgroup evolution over successive time periods.

3 Existing Methods and Framework for Tracking Communities

Online social networks evolve over time, and research has looked into the temporal aspects of social networks changing over time. It has been found that groups discovered in social networks differ in their cohesiveness or bonding, which can be in time or space. Since online social networks have time inherent in their structure, cohesive subgroups can be defined as those that are similar over time based on Social Identity Theory [57] where group members feel closer if they are similar to each other. Cohesive subgroups can also be considered as optimum subgroups by calculating the optimum number of clusters, modularity, or optimizing graphs. Similarity measures are used in research to assess the cohesiveness of subgroups. Since they explicitly consider changes in subgroupings over time, similar measures take into account network membership dynamics in the social network. Different types of similar measures can be constructed depending on the particular network dynamics observed. This section addresses the problem of tracking community in social networks inferred from online interactions, by expanding on the problem of finding subgroups initially explored through the SCAN method [56] and addressing the limitations of the SCAN method.

3.1 Social Cohesion Analysis of Networks (SCAN)

Social network analysis is a new research field in data mining. Finding subgroups within social networks is important for understanding and possibly influencing the formation and evolution of online communities. The SCAN (Social Cohesion Analysis of Networks) [56] methodology involves three steps: selecting the possible members (Select), collecting those members into possible subgroups (Collect) and choosing the cohesive subgroups over time (Choose). Social network analysis, clustering and partitioning, and similarity measurement are then used to implement each of the steps.

The Social Cohesion Analysis of Networks (SCAN) [56] method was developed for automatically identifying subgroups of people in social networks that are cohesive over time. The SCAN method is to be applied based on the premise that a social graph can be obtained from the online community interactions where the links are untyped (i.e., there are no associated semantics). In the social graph, each link represents an interaction between two individuals where one individual has responded to the other's post in the online community. The SCAN method has been designed to identify cohesive subgroups on the basis of social networks inferred from online interactions around common topics of interest. The SCAN method consists of the following three steps:

1. Select: Selecting potential members of cohesive subgroups from the social network.
2. Collect: Grouping these potential members into subgroups.
3. Choose: Choosing cohesive subgroups that have a similar membership over time.

3.2 DISSECT (Data-Intensive Socially Similar Evolving Community Tracker)

The DISSECT method addresses the following shortcomings of the SCAN method:

1. The SCAN method only focused on betweenness centrality; other centrality measures may be useful.

2. The SCAN method only looked into two types of similarity measures (constant membership and members entering the network); there is a need to examine for other types.
3. The time periods used in the SCAN method were defined ad hoc as a matter of convenience, without any systematic evaluation.
4. The SCAN method fails if semantic properties determine subgroup membership.

3.3 Framework for DISSECT Method

The Framework for DISSECT method involve following steps for tracking online community in a social network:

1. Find the initial time periods for analysis.
2. Label subgroups of people from the network dataset using content analysis and semantic properties. If possible, individuals are also labelled so as to facilitate later similarity analysis between subgroups at different time periods.
3. Select the possible members of known subgroups to be tracked using the Select step from the SCAN method.
4. Carry out hierarchical cluster analysis of interaction data taken at snapshots in time and involving known subgroups of people (using the Collect step from the SCAN method).
5. Repeat steps 3 and 4 for different values of betweeness centrality (note that the DISSECT approach is agnostic in terms of which of the many available measures of centrality should be used).
6. Calculate similarity of subgroups for the designated time periods from step 1 using the clustering results of the previous step. In this case, the similarity measure can be augmented to take into account semantic labels assigned to different people.
7. Repeat steps 2 through 6 for different time period intervals and combinations.
8. Construct a chronological view of each subgroup showing how it changes over time (as the assigned semantic labels change).

The ultimate goal in a DISSECT analysis is to trace the evolution of subgroups into communities. The DISSECT method does not stand as a theory of how communities form. However, logically it would seem that if communities do emerge out of online interactions then they are likely to evolve, initially, from smaller subgroups.

4 Conclusion

This Chapter reviews the measures of social networks, formal methods and framework for tracking online community in social network. The researchers adopt different approaches for handling problem of tracking community over time. Subgroup identification, centrality, cluster analysis, clique analysis and similarity are basic measures & methods for finding structure of community. SCAN & DISSECT are two methods or frameworks for finding cohesive group & discussed a framework for tracking community evolution in an online Community called DISSECT or Data-Intensive Socially Similar Evolving Community Tracker. This framework is an expanded and enhanced version of the SCAN method, for finding cohesive subgroups in online interactions. The framework is designed to be a step-by-step process to track the evolution of community members. This chapter proposes future direction for improving existing methods & framework of tracking community over time.

References

1. Scott: Social Network Analysis. A Handbook. Sage (2000)
2. Shuie, Y.-C.: Exploring and Mitigating Social Loafing in Online Communities. Computers and Behavior 26(4), 768–777 (2010)
3. Wasserman, S., Faust, K.: Social Network Analysis: Methods and Applications. Cambridge University Press (1995)
4. Kumar, R., Raghavan, P., Rajagopalan, S., Tomkins, A.: Trawling the web for emerging cyber-communities. Computer Networks (1999)
5. Flake, G.W., Lawrence, S., Giles, C.L., Coetzee, F.M.: Self organization and identification of web communities. IEEE Computer 35(3), 66–71 (2002)
6. Chau, M., Shiu, B., Chan, I., Chen, H.: Automated identification of web communities for business intelligence analysis. In: Proceedings of the Fourth Workshop on E-Business (WEB). ACM (2005)
7. Gruzd, A., Haythornthwaite, C.: Automated discovery and analysis of social networks from threaded discussions. Paper presented at the International Network of Social Network Analysts (2008)
8. Gibson, D., Kumar, R., Tomkins, A.: Discovering large dense subgraphs in massive graphs. In: Proceedings of the 31st international conference on Very large data bases, VLDB 2005, pp. 721–732. VLDB Endowment (2005)
9. Tantipathananandh, C., Berger-Wolf, T.Y., Kempe, D.: A framework for community identification in dynamic social networks. In: KDD 2007: Proceedings of the 13th ACM SIGKDD International Conference on Knowledge Discovery and Data Mining, pp. 717–726. ACM, New York (2007)
10. Zhao, Q., Liu, T.-Y., Ma, W.-Y.: Predicting community members based on evolution of heterogeneous networks (patent number us 2007/0239677 a1). Microsoft Corporation (2007)
11. Joachims, T.: Making large-scale svm learning practical. In: Scholkopf, B., Burgess, C., Smola, A. (eds.) Advances in Kernel Methods - Support Vector Learning (1999)
12. Nie, Z., Zhang, Y., Wen, J.-R., Ma, W.-Y.: Object-level ranking: bringing order to web objects. In: WWW 2005: Proceedings of the 14th International Conference on World Wide Web, pp. 567–574. ACM, New York (2005)
13. Fisher, D.: Using egocentric networks to understand communication. IEEE Internet Computing 9(5), 20–28 (2006)
14. Frivolt, G., Bielikov, M.: An approach for community cutting. In: Svatek, V., Snasel, V. (eds.) Proc. of the 1st Int. Workshop on Representation and Analysis of Web Space, RAWS 2005, pp. 49–54 (2005)
15. Chin, A., Chignell, M.: A social hypertext model for finding community in blogs. In: Proceedings of the 17th International ACM Conference on Hypertext and Hypermedia: Tools for Supporting Social Structures, pp. 11–22. ACM (2006)

16. Frivolt, G., Bielikov, M.: an approach for community cutting. In: Svatek, V., Snasel, V. (eds.) Proc. of the 1st Int. Workshop on Representation and Analysis of Web Space, RAWS 2005, pp. 49–54 (2005)
17. Ma, H.-W., Zeng, A.-P.X.: The connectivity structure, giant strong component and centrality of metabolic networks. Bioinformatics 19(11), 1423–1430 (2005)
18. Donetti, L., Munoz, M.A.: Detecting network communities: a new systematic and efficient algorithm. Journal of Statistical Mechanics: Theory and Experiment, 2004, 10, P10012 (2004)
19. Girvan, M., Newman, M.E.: Community structure in social and biological networks. Proc. Natl. Acad. Sci. USA 99, 7821 (2002)
20. Gloor, P.A., Laubacher, R., Dynes, S.B.C., Zhao, Y.: Visualization of communication patterns in collaborative innovation networks - analysis of some w3c working groups. In: CIKM 2003: Proceedings of the Twelfth International Conference on Information and Knowledge Management, pp. 56–60. ACM Press (2003)
21. Costenbader, E., Valente, T.W.: The stability of centrality measures when networks are sampled. Social Networks 25, 283–307 (2003)
22. Crucitti, P., Latora, V., Porta, S.: Centrality measures in spatial networks of urban streets. Physical Review E 73, 036125 (2006)
23. Estrada, E., Rodriguez-Velazquez, J.A.: Subgraph centrality in complex networks. Physical Review E 71, 056103 (2005)
24. Newman, M.E.: Modularity and community structure in networks. Proceedings of the National Academy of Sciences 103(23), 8577–8582 (2006)
25. Memon, N., Harkiolakis, N., Hicks, D.: Detecting high-value individuals in covert networks: 7/7 london bombing case study. In: IEEE/ACS International Conference on Computer Systems and Applications, AICCSA 2008, pp. 206–215 (2008)
26. Memon, N., Larsen, H.L., Hicks, D., Harkiolakis, N.: Detecting hidden hierarchy in terrorist networks: Some case studies. In: Yang, C.C., Chen, H., Chau, M., Chang, K., Lang, S.-D., Chen, P.S., Hsieh, R., Zeng, D., Wang, F.-Y., Carley, K.M., Mao, W., Zhan, J. (eds.) ISI Workshops 2008. LNCS, vol. 5075, pp. 477–489. Springer, Heidelberg (2008)
27. Freeman, C.L.: Centrality in social networks: Conceptual clarification. Social Networks 1, 215–239 (1978)
28. Kahng, G., Oh, E., Kahng, B., Kim, D.: Betweenness centrality correlation in social networks. Phys 67, 017101 (2003)
29. Newman, M.: A measure of betweenness centrality based on random walks. Social Networks 27(1), 39–54 (2005)
30. Ruhnau, B.: Eigenvector-centrality node-centrality? Social Networks 22(4), 357–365 (2000)
31. Fortunato, S., Latora, V., Marchiori, M.: Method to find community structures based on information centrality. Phys. Rev. E (Stat Nonlinear, Soft Matter Phys.) 70(5), 056104 (2004)
32. Brandes, U.: A faster algorithm for betweenness centrality. Journal of Mathematical Sociology 25(2), 163–177 (2001)
33. Borgatti, S.P., Everett, G.M., Freeman, C.L.: Ucinet for windows: software for social network analysis. Analytic Technologies, Harvard, USA Science BV, Amsterdam, The Netherlands, pp. 107–117 (2002)
34. Costenbader, E., Valente, T.W.: The stability of centrality measures when networks are sampled. Social Networks 25, 283–307 (2002)
35. Duda, R.O., Hart, P.E., Stork, D.G.: Unsupervised Learning and Clustering. Wiley, New York (2001)

36. Alba, R.D.A.: graph-theoretic definition of a sociometric clique. Journal of Mathematical Sociology 3, 113–126 (2003)
37. Balasundaram, B., Butenko, S., Hicks, I., Sachdeva, S.: Clique relaxations in social network analysis: The maximum k-plex problem. Tech. rep., Texas A and M Engineering (2007)
38. Chin, A., Chignell, M.: Identifying subcommunities using cohesive subgroups in social hypertext. In: HT 2007: Proceedings of the Eighteenth Conference on Hypertext and Hypermedia, pp. 175–178. ACM, New York (2007)
39. Brooks, C.H., Montanez, N.: Improved annotation of the blogosphere via autotagging and hierarchical clustering. In: WWW 2006: Proceedings of the 15th International Conference on World Wide Web (2006), pp. 625–632. ACM Press (2006)
40. Li, X., Liu, B., Yu, P.S.: Mining community structure of named entities from web pages and blogs. In: AAAI Spring Syposium-2006. AAAI (2006)
41. Gömez, V., Kaltenbrunner, A., Löpez, V.: Statistical analysis of the social network and discussion threads in slashdot. In: WWW 2008: Proceeding of the 17th International Conference on World Wide Web, pp. 645–654. ACM (2008)
42. Johnson, S.C.: Hierarchical clustering schemes. Psychometrika 32
43. Hartigan, J.: Clustering Algorithms. John Wiley and Sons, New York (1975)
44. Orford, J.D.: Implementation of criteria for partitioning a dendrogram. Mathematical Geology 8(1), 75–84 (1976)
45. Noack, A.: Modularity clustering is force-directed layout (2008)
46. Radicchi, F., Castellano, C., Cecconi, F., Loreto, V., Parisi, D.: Defining and identifying communities in networks. Proceedings of the National Academy of Sciences of the United States of America 101(9), 2658–2663 (2004)
47. van Duijn, M.A.J., Vermunt, J.K.: what is special about social network analysis? Methodology 2, 2–6 (2005)
48. Elmore, K.L., Richman, M.B.: Euclidean distance as a similarity metric for principal component analysis. Monthly Weather Review 129(3), 540–549 (2001)
49. Santini, S., Jain, R.: Similarity measures. IEEE Transactions on Pattern Analysis and Machine Intelligence 21(9), 871–883 (1999)
50. Tversky, A.: Features of similarity. Psychological Review 84(4), 327–352 (1977); [160] Tyler, J. R., Wilkinson, D. M., Huberman, B.A.: E-mail as spectroscopy: Automated discovery of community structure within organizations. The Information Society 21(2), 143–153 (2005)
51. Jaccard, P.: Distribution de la flore alpine dans le bassin des dranses et dans quelques rgions voisines. Bulletin del la Socit Vaudoise des Sciences Naturellese 37, 241–272 (1901)
52. Falkowski, T., Bartelheimer, J., Spiliopoulou, M.: Community dynamics mining. In: Proceedings of 14th European Conference on Information Systems (ECIS 2006), Gteborg, Sweden (2006)
53. Leskovec, J., Lang, K.J., Dasgupta, A., Mahoney, M.W.: Statistical properties of community structure in large social and information networks. In: WWW 2008: Proceeding of the 17th International Conference on World Wide Web, pp. 695–704. ACM, New York (2008)
54. Hirsch, B.J.: Psychological dimensions of social networks: A multimethod analysis. American Journal of Community Psychology 7(3), 263–277 (1979)
55. Sarason, I.G., Levine, H.M., Basham, R.B., Sarason, B.R.: Assessing social support: The social support questionnaire. Journal of Personality and Social Psychology 44, 127–139 (1983)

56. Chin, A., Chignell, M.: Automatic detection of cohesive subgroups within social hypertext:A heuristic approach. New Rev. Hypermed Multimed 14(1), 121–143 (2008)
57. Tajfel, H., Turner, J.C.: The social identity theory of inter-group behavior. In: Worchel, S., Austin, L.W. (eds.) Psychology of Intergroup Relations (1986)
58. Chin, A., Chignell, M., Wang, H.: Tracking cohesive subgroup over time in inferred social network. New Review of Hypermedia and Multimedia / Hypermedia 16(1&2), 113–139 (2010)

Social Network Analysis Approach for Studying Caste, Class and Social Support in Rural Jharkhand and West Bengal: An Empirical Attempt

Anil Kumar Choudhuri and Rabindranath Jana

Abstract. In Jharkhand, without having any effective measure of land reforms and the Panchayats as well as the absence of peasant mobilization, Total Literacy Campaign or organized women's movement, major portion of the people having dependence upon the market forces have, no doubt, extended their livelihoods to various distant urban-industrial job markets. Unlike in Jharkhand, in West Bengal, economic and political / organizational changes have been taken place. Redistribution of land through land reforms, increase of wage rate and rise of Gram Panchayat have been as a source for the rural poor in Bengal. As a result, there is competition among the land owners to retain labourers and the land-owners' authority has been weakened. Under the circumstances, in the present article, an attempt has been made to study the pattern of social networks of people in the two regions concerning social stratification by caste/ community composition, occupational class, ownership of land and their inter-face among themselves. For the study, four villages from Jharkhand and two from West Bengal were purposively selected and applying complete enumeration method and considering Head of a household or his representative as respondent, data on structural variables as well as few composite variables were collected. Besides household survey, methods of case history and group discussion with the people were also undertaken for collecting other information useful for the study. Then social network analysis techniques were adopted to analyze the data. Social Network Analysis has brought out that in Jharkhand the domain of articulation of ties of social help and support among the villagers against vulnerabilities, at the time of urgency or crisis, is primarily based upon traditional primordial relationship, kinship and, that way,

Anil Kumar Choudhuri
Sociological Resaerch Unit, Indian Statistical Institute, Giridih, Jharkhand, India
e-mail: anil@isical.ac.in

Rabindranath Jana
Sociological Research Unit, Indian Statistical Institute, Kolkata, West Bengal, India
e-mail: rabindranathjana65@gmail.com

M. Panda, S. Dehuri, and G.-N. Wang (eds.), *Social Networking*
Intelligent Systems Reference Library 65,
DOI: 10.1007/978-3-319-05164-2_7,

spreads among the members of the same caste or community. In the Rarh region of West Bengal, its base is, on the other hand, secular. It circulates primarily along the ties of neighborhood and friendship. Recently, new secular relationships have been added to this list. But these are sort of contractual, exchange-oriented relationships as these are related to the employers, leading persons in different institutions or organizations, "experts", influential persons and so on. The findings from the study may be much useful to the planners of development and methodologically, a triangulation of methods of case history, groups discussion with the concerned people and household survey may be effective strategy for future research study in the allied area.

Keywords: Social stratification, social help, reciprocity, isolatedness, reachability, connectedness, within and between caste/ and class interaction.

1 Introduction

During the last three decades socio-economic and administrative measures in West Bengal (namely, implementation of land reforms, Patta distribution, minor irrigation schemes, Panchayats, Total Literacy Campaign, minimum wage rate, share cropper registration, etc.) as well as market factors (such as opening of facilities for jobs and small business, mini-entrepreneurial opportunities, construction and improvement of road and transportation communication, etc.) have created noteworthy impact upon pattern of articulation of social relation in the villages [2]. Specially noteworthy is the articulation of new power structure and rise of new centers of power from within the lower rung of the village society [5], [13].

In case of Jharkhand the striking feature is the absence of the above mentioned socio-economic and administrative measures. Under such conditions, the proposed study will investigate to what extent in the absence of officially induced socio-economic and administrative measures as above, only the mechanism of spontaneous social interaction and market forces as such can be conducive to changes in the pattern of articulation of social relations and emergence of any new structure of power relations in the villages. The situation is all the more aggravated since Jharkhand is, on the one hand, one of the most poverty stricken regions in India, while, on the other hand, is sharply polarized by traditionally continuing caste-community based power structure which creates a strong barrier to decentralization and empowerment of the socio-economically weaker sections in the villages. Because of socio-economic and administrative measures implemented officially, West Bengal becomes a typical case whereas, left to traditionally continuing mechanism of market-based spontaneity, Jharkhand becomes an atypical example which may have been replicated in other states as well. In a way our findings from Jharkhand, in this sense, can be more representative in so far as the county as a whole is concerned. That adds to the importance of our study in Jharkhand.

Against the above two different macro-scenarios, the villages within 20 miles circle around the town of Giridih in Jharkhand and the villages in Md. Bazar Block in Birbhum district of West Bengal were considered to be suitable for a comparative sociological study. Firstly, the similarity of their natural and physical characteristics adds to the merit of a comparative study. Both are mostly rain-fed mono-crop (namely, traditional cultivation of Aman rice). Both the regions are situated towards the tail-end of the east India plateau of Chhotanagpur. According to indigenous agricultural expertise, the distribution of agricultural land by local classification of soil situation in the two regions is also more or less similar. Furthermore they are largely exposed to similar urban-market contacts.

Equally noteworthy is the composition of their castes and communities. The two villages of West Bengal included for our comparative study are situated close to the border of Dumka district of Jharkhand in the tail-end of rainfed east Indian plateau and the fringe region of historically existing two socio-cultural settings. In fact, the other four villages around Giridih town in Jharkhand are also located in the milieu of the same plateau and a somewhat similar socio-cultural but in a different direction. Thus, according to geographical and agricultural practices based on land-situation as well as earlier social composition, the two villages of Birbhum in West Bengal are similar to the four others in Jharkhand. It is also observed that, as usual practice in rural Jharkhand, residential areas of the study villages are mostly adjacent to both sides of the roads through the villages.

As one can see from the maps, village habitation still remains segregated by castes and communities into different neighbourhoods which are locally called a Tolah or a Mohallah in Jharkhand and a Para in West Bengal. In the four villages of Jharkhand unlike the two of West Bengal, a single caste or community inhabits most of the neighbourhoods except one in Harsingradih situated by the side of the metalled road adjacent to Giridih town. This neighbourhood is, in fact, a relatively "recent" one where the households primarily meet the demands of those moving to or from the town. They belong to Gowala, Rajak, Kahar and Ghatwar caste who have set up shops to sell stationeries and groceries. The Muslims are also engaged in these occupations while those of Lohar, Kumhar and Chamar castes pursue traditional family crafts of making and selling iron, earthen and leather skin products for the near urban market.

The study has been undertaken in two phases during 2001 2004. The central thrust of the study is as follows. Firstly, we would study some parameters of social stratification such as caste / community composition, occupational class and land ownership, on the one hand and their inter-face, on the other hand with the articulation of ties of social network of help and support, at the time of an emergency, among the villagers. Our objective is an endeavour to explore their inter-face.

We have also intended to find out whether macro-factors like the urban-market influence or road connections have made any effect on the pattern of distribution of ties of social interaction in a village and whether the village development programmes have been effective to ameliorate the conditions of uneven distribution of power.

We first conducted an exploratory study in two villages for collation of data concerning formulation of probable hypothesis. We have also replicated the study for their verification by further in-depth investigation in the next phase in two other comparable villages. Thus, two villages were selected in each phase, one with good urban-market connection and another remote agricultural village. These were selected from among the villages within 20 miles around Giridih town in Jharkhand. Altogether four villages were thus surveyed in Jharkhand. Since West Bengal provided a socio-economic and administrative contrast, two additional villages were similarly selected from out of 21 villages studied earlier in Md. Bazar Block, Birbhum District [2]. The purpose was to compare and examine broadly whether socio-economic and administrative context as stated earlier made any difference.

2 Methodology

2.1 Village Selection for the Study

In the first phase of the project, two villages selected for the study in Jharkhand were Harsingradih (Thana No.-281), in Daridih Gram Panchayat, under Police Station and Block in Giridih and Chitmadih (Thana No. -118) in Nawahar Gram Panchayat under Police Station and Block Bengabad . These two were supplemented, as stated above, in the second phase of the project by the following four villages - two from Jharkhand and two from West Bengal: Baghra (Thana No. -38G) in Madhwa Gram Panchayat under Bengabad Block and Police Station and Mahacho (Thana No. -8G) in Ojhadih Gram Panchayat under Police station and Block Bengabad. The two villages of West Bengal selected from Birbhum district were Maladanga (J.L. No. 112) in Bhutura Gram Panchayat and Raspur (J.L. No. 174) in Charicha Gram Panchayayt under Md. Bazar Block and Police Station [6].

2.2 Data Collection and Level of Analysis

All the households in each of the aforesaid six villages have been enumerated. A complete enumeration of each village was done because, otherwise, many of the parameters of social network cannot be analysed from a sample [10], [12], [15]. In the study, both the quantitative and qualitative methods of obtaining data and analysis have been used. Structural data have been mostly collected by quantitative survey, interviewing respondents with questionnaires. This was supplemented by qualitative data collected from case studies, group discussions held in the villages, historical narratives told by locally knowledgeable "aged villagers" and so on. This part of investigation was focused to delineate villagers' perception and evaluation of a phenomenon. Similarly, in addition to conventional tools of analysis of quantitative survey data, methods of qualitative data analysis such as obtaining standardised entropy or index of qualitative variation were applied [16].

2.3 *Limitations of Analysis by Exact Caste Community (CC) and Source of Livelihood (SL)*

As caste / group affiliation of households and sources of their livelihood differ largely from village to village, choice of units for our analysis has got to be a standardised one so that these can be compared over all the six villages. Given below is a simple example to illustrate the limitation of analysis by exact community in our situation. It has already been pointed out that the village population belongs to thirty-three different castes and communities. Among these only two, the Muslims and the Gowalas are found in each of the six villages. On the other hand, sixteen of the remaining of thirty-one caste/communities are found in one village out of six villages, while ten in two villages and five in three villages. This is relevant not only for the dominant castes, but also for other castes, like functionally important castes , such as, Barhi (Carpenter), Kamar (Blacksmith), Rajak (Washerman), Muchi (Cobbler) and so on.

For similar reasons, exact source of livelihood as such cannot be used for analysis. It becomes more constrained than in case of "castes / communities". A source of livelihood (SL) refers to the source from which the livelihood of a household is derived. If it obtains its livelihood from more than one sources, the one from which the major part, whether in cash or kind or both, is obtained, is to be considered unless it is specified otherwise. In order to grasp the nuances of economic reality of the households, an exact description of a SL includes its economic sector, nature of occupation and job performed and employment status of the person engaged. Thus altogether seventy-seven different SL-s (Appendix-A) were identified which were pursued by households in six villages. Among these only three (Cultivator of own land by family labour, Day labourer and Mason) were found in all the six villages. The distribution of SL-s by number of villages is as follows: 130, 226, 315, 48, 57, 66 (1 = Harsingraidih, 2 = Chitmadih, 3 = Bagrah, 4 = Mahacho, 5 = Maladanga, 6 = Raspur). Therefore, on the basis of exact caste / community or source of livelihood affiliation a comparative analysis by exact one becomes quite complex and may reduce its validity, because, for instance, locally dominant castes such as the Bavans and the Maharis are found in Harsinghradih and Chitmadi respectively, but not in the other villages.

Furthermore, frequency distribution, by exact caste / community or by exact SL show that most of the cells remain unrepresented, i.e., either empty of any frequency, or very low (less than 5) frequency in case of the six villages under study in our study. Hence, exact" caste / community or source of livelihood cannot provide any meaningful interpretation of the results of data analysis. Obviously, for the above reasons, we had to classify the "exact" categories into a few meaningful and analytically appropriate categories. We would next describe briefly their underlying rationale.

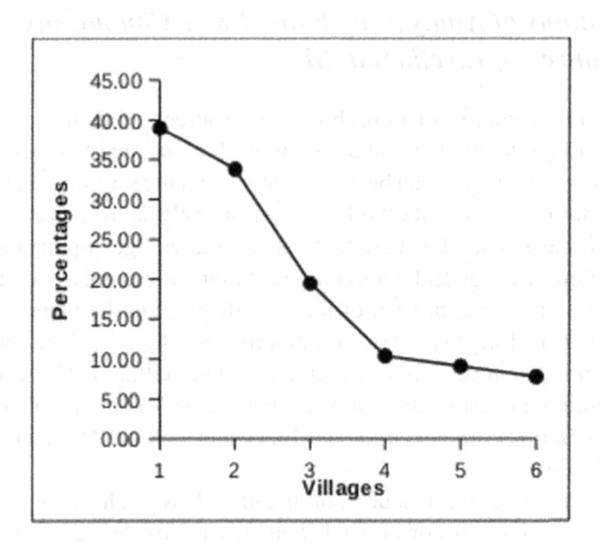

Fig. 1 Distribution of occupations over the villages

2.4 *Formation of Categories of CC and SL*

Categories of castes / communities were formed by collecting locally prevalent notion of status differentiation considering social, ritual, economic and cultural characteristics. These categories were deduced from the perception of knowledgeable villagers as well as from experts in Giridih town and around when a large scale survey was earlier undertaken in Giridih town and in the surrounding villages. Collectively their perception provided a broad consensus regarding classification of castes and communities into four distinct groups, as follows (a) the Hindu upper castes consisting of castes normally regarded as high and middle ranks such as the Brahmin, Bavan, Kayastha, Rajput, Mahuri, Napit, Gowala, Sadgop, Koiris, Kahars,etc.; (b) the Hindu - low including scheduled castes and other Hindu castes of ritually low rank, e,g., Chamar, Bagdi, Turi, Hari etc.; (c) the scheduled tribes (the Santhal, Mal, Orang, Kora, etc.); and (d) the Muslims. Even with the help of this classification, it is found that some of the cells in the Table have zero or very low frequency. For example, the Hindu - high caste people are not found in the village of Bagrah and Mahacho. In the same way, no scheduled tribe lives in the selected villages of Jharkhand. For bringing a balance in our analysis, we put (c) and (d) together. Otherwise, we have treated such individual non-representation of a category as a matter of structural zero.

3 Brief Historical Background of the Villages

3.1 Location and other Facilities

Harsingradih and Bagrah enjoy comparatively greater access to better road and other infra-structural facilities compared to Chitmadih and Mahacho, because the former two are situated adjacent to the town of Giridih as well as very close to major roads connecting Giridih town with other important areas. Hence, market exposure and urban connections are high in Harsinghradih and Bagrah. Chitmadih is a very remote agricultural village and so also is Mahacho. A broad outline of locational characteristics and socio-economic composition of the villages are given in Panel-A, B & C of Table 1.

Table 1

Panel A: Location of the villages

State	Village (alphabetic code).	Total house-holds	Area in acres	Distance from District H.Q. (Km)		Distance from the nearest wholesale market (Km)	
				By road	Walking	By road	Walking
(0)	(1)	(2)	(3)	(4)	(5)	(6)	(7)
Jharkhand	Harsingradih (Hd)	313	485	1.5	1.5	1.5	1.5
	Chitmadih (Cd)	175	602	35	32	3	1
	Bagrah (Bg)	104	130	7	5	7	5
	Mahacho (Mc)	89	377	23	15	23	15
West Bengal	Maladanga (Md)	341	550	8	8	0	0
	Raspur (Rp)	174	575	15	15	8	8

Panel B : Location of the villages [1]

Village	Distance from (in Km.)												
	PS	RS	BO	BA	BR	CS	PO	GP	HO	PHC	CO	PR	HS
(1)	(2)	(3)	(4)	(5)	(6)	(7)	(8)	(9)	(10)	(11)	(12)	(13)	(14)
Hd	4	4	8	0	0	5	0	0	6	0	0	0	3
Cd	15	35	15	5	2	15	2	2	35	2	27	0	0.5
Br	10	7	18	7	0	7	8	2	7	4	7	0	7
Mc	10	23	10	7	2	10	2.5	2	23	2.5	23	0	10
Md	3	10	5	3	0	3	0.5	0.5	8	0.5	8	0	0.5
Rp	12	15	15	12	3	12	3	6	15	4	15	0	6

[1] **BA**=Bank, **BO**=Block Office, **BR**=Bus Route, **CO**=Collage, **HO**=Hospital, **HS**=High School, **PO**=Post Office, **PR**=Primary School, **PS**=Police Station, **RS**=Railways Station, **CS**=Co-operative society, **GP**=Gram Panchayat, **PHC**=Primary Health Centre.

Table 1 (*continued*)

Panel C : Summary of socio-economic composition of villages[2]

Village alphabetic code	Principal sector of source of livelihood of a household (% in bracket)			Distribution of households by caste/community categories in each village (Exact number. of castes/communities in a category within brackets)							
	Agriculture	Non agriculture	Total	H1	H2	H3	ST	M	Total	% (H1 + H2)	% (H3 + ST + M)
(1)	(2)	(3)	(4)	(5)	(6)	(7)	(8)	(9)	(10)	(11)	(12)
Hd	15 (4.79)	298 (95.21)	313 (100.00)	54 (4)	39 (5)	186 (4)	00 (00)	34 (1)	313 (14)	29.71 (9)	70.29 (5)
Bg	21 (20.19)	83 (79.81)	104 (100.00)	00 (00)	67 (4)	23 (1)	00 (00)	14 (1)	104 (6)	64.42 (4)	35.58 (2)
Cd	45 (25.71)	130 (74.29)	175 (100.00)	51 (2)	68 (5)	3 (2)	00 (00)	53 (1)	175 (10)	68.00 (7)	32.00 (3)
Mc	50 (56.18)	39 (43.82)	89 (100.00)	00 (00)	28 (2)	51 (2)	00 (00)	10 (1)	89 (5)	31.46 (2)	68.54 (3)
Md	108 (31.67)	233 (68.33)	341 (100.00)	21 (2)	72 (5)	166 (6)	60 (2)	22 (1)	341 (16)	27.27 (7)	72.73 (9)
Rp	101 (58.05)	73 (41.95)	174 (100.00)	1 (1)	44 (4)	56 (3)	64 (3)	9 (1)	174 (12)	25.86 (5)	74.14 (7)

Briefly speaking, Harsingradih, Baghra, Mahacho and Raspur are un-irrigated rain-fed, mono-crop agricultural villages. Village Maladanga is an agricultural, but multi-crop village. It enjoys canal irrigation facilities. The villages Harsingradih, Baghra and Maladanga are very close to wholesale markets. Their urban exposure is also very high. The remaining three villages Raspur, Chitmadih and Mahacho, are interior villages without much road and transport connection with any urban or market center. In this context, it is also noteworthy that although both agriculture and non-agriculture sectors provide sources of livelihood to the villagers in all the selected villages, the latter is particularly the major source in all the villages (Table 1: Panel C). Our summary data regarding the villages also show the presence of different categories of Hindu castes and other communities in each village. Again, Hindu high and middle caste households own most of the land in all the villages except Raspur where the low castes, tribals and Muslims own major share. We can now discuss to explain how these socio-economic features have come up.

[2] ** Traditionally followed ritual categories; Hindu high (H1), middle (H2), low castes including SC (H3), Tribal (ST) and Muslim (M).

3.2 Settlement Patterns of the Villages

Through a quick look at locally collected oral histories of settlement pattern in each of these villages, it will give us a meaningful backdrop of what we find now (Appendix B). Unfortunately, we could not gather much in this regard in case of the two West Bengal villages. Yet we are inclined to summarise what we have learnt from the villagers through group discussions. This is given below.

Structural composition, both social and economic, of a village has a certain development cycle with its own specificities. But still there will have an overall commonality among them. The cycle of a village has passed through a sort of similar stages across few generations in the past since the inception of the first settlement upto village formation. In fact, local histories also provide an explanation based on ground reality as to why habitation pattern has developed like a mosaic of castes and communities in the villages around Giridih unlike that of a mixed bag type in Birbhum. In this context, Fox [8] has described nicely for getting methodological idea of constructing village development cycle and its stages from local historical materials.

The history of origin of Giridih town does not go back beyond the middle of nineteenth century. Earlier the area was a largely uninhabited area full of forests and wild animals. For various physical and natural constraints, such as, lack of water resources, acute water crisis during summer, high risks of rice cultivation due to irregular rainfall, uncertainty of rainfall during September-October (Hatia rainfall), combined with land situation and quality of agricultural land, restricted any extensive cultivation of rice or wheat. The population, then, mostly settled around the riverine region in "the North" or "in the East" and nearby places where they were growing crops suitable for cultivation in riverine areas, but not those crops which could be sustainable here [4].

However, increasing population pressure in the native village or some specific reason of personal or familial nature would push someone or a small group of kins to move out of the original native village to this region of Giridih in search of some land suitable for agricultural purpose or for cattle grazing. A suitable opportunity would prompt them to clear the jungle, make land somewhat amenable for the purpose and settle down. This was the first stage of the cycle.

Such initial migrant settlers were usually offshoots of some traditionally agricultural or cattle grazing castes. That is how small kin group of the Koiris, Gowalas, Yadavs, Bavans, etc., became the first original settlers in these villages of Giridih. In addition, the landlords of one village residing elsewhere made a few Rajputs supervisory farmers and they began to stay in the village. Similarly, in another village, few Turis were allowed by the landlords (staying elsewhere) to clear jungle and begin cultivation. The Koiris or Yadavs settled in the village later.

The second stage of the cycle was the period when the agriculturists brought people belonging to castes such as Chamar, Turi, Dusadh or even Muslims as either guards against wild animals or bandits, or, as their labourers, both agricultural and non-agricultural. These castes were, in most of the cases, given dwelling land sometime with an adjacent piece of land for kitchen garden. The land to build their dwellings was allotted by caste in different parts of the village according to the convenience of their employers. As in case of the first settlers, those who settled next also formed small kin groups. This stage is the beginning of the village formation.

The functional requirements of a village has later attracted the castes rendering various services, like barbers, tailors, carpenters, washer men and so on as well as supplying groceries or other materials, manufacturing own products for domestic consumption or household / agricultural production,, earthen wares, weaving hand loom products, etc. This was the third stage of development cycle when the village has become more or less self-contained.

Subsequently, others were added to the picture, as if as tail-enders. Hence, a broad sketch of the process of village settlement can be modelled as consisting of three stages as above. Oral histories of these villages have been appended in the form of a somewhat extended summary (Appendix-B). One of the major implications of this pattern of village settlement was a steep segregation of ownership of village-land by caste and community contributing to build a sort of in-built mechanism having "good" land by a few privileged castes of high / middle ritualistic ranks who had settled early.

Discussion with the local experts as well as the knowledgeable villagers in Maladanga and Raspur, the two villages of Birbhum, did not bring out much relevant material in this regard except that the Santals migrated to these villages after their rebellion was suppressed by the British army in the middle of the nineteenth century. Others (except the Sadgops) have already settled in the villages by the side of the old roads of " pre-British Badsahi period" linking Mallarpur, Angargaria, Rajnagar and Dumka, which were functionally very prosperous as administrative and trading centres. Hence, the dwellings were not segregated caste-wise in these villages. Rather, they eventually became much more mixed, being adjacent to one another.

4 Social Stratification of the Villages

In this chapter, we would discuss caste - community - composition based stratification of the villages. Except the Muslims and the Yadavs no other caste is present in each of the villages (Appendix C). From the Appendix C, it is clear that the Hindu-low- caste people are very much in presence (40.55%); though in Chitmadih only three households were found belonging to Hindu-low-caste. In Bagrah and Mahacho there is none belonging to Hindu-high-caste, whereas in Harsingradih, Chitmadih, Bagrah and Mahacho no tribal people live.

4.1 By Caste Community

The entire population in each village has been divided into two categories, CCI and CCII according to broadly ascribed status based on the basis of caste - community ranking in the local social milieu. On many instances it is conceptually more meaningful, as it is said, to see the "forest" rather than each "tree" by itself.

In CCI belong the Hindu High (H1) and Middle castes (H2). Incidentally, the Hindu Middle castes occupy most of the land in the villages of Bagrah, Mahacho and Raspur and they enjoy political and social power as well. Thus, they are no less "dominant" than the High - castes. Occupationally, they are engaged in cultivation, business and services. The only social difference between them concerns customs, which, however, are noticeably narrowing down. Our observations during the course of field study indicated that the economic condition of the Hindu High and Middle castes was, more or less, the same.

In the second category CCII, we include the remaining Hindu - Low (H3) (including Scheduled Castes), Scheduled Tribes and the Muslims. However, as and where found possible, the category H3 has been treated separately, because the people belonging to the castes in this category happen to be placed socially and culturally in the same line. Moreover, living condition, occupational pattern and distribution of land owned are similar. For their livelihood they often depend upon non -agricultural activities, mainly labour of various types.

Table 2

Panel - A: Distribution of households by caste / community categories in Jharkhand villages (for 1960 & 2001-02) and West Bengal villages (for 2001-02) [% in first bracket].

Caste/ com- munity category	Jharkhand villages								West Bengal villages	
	Hd		Cd		Bg		Mc		Md	Rp
	1960	2001-02	1960	2001-02	1960	2001-02	1960	2001-02	2001-02	2001-02
(1)	(2)	(3)	(4)	(5)	(6)	(7)	(8)	(9)	(10)	(11)
H1	23 (14.65)	54 (17.25)	29 (43.94)	51 (29.14)	0 (0.00)	0 (0.00)	0 (0.00)	0 (0.00)	21 (6.16)	1 (0.58)
H2	26 (16.56)	39 (12.46)	19 (28.79)	68 (38.86)	38 (66.67)	67 (64.42)	16 (45.71)	28 (31.46)	72 (21.11)	44 (25.29)
H3	84 (53.50)	186 (59.43)	2 (3.03)	3 (1.17)	12 (21.05)	23 (22.12)	18 (51.43)	51 (57.30)	166 (48.68)	56 (32.18)
S.T.	0 (0.00)	0 (0.00)	0 (0.00)	0 (0.00)	0 (0.00)	0 (0.00)	0 (0.00)	0 (0.00)	60 (17.60)	64 (36.78)
Muslim	24 (15.29)	34 (10.86)	16 (24.24)	53 (30.29)	7 (12.28)	14 (13.46)	1 (2.86)	10 (11.24)	22 (6.45)	9 (5.17)
Total	157 (100)	313 (100)	66 (100)	175 (100)	57 (100)	104 (100)	35 (100)	89 (100)	341 (100)	174 (100)
CC1: H1+ H2	49 (31.21)	93 (29.71)	48 (72.73)	119 (68.00)	38 (66.67)	67 (64.42)	16 (45.71)	28 (31.46)	93 (27.27)	45 (25.86)
CC2: H3+ S.T. + M	108 (68.79)	220 (70.29)	18 (27.27)	56 (32.00)	19 (33.33)	37 (35.58)	19 (54.29)	61 (68.54)	248 (72.73)	129 (74.14)

Table 2 (*continued*)

Panel - B: Standardised Entropy* for 2001-02

Caste/community Category	Villages of Jharkhand			Villages of West Bengal
	Hd + Cd + Bg + Mc	Hd + Mc	Cd + Bg	Md + Rp
(1)	(2)	(3)	(4)	(5)
H1 + H2	307 (45.08)	121 (30.10)	168 (66.67)	138 (26.80)
H3	263 (38.62)	237 (58.96)	26 (9.33)	222 (43.10)
ST + M	111 (16.30)	44 (10.94)	67 (24.01)	155 (30.10)
Total	681 (100.00)	402 (100.00)	279 (100.00)	515 (100.00)
Standardized entropy	0.9305	0.8328	0.7591	0.9804

Panel - C: Standardised Entropy[3] for 1960

Caste/community Category	Villages of Jharkhand		
	Hd + Cd + Bg + Mc	Hd + Mc	Cd + Bg
(1)	(2)	(3)	(4)
H1 + H2	151 (47.94)	87 (40.65)	64 (63.37)
H3	116 (36.82)	96 (44.86)	20 (19.80)
ST + M	48 (15.24)	31 (14.49)	17 (16.83)
Total	315 (100.00)	214 (100.00)	101 (100.00)
Standardized entropy	0.9167	0.9152	0.8280

On the whole as Panel B of Table-2 summarises two among the four villages of Jharkhand, namely, Chitmadih (Cd) and Bagrah (Bg) are numerically dominated by Hindu high and middle castes (CCI) whereas the other two by CCII (Hindu-low, SC, ST and Muslim communities). If we consider H3 as the mid-point, the former two villages (Cd, Bg) are highly skewed with (H1 +H2) at the left of the axis and the ST, M, on the right, as a sort of outliers. In the other two Harsingradih and Mahacho (Hd, Mc), the distribution is mesokurtic. Thus their social compositions

[3] $St.En.(J) = -\sum^k p_k log(p_k)/logk, [0 \leq J \leq 1]$.

are considerably stratified, only shapes are different. Their standardised entropy is close being 0.7591 and o.8328 respectively. The shape of stratification of the two West Bengal villages is quite different, notably flatter, platykurtic. The value of standardised entropy measure is relative quite high, 0.9804. Oral histories of settlement of caste / communities in different villages presented earlier throw some light to explain currently observed qualitative differentials in shape of stratification.

4.2 Stratification by Source of Livelihood (SL) Categories and Ownership of Land

From the standpoint of occupational affiliation, all the villages show a significant level of stratification. From Panel C of Table 1, it is seen that the villagers are mostly dependent on non-agricultural activities except in Raspur and Mahacho where more than 50% of the households depend upon agriculture, cultivation in particular.

Table 3

Panel A: Distribution of households by family principal source of livelihood (SL) in Jharkhand villages (for 1960 & 2001-02) and West Bengal villages (for 2001-02) [% in first bracket].

Source of Livelihood	Jharkhand villages								West Bengal villages	
	Hd		Cd		Bg		Mc		Md	Rp
	1960	2001 - 02	1960	2001 - 02	1960	2001 - 02	1960	2001 - 02	2001 - 02	2001 - 02
(1)	(2)	(3)	(4)	(5)	(6)	(7)	(8)	(9)	(10)	(11)
Farmer+ cultivator+ share cropper (FC)	16 (10.19)	15 (4.79)	59 (89.39)	45 (25.71)	29 (50.88)	21 (20.19)	28 (80.00)	50 (56.18)	108 (31.67)	101 (58.05)
Profession + Service (PS)	31 (19.75)	48 (15.34)	4 (6.05)	15 (8.57)	10 (17.54)	5 (4.81)	1 (2.86)	8 (8.99)	16 (4.69)	4 (2.31)
Trade & Small business (TB)	0 (0.00)	31 (9.90)	1 (1.52)	4 (4.57)	0 (0.00)	3 (2.88)	0 (0.00)	0 (0.00)	9 (2.64)	20 (11.49)
Petty business + Craft & Artisan (PB/CA)	16 (10.19)	15 (4.79)	1 (1.52)	13 (7.44)	2 (3.51)	4 (3.85)	0 (0.00)	5 (5.62)	14 (4.11)	7 (4.02)
Labourers (L)	93 (59.24)	195 (62.30)	1 (1.52)	94 (53.71)	16 (28.07)	70 (67.31)	6 (17.14)	26 (29.21)	185 (54.25)	37 (21.26)
Others (O)	1 (0.63)	9 (2.88)	0 (0.00)	0 (0.00)	0 (0.00)	1 (0.96)	0 (0.00)	0 (0.00)	9 (2.64)	5 (2.87)
Total	157 (100)	313 (100)	66 (100)	175 (100)	57 (100)	104 (100)	35 (100)	89 (100)	341 (100)	174 (100)

Table 3 (*continued*)

Panel - B: Standardised Entropy for 2001-02

Source of Livelihood	Villages of Jharkhand			Villages of West Bengal
	Hd + Cd + Bg + Mc	Hd + Mc	Cd + Bg	Md + Rp
(1)	(2)	(3)	(4)	(5)
Farmer + cultivator + share cropper	131 (19.24)	36 (8.64)	95 (35.99)	209 (40.58)
Profession+Service+ Trade & Small business	118 (17.33)	87 (20.86)	31 (11.74)	49 (9.52)
Petty business + Craft & Artisan + Labourers + Others	432 (63.43)	294 (70.50)	138 (52.27)	257 (49.90)
Total	681 (100.00)	417 (100.00)	264 (100.00)	515 (100.00)
Standardized entropy	0.8280	0.7145	0.8724	0.8527

Panel - C: Standardised Entropy for 1960

Caste/community Category	Villages of Jharkhand		
	Hd + Cd + Bg + Mc	Hd + Mc	Cd + Bg
(1)	(2)	(3)	(4)
Farmer + cultivator + share cropper	132 (41.91)	45 (21.03)	87 (86.14)
Profession + Service + Trade & Small business	47 (14.92)	41 (19.16)	6 (5.94)
Petty business + Craft & Artisan + Labourers + Others	136 (43.17)	128 (59.81)	8 (6.92)
Total	315 (100.00)	214 (100.00)	101 (100.00)
Standardized entropy	0.9202	0.8665	0.4524

Panel B of Table 3 shows that all the six villages are steeply stratified w.r.t. occupational class. But, on the whole, it is skewed in Jharkhand while polarised in the two villages of Birbhum. Agricultural developmental measures of the State

Government and land reforms have strengthened the farming activities of particularly the middle and lower rung of the farmers in West Bengal, which have stabilised them as agriculturists - hence, a polarised trend has taken shape.

The Table 4 (given below) exhibits that, in case of ownership of land, there lies a great degree of stratification. It is very significant that more than 50% of the people in both Harsingradih and Maladanga are bereft of land. Both these villages are relatively more exposed to commercial activities in urban areas. However, on the whole, we find a steeply stratified skewed stratification of agricultural population by land ownership in villages of Jharkhand unlike that in West Bengal, where it tends to be a flat platykurtic distribution of land among middle and lower rung of farmers. Among the farmers, again, we find acutely skewed stratification by land ownership in Hd and Bg, the "urban villages", to say, while it is flattened, in Cd and Mc.

Table 4

Panel-A: Distribution of households in a village by class on land ownership (in bigha) for 2001 (% in bracket)

Villages	Landholding in Bigha (L)						Total
	Nil	$0 < L \leq 1.5$	$1.5 < L \leq 3.0$	$3.0 < L \leq 7.0$	$7.0 < L \leq 15$	$15 < L$	
(1)	(2)	(3)	(4)	(5)	(6)	(7)	(8)
Hd	204 (65.18)	59 (18.85)	31 (9.90)	9 (2.88)	7 (2.23)	3 (0.96)	313 (100.00)
Cd	32 (18.29)	64 (36.57)	20 (11.43)	27 (15.43)	29 (16.57)	3 (1.71)	175 (100.00)
Bg	26 (25.00)	40 (38.47)	21 (20.19)	13 (12.50)	2 (1.92)	2 (1.92)	104 (100.00)
Mc	4 (4.50)	43 (48.32)	30 (33.71)	8 (8.98)	4 (4.49)	0 (0.00)	89 (100.00)
Md	172 (50.44)	46 (13.49)	51 (14.96)	43 (12.61)	23 (6.74)	6 (1.76)	341 (100.00)
Rp	21 (12.07)	37 (21.26)	45 (25.86)	53 (30.46)	12 (6.90)	6 (3.45)	174 (100.00)
Total	459 (38.38)	289 (24.16)	198 (16.56)	153 (12.79)	77 (6.44)	20 (1.67)	1196 (100.00)

Panel-B: Standardised entropy

Landholding (L) in Bigha	Villages of Jhankhand			Villages of West Bengal
	Hd + Cd + Bg + Mc	Hd + Bg	Cd + Mc	Md + Rp
(1)	(2)	(3)	(4)	(5)
$3.0 \leq L$	308 (74.22)	381 (91.37)	193 (73.10)	179 (55.59)

Table 4 (*continued*)

$3.0 < L \leq 15.0$	99 (23.86)	31 (7.43)	68 (25.76)	131 (40.68)
$15.0 < L$	8 (1.92)	5 (1.20)	3 (1.14)	12 (3.73)
Total	415 (100.00)	417 (100.00)	264 (100.00)	322 (100.00)
Standardized entropy	0.5817	0.2992	0.5730	0.7418

4.3 What Patterns Observed

Based on the aforesaid scenario, we would look forward to discuss some salient aspects of life and living. It is evident from oral histories gathered in the villages that traditionally agriculture and the related activities have been the chief sources of livelihood in all the six villages. But the occupational pattern, to a great extent, has now changed and the non-agricultural activities have become the principal sources of occupation. The collapse of mica industry, sickening of coal industry, and uncertainty of cultivation in Giridih area and fragmentation of land due to increase in number of owners over generations are the primary factors for which the villagers were compelled to take resort to non-agricultural activities such as coal carrying, hawkery, rickshaw-pulling, brick labour etc. On the other hand, a huge section of population from each of the Jharkhand villages have migrated initially for menial jobs in Kolkata, Ranigunj and Asansol region in West Bengal and recently for being employed in crafts, as artisans, and in shops in Pune, Meerut, Surat, Delhi, Rajasthan etc. The inflow of money from the migrants has acted as a buffer to the deteriorating economic condition in these villages of Jharkhand. The economic condition of these villages has rather, somewhat, improved. It is notable that the migration taking place recently is mainly from the rural agricultural sector to urban crafts and commercial sector and mostly from among the middle / low Hindu castes and the Muslims. The process has opened up slowly an access to power - base for the weaker sections in these villages.

The High-caste Hindus, like the Bavans and the Brahmins, who were powerful in the past, are still trying to maintain the same status in the village Harsinghradih. But now-a- days their "powerfulness" has suffered a considerable setback with the Turis, the Rajaks, and the Gowalas challenging them socially and even legally, as and when required, since the level of their political awareness is now high and they enjoy advantageous contacts with the outside world. In Chitmadih also, the power status of the Rajputs is getting degraded. The Koiris and the Muslims are becoming the power pioneers. The Middle caste groups - the Koiris and the Gowalas socially and economically dominate in Bagrah. The Gowalas who were previously powerful are running far behind in power context in the village Mahacho under the present circumstances. It is the turn for the Hindu-low-castes like the Dushadhs and the Chamars to get hold of village power base.

In Maladanga the Brahmins, Sadgops and Gowalas still continue to exercise their control over the village. Observation confirms that some households of the down-trodden castes have got themselves socially uplifted. Maladanga, out of all the villages under the present study, is the sole village to enjoy noteworthy irrigational facilities. However, in spite of irrigational facilities, the most of the villagers are earning their livelihood out of non-agricultural activities as the village is endowed with many positional amenities like its proximity to high-way junction, market centre, etc. The villagers in Raspur do not enjoy such locational facilities and, by and large, are dependent on agriculture.

The various institutions and organizations which implement policies meant for development like the Panchayats, the Krishak Sabha, the Mahila Samiti etc. and, above all, the measures of land reforms have given a new socio- economic dimension which has changed the level of consciousness of the villagers. On the basis of these new interventions, the villagers get their collective efforts mobilised by village or neighbourhood or class, but not by caste. In the villages under our study in West Bengal, the caste based co-operation, in the present scenario, is rarely seen. On the other hand, the castes in the lower rung, with the help of existing administrative advantages have come up at a rapid pace to secure social mobility. This upliftment has resulted in altering their power equations with the dominant castes.

Lastly, we briefly discuss recent longitudinal changes with reference to social composition by caste / community categories and principal sources of livelihood (SL). We are inclined to have a comparative analysis based on the present data and that available from a survey of all villages (1875) within 20-mile radius from Giridih town earlier undertaken by Sociological Research Unit of Indian Statistical Institute, Kolkata in 1960 [3].

A comparison of the summary social composition of Jharkhand villages between 1960 & 2001-02 (Panel B with Panel C of each of Table-2 and Table-3) shows quite clearly no change in caste/community composition except the village Mahacho and a noteworthy change in the pattern of occupational composition.

If the data on exact caste / community in each village of both 2001-2002 and 1960 (Panel A, B & C of Table-2) are placed side by side we observe that in Harsinghradih the castes like the Dushads, the Mahuris, the Laheris and the Pasis have left the village. On the other hand, there has been the settlement of the Kayastha as the new entrants in the village. In the village Chitmadih the Lohars are the new comers to have settled down.

The oral history of the two villages of West Bengal reveals that the village population, on the whole, is on the rise. In Maladanga one Pasi household has become a new resident. One Rajput household has been the new dweller in Raspur in terms of marriage. Unlike the villagers of Jharkhand the level of social perception as well as behaviour of the villagers of West Bengal no longer remains confined within caste hierarchy.

On the contrary to social composition, the adjoining tables show a noteworthy change in the pattern of occupational affiliation in 2001 -2002 compared to that of 1960 (Panel B & C of Table-3) in the villages under study in Jharkhand. The villagers, who were earlier pursuing agricultural activities, have now been forced to change their sources of livelihood. The overall pattern of shift is to petty business, craft work / artisanry, working as labourers, etc. But there are two distinct shapes of change w,r.t. occupational class structure in different villages. It was already skewed in Harsingradih and Bagrah (Hd, Bg) in 1960 and has become much more steeply skewed in 2001. This is due to their proximity to Giridih town where jobs are available within the working distance "from home in the village", on the one hand, and Mica factories have mostly closed down forcing the households to go for urban jobs. In fact, Bg (Bagrah) was locally known as a "Mica village", so to say. On the contrary, the other two villages, Chitmadih and Mahacho (Cd, Mc), where the overwhelming number of households were farmers or own - account cultivator, thus, being flat in shape, have become polarised. The relevant data are presented in summary part in Panel B of Table.

It may also be noted that additional income remitted by migrants has helped the growth of petty business, craft etc. in the latter villages. The patterns of stratification on occupational classes in villages of Md. Bazar have been studied by Bandyopadhyay and von Eschen [2].

This change in economic scenario has opened up the gate for migration to a large extent and that, too, to industrial sectors. The closure of Mica establishments and the sickening of coal industries and continuous partitioning of land are major factors responsible for this change in livelihood pattern. The needy families in the villages are still dependent on the affluent households in one way or the other. On the other hand, West Bengal with implementation of land reforms and schemes of minor/medium irrigation (either individually through the bank loans or with the help of official programmes), decentralisation of power structure and implementation of the Minimum Wage Act have been able to overcome, to a considerable extent, the trends of inequality in the state. These administrative steps have stabilized the sources of livelihood of the villagers in West Bengal which was traditionally based on agriculture and its related activities. Besides, a large mass of them have got themselves engaged in local non-agricultural industries such as stone - quarry, furniture making, pottery, brick making, china - clay etc. The opening of occupational change has created a new outlook among the people. They are free to opt for any livelihood they would like and, thus, many of them do not depend on the affluent families. Moreover, during the last two decades along with the market - growth, improvement of transport and communication facilities, functioning of the Panchayats, a change in political awareness has also occurred. Hence, there was no new noteworthy out-migration for jobs.

It is also notable that the impact of emergence of market - oriented non-agricultural sector has also "changed" occupational profile in Jharkhand, but the absence of measures like those implemented in West Bengal has failed to ameliorate the inequal features of social stratification of the village communities there. It has rather made the system of stratification by class much steeper and unfavourable for the lower stratum of the village communities. Moreover, the two systems of stratification, caste and class, being confounded with one another, have produced a virulently aggressive and exploitative system of stratification. In West Bengal, the class-relation' has tended to replace caste-relation'.

4.4 *Inter-relationship of CC and SL Categories in Villages*

We would now look at inter-relationship of different categories of castes and communities with reference to occupational patterns and land-ownership. It is already stated that in the state of Jharkhand the caste based identity is intense whereas, as we have stated earlier, its prevalence is quite less in the West Bengal villages.

Table 5

Distribution of households by caste/community and principal source of livelihood (% in bracket)

Caste/community categories	Principal sources of livelihood [4]						Total
	1	2	3	4	5	6	
(1)	(2)	(3)	(4)	(5)	(6)	(7)	(8)
Village: Harsingraidih (2001-02)							
H1	13 (24.07)	15 (27.78)	4 (7.41)	0 (0.00)	18 (33.33)	4 (7.41)	54 (100.00)
H2	1 (2.56)	7 (17.95)	10 (25.64)	5 (12.82)	15 (38.47)	1 (2.56)	39 (100.00)
H3	0 (0.00)	16 (8.60)	9 (4.84)	9 (4.84)	150 (80.65)	2 (1.07)	186 (100.00)
S.T.	0 (0.00)	0 (0.00)	0 (0.00)	0 (0.00)	0 (0.00)	0 (0.00)	0 (0.00)
M	1 (2.94)	10 (29.41)	8 (23.54)	1 (2.94)	12 (35.29)	2 (5.88)	34 (100.00)
Total	15 (4.79)	48 (15.34)	31 (9.90)	15 (4.97)	195 (62.30)	9 (2.88)	313 (1.00)
CCI: H1 + H2	14 (15.05)	22 (23.66)	14 (15.05)	5 (5.38)	33 (35.45)	5 (5.38)	93 (100.00)
CCII: H3 + S.T. + M	1 (0.45)	26 (11.82)	17 (7.73)	10 (4.54)	162 (73.64)	4 (1.82)	220 (100.00)
Village: Chitmadih (2001-02)							
H1	16 (31.38)	3 (5.88)	3 (5.88)	3 (5.88)	26 (50.98)	0 (0.00)	51 (100.00)
H2	17 (25.00)	7 (10.29)	4 (5.88)	7 (10.29)	33 (48.54)	0 (0.00)	68 (100.00)
H3	0 (0.00)	0 (0.00)	0 (0.00)	1 (33.33)	2 (66.67)	0 (0.00)	3 (100.00)

[4] 1 = Farmer + cultivator, 2 = Profession + service, 3 = Trade and Small business, 4 = Petty business, craft-artisan, 5 = Labourers, 6 = Others.

Table 5 (*continued*)

S.T.	0 (0.00)	0 (0.00)	0 (0.00)	0 (0.00)	0 (0.00)	0 (0.00)	0 (0.00)
M	12 (22.65)	5 (9.44)	1 (1.89)	2 (3.77)	33 (62.26)	0 (0.00)	53 (100.00)
Total	45 (25.71)	15 (8.57)	8 (4.57)	13 (7.44)	94 (53.71)	0 (0.00)	175 (100.00)
CCI: H1 + H2	33 (27.74)	10 (8.40)	7 (5.88)	10 (8.40)	59 (49.58)	0 (0.00)	119 (100.00)
CCII: H3 + S.T. + M	12 (21.43)	5 (8.93)	1 (1.78)	3 (5.36)	35 (62.50)	0 (0.00)	56 (100.00)
Village: Bagrah (2001-02)							
H1	0 (0.00)	0 (0.00)	0 (0.00)	0 (0.00)	0 (0.00)	0 (0.00)	0 (0.00)
H2	20 (29.85)	3 (4.48)	2 (2.99)	4 (5.97)	37 (55.22)	1 (1.49)	67 (100.00)
H3	0 (0.00)	1 (4.35)	0 (0.00)	0 (0.00)	22 (95.65)	0 (0.00)	23 (100.00)
S.T.	0 (0.00)	0 (0.00)	0 (0.00)	0 (0.00)	0 (0.00)	0 (0.00)	0 (0.00)
M	1 (7.14)	1 (7.14)	1 (7.14)	0 (0.00)	11 (78.58)	0 (0.00)	14 (100.00)
Total	21 (20.19)	5 (4.81)	3 (2.88)	4 (3.85)	70 (67.31)	1 (0.96)	104 (100.00)
CCI: H1 + H2	20 (29.85)	3 (4.48)	2 (2.99)	4 (5.97)	37 (55.22)	1 (1.49)	67 (100.00)
CCII: H3 + S.T. + M	1 (2.70)	2 (5.41)	1 (2.70)	0 (0.00)	33 (89.19)	0 (0.00)	37 (100.00)
Village: Mahacho (2001-02)							
H1	0 (0.00)	0 (0.00)	0 (0.00)	0 (0.00)	0 (0.00)	0 (0.00)	0 (0.00)
H2	21 (75.00)	2 (7.14)	0 (0.00)	1 (3.57)	4 (14.29)	0 (0.00)	28 (100.00)
H3	22 (43.00)	5 (9.80)	0 (0.00)	4 (7.84)	20 (39.22)	0 (0.00)	51 (100.00)
S.T.	0 (0.00)	0 (0.00)	0 (0.00)	0 (0.00)	0 (0.00)	0 (0.00)	0 (0.00)
M	7 (70.00)	1 (10.00)	0 (0.00)	0 (0.00)	2 (20.00)	0 (0.00)	10 (100.0)
Total	50 (56.18)	8 (8.99)	0 (0.00)	5 (5.62)	26 (29.21)	0 (0.00)	89 (100.00)
CCI: H1 + H2	21 (75.00)	2 (7.14)	0 (0.00)	1 (3.57)	4 (14.29)	0 (0.00)	28 (100.00)

Table 5 (*continued*)

CCII: H3 + S.T. + M	29 (47.54)	6 (9.84)	0 (0.00)	4 (6.56)	22 (36.06)	0 (0.00)	61 (100.00)
Village: Maladanga (2001-02)							
H1	4 (19.05)	7 (33.33)	3 (14.29)	1 (4.76)	1 (4.76)	5 (23.81)	21 (100.00)
H2	50 (69.44)	6 (8.34)	5 (6.94)	3 (4.17)	7 (9.72)	1 (1.39)	72 (100.00)
H3	42 (25.30)	2 (1.20)	0 (0.00)	7 (4.22)	114 (68.68)	1 (0.60)	116 (100.00)
S.T.	9 (15.00)	1 (1.67)	0 (0.00)	0 (0.00)	50 (83.33)	0 (0.00)	60 (100.00)
M	3 (13.64)	0 (0.00)	1 (4.55)	3 (13.64)	13 (59.08)	2 (9.09)	22 (100.00)
Total	108 (31.67)	16 (4.69)	9 (2.64)	14 (4.11)	185 (54.25)	9 (2.64)	341 (100.00)
CCI: H1 + H2	54 (58.07)	13 (13.98)	8 (8.60)	4 (4.30)	8 (8.60)	6 (6.45)	93 (100.00)
CCII: H3 + S.T. + M	54 (21.78)	3 (1.21)	1 (0.40)	10 (4.03)	177 (71.37)	3 (1.21)	248 (100.00)
Village: Raspur (2001-02)							
H1	1 (100.00)	0 (0.00)	0 (0.00)	0 (0.00)	0 (0.00)	0 (0.00)	1 (100.00)
H2	16 (36.35)	3 (6.82)	11 (25.00)	2 (4.55)	10 (22.73)	2 (4.55)	44 (100.00)
H3	34 (60.71)	0 (0.00)	5 (8.93)	3 (5.36)	12 (21.43)	2 (3.57)	56 (100.00)
S.T.	47 (73.44)	1 (1.56)	1 (1.56)	1 (1.56)	14 (21.88)	0 (0.00)	64 (100.00)
M	3 (33.33)	0 (0.00)	3 (33.33)	1 (11.11)	1 (11.11)	1 (11.11)	9 (100.00)
Total	101 (58.05)	4 (2.31)	20 (11.49)	7 (4.02)	37 (21.26)	5 (2.87)	174 (100.00)
CCI: H1 + H2	17 (37.78)	3 (6.67)	11 (24.45)	2 (4.44)	10 (22.22)	2 (4.44)	45 (100.00)
CCII: H3 + S.T. + M	84 (65.12)	1 (0.77)	9 (6.98)	5 (3.88)	27 (20.93)	3 (2.32)	129 (100.00)

On the basis of Table 5, we find that in Harsinghradih, Bagrah and Maladanga the Hindu-high and Middle-caste villagers are placed in a better position as compared to the Hindu- low- caste and the Muslims, so far as occupational class position is considered. Again, the pattern of relationship between caste - community and livelihood remains more or less similar in the other three villages, namely, Chitmadih,

Mahacho and Raspur. In these three villages the occupational class position of the lower rung of caste/community stratum (CCII) is notably improved as compared to that in the former three villages. On the whole, the line of demarcation is proximity to urban-commercial connection of the village.

However, the relative amelioration of occupational profile of household in lower rung (CCII) in Maladanga as comparaed to Harsingradih and Bagrah is also to be noted. Following the feedback of land reform policy and registration of share-croppers there has been a considerable stabilisation of agricultural system marked by the increase in the number of both farmers and own - account cultivators - cum - share croppers among the Hindu-low-caste, scheduled tribe and Muslims in Maladanga and Raspur.

Table 6
Distribution of households by caste/community and land ownership in bigha (% in bracket)

Caste/community categories	Landholding (L) in Bigha						Total
	Nil	$0 < L \leq 1.5$	$1.5 < L \leq 3.0$	$3.0 < L \leq 7.0$	$7.0 < L \leq 15$	$15 < L$	
(1)	(2)	(3)	(4)	(5)	(6)	(7)	(8)
Village: Harsingraidih (2001-02)							
H1	8 (14.81)	14 (25.93)	24 (44.44)	3 (5.56)	2 (2.70)	3 (5.56)	54 (100.00)
H2	23 (58.98)	9 (23.08)	3 (7.69)	3 (7.69)	1 (2.56)	0 (0.00)	39 (100.00)
H3	150 (80.65)	28 (15.05)	3 (1.61)	1 (0.54)	4 (2.15)	0 (0.00)	186 (100.00)
S.T.	0 (0.00)	0 (0.00)	0 (0.00)	0 (0.00)	0 (0.00)	0 (0.00)	0 (0.00)
M	23 (67.65)	8 (23.53)	1 (2.94)	2 (5.88)	0 (0.00)	0 (0.00)	34 (100.00)
Total	204 (65.18)	59 (18.85)	31 (9.90)	9 (2.88)	7 (2.23)	3 (0.96)	313 (100.00)
CCI: H1 + H2	31 (33.34)	23 (24.73)	27 (29.04)	6 (6.45)	3 (3.22)	3 (3.22)	93 (100.00)
CCII: H3 + S.T. + M	173 (78.64)	36 (16.36)	4 (1.82)	3 (1.36)	4 (1.82)	0 (0.00)	220 (100.00)
Village: Chitmadih (2001-02)							
H1	6 (11.77)	22 (43.14)	5 (9.80)	11 (21.57)	6 (11.76)	1 (1.96)	51 (100.00)
H2	9 (13.24)	23 (33.82)	8 (11.76)	11 (16.18)	15 (22.06)	2 (2.94)	68 (100.00)
H3	1 (33.33)	1 (33.33)	0 (0.00)	1 (33.33)	0 (0.00)	0 (0.00)	3 (100.00)

Table 6 (*continued*)

S.T.	0 (0.00)	0 (0.00)	0 (0.00)	0 (0.00)	0 (0.00)	0 (0.00)	0 (0.00)
M	16 (30.19)	18 (33.96)	7 (13.21)	4 (7.55)	8 (15.09)	0 (0.00)	53 (100.00)
Total	32 (18.29)	64 (36.57)	20 (11.43)	27 (15.43)	29 (16.57)	3 (1.71)	175 (100.00)
CCI: H1 + H2	15 (12.60)	45 (37.82)	13 (10.92)	22 (18.49)	21 (17.65)	3 (2.52)	119 (100.00)
CCII: H3 + S.T. + M	17 (30.36)	19 (33.93)	7 (12.50)	5 (8.93)	8 (14.28)	0 (0.00)	56 (100.00)
Village: Bagrah (2001-02)							
H1	0 (0.00)	0 (0.00)	0 (0.00)	0 (0.00)	0 (0.00)	0 (0.00)	0 (0.00)
H2	4 (5.97)	26 (38.80)	21 (31.34)	12 (17.91)	2 (2.99)	2 (2.99)	67 (100.00)
H3	20 (86.96)	3 (13.04)	0 (0.00)	0 (0.00)	0 (0.00)	0 (0.00)	23 (100.00)
S.T.	0 (0.00)	0 (0.00)	0 (0.00)	0 (0.00)	0 (0.00)	0 (0.00)	0 (0.00)
M	2 (14.29)	11 (78.57)	0 (0.00)	1 (7.14)	0 (0.00)	0 (0.00)	14 (100.00)
Total	26 (25.00)	40 (38.48)	21 (20.14)	13 (12.50)	2 (1.92)	2 (1.92)	104 (100.00)
CCI: H1 + H2	4 (5.97)	26 (38.81)	21 (31.35)	12 (17.91)	2 (2.98)	2 (2.98)	67 (100.00)
CCII: H3 + S.T. + M	22 (59.46)	14 (37.84)	0 (0.00)	1 (2.70)	0 (0.00)	0 (0.00)	37 (100.00)
Village: Mahacho (2001-02)							
H1	0 (0.00)	0 (0.00)	0 (0.00)	0 (0.00)	0 (0.00)	0 (0.00)	0 (0.00)
H2	0 (0.00)	6 (21.44)	17 (60.71)	2 (7.14)	3 (10.71)	0 (0.00)	28 (100.00)
H3	4 (7.84)	34 (66.66)	6 (11.77)	6 (11.77)	1 (1.96)	0 (0.00)	51 (100.00)
S.T.	0 (0.00)	0 (0.00)	0 (0.00)	0 (0.00)	0 (0.00)	0 (0.00)	0 (0.00)
M	0 (0.00)	3 (30.00)	7 (70.00)	0 (0.00)	0 (0.00)	0 (0.00)	10 (100.00)
Total	4 (4.50)	43 (48.32)	30 (33.70)	8 (8.98)	4 (4.50)	0 (0.0)	89 (100.00)
CCI: H1 + H2	0 (0.00)	6 (21.43)	17 (60.71)	2 (4.14)	3 (10.72)	0 (0.00)	28 (100.00)

Table 6 (*continued*)

CCII: H3 + S.T. + M	4 (6.56)	37 (60.66)	13 (21.31)	6 (9.84)	1 (1.63)	0 (0.00)	61 (100.00)
Village: Maladanga (2001-02)							
H1	8 (38.10)	0 (0.00)	3 (14.28)	5 (23.81)	4 (19.05)	1 (4.76)	21 (100.00)
H2	11 (15.28)	9 (12.50)	15 (20.83)	19 (26.39)	13 (18.06)	5 (6.94)	72 (100.00)
H3	91 (54.82)	32 (19.28)	21 (12.65)	17 (10.24)	5 (3.01)	0 (0.00)	166 (100.00)
S.T.	49 (81.66)	4 (6.67)	6 (10.00)	1 (1.67)	0 (0.00)	0 (0.00)	60 (100.00)
M	13 (59.08)	1 (4.55)	6 (27.27)	1 (4.55)	1 (4.55)	0 (0.00)	22 (100.00)
Total	172 (50.44)	46 (13.49)	51 (14.96)	43 (12.61)	23 (6.74)	6 (1.76)	341 (100.00)
CCI: H1 + H2	19 (20.43)	9 (9.68)	18 (19.35)	24 (25.81)	17 (18.28)	6 (6.45)	93 (100.00)
CCII: H3 + S.T. + M	153 (61.69)	37 (14.92)	33 (13.31)	19 (7.66)	6 (2.42)	0 (0.00)	248 (100.00)
Village: Raspur (2001-02)							
H1	0 (0.00)	0 (0.00)	0 (0.00)	0 (0.00)	0 (0.00)	1 (100.00)	1 (100.00)
H2	6 (13.64)	9 (20.45)	4 (9.09)	19 (43.18)	2 (4.55)	4 (9.09)	44 (100.00)
H3	3 (5.35)	15 (26.79)	12 (21.43)	19 (33.93)	6 (10.71)	1 (1.79)	56 (100.00)
S.T.	9 (14.06)	10 (15.63)	29 (45.31)	13 (20.31)	3 (4.69)	0 (0.00)	64 (100.00)
M	3 (33.33)	3 (33.33)	0 (0.00)	2 (22.23)	1 (11.11)	0 (0.00)	9 (100.00)
Total	21 (12.07)	37 (21.26)	45 (25.86)	53 (30.46)	12 (6.90)	6 (3.45)	174 (100.00)
CCI: H1 + H2	6 (13.33)	9 (20.00)	4 (8.89)	19 (42.22)	2 (4.45)	5 (11.11)	45 (100.00)
CCII: H3 + S.T. + M	15 (11.63)	28 (21.71)	41 (31.78)	34 (26.36)	10 (7.75)	1 (0.77)	129 (100.00)

When we refer to land, it is agricultural land. Its distribution among the villagers by size of ownership indicates that the villagers are mostly small or marginal farmers (Table 6). For example, it is found that in Harsingradih 65.18

Broadly speaking, the Tables given above show that most of the Hindu-low-caste scheduled tribes and Muslims people are, in general, landless and marginal farmers as well as engaged as labours. The Hindu Middle caste people, who constitute the second largest mass, are, in general, small or marginal farmers. The Hindu High caste people are mostly engaged in whole range of occupations, viz, marginal and small farmers, big farmers, profession-s, service-s, trade, business etc.

The villages of Jharkhand are endowed with a remarkable feature that the number of marginal farmers constitutes the largest occupational class in the social composition in a village. As indicated in earlier Table 3, the partitioning of land over generations is the root cause for the rapid growth in the number of marginal farmers. The consequent effect is minimization of size of land owned by the households. All these have led to migration, in large numbers, of the villagers out of the state for survival. Those who have not migrated are accelerating conflicts over land using caste identity as a quick instrument for mobilization of own support. While gathering materials for village history, we have come across ample illustrations of such inter-caste conflicts for land possession.

In the two villages of West Bengal the scheduled caste and scheduled tribe households occur much more in numbers. But as we have gathered from local villagers, due to growth of local internal market or domestic consumption as well agriculture becoming more profitable, there has been a remarkable two-way shift between non-agriculture and agriculture. Unlike Jharkhand, the SC and ST population in the region has been drawn in both the streams according to their potential and access to resources instead of migrating out of the state, or becoming embroiled in conflicts over acquiring of land by any means.

The truth lies in the fact that in Jharkhand people are still getting on with their class interests under guise of caste identity. We, thus, find an ongoing tussle between the land - owning Bavans and the other rising aspirant land - owners like the Gowalas, the Chamars, the Rajaks and the Turis in the village Harsinghradih. In Mahacho the combined endeavour taken up by the Dushads and the Chamars in establishing their control over the social power in the village is also a good example in this connection. In Chitmadih, the Koiris and the Muslims are going on with their struggle for land by the erstwhile land owning Rajputs. Incidentally, as we have already mentioned, that is why, "the caste" and "the class" get confounded in the villages of Jharkhand. The distinction between "caste" and "class" gets easily blurred in rural Jharkhand and the villagers, according to the contingencies of place, time, and severity of conflict/competition, sometimes get themselves represented as caste, sometimes as class, and sometimes, as both. On the other hand, in West Bengal, the influence of "caste" is not felt in socio-economic relations, though it runs through the socio-cultural ethos of the region. The down-trodden as a "class" are becoming endowed with access to land and job market. The base of socio-political power is, thus, tilting in their favour. Thus, they are shifting from the control of the upper class, by decentralisation of administrative power structure through Panchayats. Hence, among the villagers of West Bengal class provides functionally more effective form of identity.

5 Parameters of Life and Living of the Villagers Required for Social Network

5.1 *Type and Timing of Needs and Urgencies*

One of the strategies of the people is their potential and ability to get mobilised and act collectively in order to fulfil their needs or necessities. On the other hand, in course of even their daily routine life, situations often occur when a requirement is faced by a household which it has to fulfil immediately even by taking help from others. In such circumstances they have to act on their own, informally, as a household or as a family member. They interact in various dimensions with one another as neighbours, friends, kins, members of same caste / community, co-villagers, etc. Thus, in course of regular interaction in any dimension of daily life, households build stable ties among themselves. We call such an articulation of social ties among a set of households a social network. Obviously, a social network (SN) has to be studied empirically since it may cut across boundaries set up by a priorily given structural-functional parameters. Rather it is flexible enough to grasp real life parameters in course of praxis. We have stated at the beginning of the report that our objective is to study networks of flow of informal help and support as and when required by the villagers. Such flows of help and support operate, in an unnoticed manner, so to say, either directly, pair - wise, or through intermediaries in the network. Thus, this process flows subtly beneath the surface, keeping a low profile, but steadily, to meet various urgent requirements of daily life and living that one cannot fulfill by one's own resources at hand or through any formal institution.

In rural society, it so happens, very often, that one family has to depend on one or more families in order to fulfill various needs. On the basis of the nature of needs, some are resolved by the "women" members of a household, while there are some for the males some may be by the "Head" of a family, some by other. For example, on sudden arrival of a relative a family may face "crisis" due to shortage of essential commodities like rice, fish or egg etc. This is met generally by the female members by requesting others for help. On the other hand, if "cash" was required to buy food or clothes for guest or, for monetary need at the time of a sudden illness, or for purchasing seeds or fertilizers or hiring a pump for irrigation, for cultivation, it is for the male members to act. Help to meet such urgent needs can be broadly divided into following types: (i) material, (ii) financial, (iii) manpower / physical. From the villages under our study, we observe, that there is a variation among the villages in type of help required on the whole. Our qualitative data show that commodities ranging from gold ornaments and money to "negligible items" for cooking food like chilly and turmeric, are requested for and taken or given in order to avoid critical situation in the household.

5.2 *Availability of Help from Formal Institutions to Satisfy Urgent Needs*

It is a matter of very common experience that in the villages, people cannot get timely help from the organised institutions in case of urgency. The institutions have to go by their particular rules and regulations. For illustration, it may be almost an absolutely necessary condition to utilise properly a loan taken earlier from that institution and repaying it. Again, it may be noted that banks cannot give small monetary loans on "as and when required" basis to a villager. Even when they can, procedural rules have to be followed and formalities to be gone through before sanctioning the loan. Moreover, a bank cannot give a loan in kind. As noted earlier, in West Bengal requests have been sent mostly for material help which may not be dealt with by formal organisations.

The picture of Jharkhand in this matter, as gathered from the observation of four villages, is somewhat, different. The Panchayati Raj and its corresponding functions are poised, to a large extent to be stagnant. The situation has been aggravated, since, for more than 25 years, there was no election for formation of the Panchayats. The peasant organisations are not organised properly till date. The political activities and level of public awareness are wholly caste based. Therefore, while the functioning of formal institutions in West Bengal have got accelerated and oriented to provide help and supports, as quickly and as far as possible, to the common villagers, reverse is the case with Jharkhand.

5.3 *How One Decides to Approach Whom for Help or Support*

For a household it is quite an important feature during a crisis situation as to how does it decide to whom to go for help. Socially men have to shape their lives by interaction with "other" in various dimensions at different levels. That way they have to make compromise and build an understanding with neighbours, co-villagers, neighbouring villagers, kins, friends, caste, religion etc. On the basis of all these things, they classify acquaintance circles by resource base and probability of getting a request fulfilled. The last element in decision-making is availability on the day. The whole process involves stages of deliberate decisions, which should not be mistaken as a random process.

We would now discuss the idea of the word help' among the villagers. This information has been gathered through group discussions with various cross-sections of people in the villages. We have already narrated earlier different types of help. Let us consider some of the facts we have gathered in the light of the perception nourished by villagers.

Different meanings imputed to the same action by two actors-ego (at the receiving aid) and the alter (at the giver's end): For instance, suppose, a poor family finds it almost impossible to buy fish or sweet for lack of money on the day of marriage of the girl. Under such circumstances, the neighbour or co - villagers are used to lend their hands, generally, to help the poor family. The head of the poor family, in this case, considers the "help" as a matter of benevolence of others, while the others view differently. They consider the act as carrying out a social duty.

Considering the act as a matter of upholding collective prestige or status, some affinal kins (kutums) have come to visit a family from another village. In order to treat them, a family needs good quality of utensils which the host family does not possess, or, say lack enough of them to treat the guests. In this situation, another family may come forward to help the host, though it is not claimed that they are "helping". On the contrary, according to them they have done so just to keep up the prestige of the village or muhallah or caste / community.

Viewing the act as a "natural" phenomenon of daily life and living: A farmer is short of seedlings while his sowing his piece of land. But he finds that another farmer has a seedbed containing a considerable number of seedlings a part of which is "surplus" and will not be used, but thrown away. The farmer approaches the latter and requests him to give the surplus" seedlings. Subsequently, the latter farmer may be in need of some tools which he does not possess, but the farmer does not possess and gets them from him. The villagers never see any act of help' in it, but, in fact it is.

Villagers usually regard such acts as a matter of routine behaviour in daily life. But that way, unconsciously, they perform social acts as means to serve duty. Again in Jharkhand, the grocers lend essential commodities to the villagers when needed. The grocers think that they are "helping" the villagers by this kind of lending. But, on the contrary, the villagers opine that the grocers are not providing any "help" to them since the grocers lend them in exchange of higher price against each commodity in latter course. The view in this respect is quite different in West Bengal villages. The villagers would report that the grocer is providing "help" because they are providing the commodities at the time when one needs it most.

Another illustration of unconsciously given help is the following. An experienced farmer advises his fellow farmer while he is passing by the latter's paddy field about how to apply pesticides. Would this spontaneous advice be accounted for help'? In fact, he has helped the latter, even though his suggestion may have been triggered by the fear that the pests may spread to all the neighbouring paddy fields including his own, from the already pest attacked fields.

Socially speaking, help or support is not a routinized behaviour nor ritualistic or casual in nature. The relevant question we began with how a man decides from whom to take "help" in a situation of emergency. It seems this selection is guided by the following parameters: Time when help is required, Type of help including quantity and value of help, place where it is required, and who requires it. This selection is not random, casual and directionless. There are certain circles consisting of selected people called potential forces to act as the centres of help', which change with time, place and situation. A person is accustomed to his different centres of help' as per his social status, personality and social inter-connectivity.

5.4 Why Help Is Provided

We thus find that help may be provided in various forms by different sections of population - such as kin, neighbours, friends, employers, local leaders, experienced persons, shop-owners etc. One question, now, can be raised as to why these persons come out for lending help. In this regard two major factors come into operation: social values' and self interest'. The social values in many cases, especially, in crisis-ridden situations, pave the way for help. There are various sources of value-oriented motivation to provide help. It may be a matter of traditional respect; normative duty; extending power base; spreading influence and popularity; and so on. It may be a matter of common humanitarian consideration as well. For example, taking a sick people away to hospital at night, giving rice to an unfed family, assisting a family to carry out rituals when a person has died and so on. The self interest factor comes out mainly on calculative futuristic approach. In this case, the person gives assistance or help to the seeker on condition of getting the help from the latter when the former needs the same. For illustrations, employers give help to the employees, relatives help relatives, etc. Our study also depicts the fact that "help" operates in the society, on the basis of mutual give and take process. Lastly, culturally identified feeling of solidarity within a caste or community can act as a powerful source to motivate one to help another. This cultural identity may be a follow-up of village settlement pattern which we have discussed earlier.

A large proportion of the high and middle caste villagers have migrated separately by themselves from north Bihar. These are Gowalas, Koiris, Rajputs, Bavans, Brahmins, etc. apart from the tribal and Hindu low caste autochthones such as Turis, Dusadhs, Santals and so on. The settlement, as per our study, was based on kin groups with a few "primary settlers", migrating generations back. With the growth of population, the primary settlers form a large section of village mass and identify themselves as strong caste groups.

In the villages of West Bengal, however, the type of settlement has not followed this pattern. Consequently, "help" in the Jharkhand villages is mostly confined within one's kin-circle and that way, within one's caste group. But in West Bengal the "help giver" is often a neighbour who may be a kin or not, or who may belong to the persons of the same caste or not.

5.5 Distance of Spread of Sources of Helps and Assistance

Earlier the boundary of flows help was restricted mostly within the village. But now with the better means of communication and improvements in the system of transportation, the villagers are now used to send requests for to help those who have migrated outside the village, even someone who reside 100 or more kilometres away. Distance has not dampened the efficacy of social network, nor it has led to more and more atomisation of the rural people. Social networks are rather working in a new form.

Now, before going to the data gathered from SNA, we respond to few natural queries regarding a household requesting another for help or support. Such as: How one decides to request whom for help (or, in other words, how does one decide to go to whom for help)? Why does one provide help or support at all to someone? How for the network spreads? And so on.

6 Findings from Social Network Analysis

In measuring reciprocity, Rao and Bandyopadhyay [14] have studied different measures of reciprocity. It is also available in several studies [9], [14] that the s_3-measure with given out-degrees is more suitable for standardizing the observed reciprocal pairs of a social network. Hence, for standardizing the number of reciprocal pairs, we adopt the s3-measure of reciprocity. Besides, the other measures adopted here have been discussed in details by several researchers [1], [11]. Now we start the network analysis by giving summary measures of the relevant characteristics of the social networks of the six villages as in Table 7 given below.

Table 7
Summary measures of the social networks of the six villages

Measure	Four villages in Jharkhand				Two villages in West Bengal	
	Hd	**Cd**	**Bg**	**Mc**	**Md**	**Rp**
(1)	(2)	(3)	(4)	(5)	(6)	(7)
No. of hhs. (n)	313	175	104	89	341	174
No. of ties inside village in %	51.12	75.00	42.72	53.92	43.01	65.50
No. of ties going to outside village in %	48.88	25.00	57.28	46.08	56.99	34.50
Total Ties (inside vill. + outside vill.)	100.00 (n_1=1023)	100.00 (n_1=464)	100.00 (n_1=309)	100.00 (n_1=217)	100.00 (n_1=744)	100.00 (n_1=458)
Ties within castes (inside vill.) in %	78.78	56.90	78.79	70.09	36.56	48.00
Ties between castes (inside vill.) in %	21.22	43.10	21.21	29.91	63.44	52.00
Total Ties (only inside vill.)	100.00 (n_2=523)	100.00 (n_2=348)	100.00 (n_2=132)	100.00 (n_2=117)	100.00 (n_2=320)	100.00 (n_2=300)
Ties within castes (outside vill.) in %	81.40	74.14	75.14	62.00	23.35	39.87
Ties between castes (outside vill.) in %	18.60	25.86	24.86	38.00	76.65	60.13
Total Ties (only outside vill.)	100.00 (n_3=500)	100.00 (n_3=116)	100.00 (n_3=177)	100.00 (n_3=100)	100.00 (n_3=424)	100.00 (n_3=158)
Average out-degree : within village	1.67	1.99	1.27	1.3	0.94	1.72
Total	3.27	2.65	2.97	2.44	2.18	2.63
No. of reciprocal pair	148	34	14	9	11	10

Table 7 (*continued*)

s_3- measure of Reciprocity (in %)	18.39	19.54	21.21	15.52	6.88	6.67
No. of isolates (within village ties)	7.35	2.86	10.58	13.48	22.29	4.02
No. of isolates.	0.64	0.00	0.00	3.37	2.64	0.57
Percentage of reach. Pairs (dir.)	10.62	7.06	6.74	15.67	0.75	4.34
No. of strong components (p)	223	137	74	68	330	159
Size of largest strong comp.	27	6	19	16	2	4
Average finite distance(dir)	8.2	2.99	3.16	4.49	0.96	2.37
Maximum finite distance (dir)	28	10	10	14	6	9
Percentage of reach. Pairs (undir.)	68.11	94.38	49.26	69.40	39.97	92.14
No. of weak components (q)	33	6	17	14	94	8
Size of largest weak comp.	258	170	72	74	215	167
Average finite distance (undir)	5.78	3.22	5.99	4.86	4.60	4.33
Maximum finite distance (undir)	16	11	17	13	12	11
Connectedness: strong [(n-p)/(n-1)] in %	28.85	28.84	29.13	21.00	3.24	8.67
Connectedness: weak [(n-q)/(n-1)] in %	89.74	97.13	84.47	85.23	72.65	95.95

What observed from the above Table 7 is described as follows. There is a considerable average village size of households (199) with the minimum (=89) in Jharkhand village Mahacho (Mc) and maximum (=341) in West Bengal village Maladanga (Md). It is already mentioned earlier that, according to village selection scheme, out of the four Jharkhand villages, Harsingraidih (Hd) and Bagrah (Bg) are nearer to Giridih town and metal road than both Chitmadih (Cd) and Mahacho (Mc). In case of West Bengal, in the same way Maladanga (Md) is nearer to block town and metal road than Raspur (Rp). In case of the villages far from the town, the interaction among the people inside the respective village is greater (more than 50% in each village) than the people outside the village, whereas, in case of the villages nearer to the town, it is distributed about fifty-fifty to inside village people and outside village people. It indicates that the people of the villages nearer to town have greater tendency of establishing connection to their outer world than the people of the villages far from the town. In Jharkhand, ties within a village are more concentrated to within the same caste than between the castes, whereas, in West Bengal, ties within a village flow between the castes little more than within the same caste. It shows that caste-based interaction for getting help is more in Jharkhand villages and, on the other hand, its base is secular in case of West Bengal. Even the same picture is observed in case of articulation of ties going to outside village(s). The settlement pattern of the Jharkhand villages may be one of the reasons for becoming caste-based ties.

Table 8
Percentage distribution of inside village ties of help by class in each village

Class	Villages of Jharkhand				Villages of West Bengal	
	Hd	Cd	Bg	Mc	Md	Rp
(1)	(2)	(3)	(4)	(5)	(6)	(7)
Panel A: Class by source of livelihood						
Within class	53.54	33.62	55.30	57.26	29.06	42.33
Between class	46.46	66.38	44.70	42.74	70.94	57.67
Total	100.00 (n_2=523)	100.00 (n_2=348)	100.00 (n_2=132)	100.00 (n_2=117)	100.00 (n_2=320)	100.00 (n_2=300)
Panel B: Class by land ownership						
Within class	60.04	29.89	55.30	48.72	20.00	20.67
Between class	39.96	70.11	44.70	51.28	80.00	79.33
Total	100.00 (n_2=523)	100.00 (n_2=348)	100.00 (n_2=132)	100.00 (n_2=117)	100.00 (n_2=320)	100.00 (n_2=300)

Again, when we go through the distribution of inside village ties across the classes, by family main source of livelihood or land ownership, the almost similar picture is observed (Table 8). Here also, in case of Jharkhand villages, ties have higher tendency to be distributed within same class, whereas it is greater between the classes than within class in case of West Bengal villages. It indicates that people of Jharkhand villages depend more on the people belonging to same class for getting help, whereas, in West Bengal villages, it is higher on the people belonging to other classes. The average out-degree indicating individual expansiveness or the percentage of isolation reflecting the self-dependence of households is not high throughout all the six villages. Even the measure (s3) of reciprocity illuminating the solidarity/cohesiveness among the households is not also high. On the other hand, reachability (i.e. possibility of a household approaching another, either directly or indirectly through intermediaries for getting help) is not high, but moderate. If we ignore directions (i.e. if we assume that A can go to B and B can go to A whenever at least one of them goes to the other), reachability becomes increased tremendously with minimum 39.97

Now, to make an in-depth analysis, we concentrate to the collected data on: to whom (i.e. relation) people go for help; the amount/quantity and type of help sought; and what purposes served by help received.

6.1 From Whom Help Was Sought

Sources to which requests for help at different times of emergency were made by the villagers and received, were gathered and classified. This is shown in Table 9. It is clearly understood from the Table 9 (given above) that non-formal help from kin, neighbours, employers etc. are much in abundance in village society under our study. However, in the villages of Jharkhand the kin's (hence, "caste") provide the bulk of informal support while in West Bengal villages the neighbours give the lion's share of informal support.

Table 9

Percentage distribution of ties by sources of help in each village

Sources of help	Villages of Jharkhand				Villages of West Bengal	
	Hd	Cd	Bg	Mc	Md	Rp
(1)	(2)	(3)	(4)	(5)	(6)	(7)
Panel A: On inside village ties						
Kin	66.35	65.98	73.48	72.25	20.63	18.67
Neighbour	22.75	22.70	12.88	20.06	60.88	70.67
Friend	4.02	3.45	4.55	5.13	3.93	6.00
Employer-Employee	1.91	1.43	2.00	1.06	10.44	2.33
Others	4.97	6.44	7.09	1.50	4.12	2.33
Total	100.00 (n_2=523)	100.00 (n_2=348)	100.00 (n_2=132)	100.00 (n_2=117)	100.00 (n_2=320)	100.00 (n_2=300)
Panel B: On outside village ties						
Kin	71.40	73.45	69.41	68.00	18.07	19.62
Neighbour	3.20	2.59	5.08	0.00	60.25	62.97
Friend	12.80	10.34	7.35	14.00	11.12	8.80
Employer-Employee	5.00	11.21	12.99	11.00	5.36	4.27
Others	7.60	2.41	5.17	7.00	5.20	4.34
Total	100.00 (n_3=500)	100.00 (n_3=116)	100.00 (n_3=177)	100.00 (n_3=100)	100.00 (n_3=424)	100.00 (n_3=158)

6.2 Amount/Quantity and Type of Help Sought

We have classified the requests for help by its type, as financial, material, and physical (man-power, advice etc.) in Table 10 (given below). It is obvious from this

Table that non-formal help ties for money are much in the villages of Jharkhand, whereas, this is high for material exchange in case of West Bengal villages. Again, there is, in Jharkhand villages, non-formal advice or man-power help to some extent, but it is almost nil in West Bengal villages.

Table 10
Distribution of ties (inside + outside Vill.) by types ("at all") of help in each village [% in bracket]

Type of help	Villages of Jharkhand				Villages of West Bengal	
	Hd	Cd	Bg	Mc	Md	Rp
(1)	(2)	(3)	(4)	(5)	(6)	(7)
Financial	701 (68.52)	287 (61.85)	203 (65.70)	144 (66.36)	135 (18.15)	156 (34.06)
Material	211 (20.63)	88 (18.97)	57 (18.45)	51 (23.50)	629 (84.54)	305 (66.59)
Physical	422 (41.25)	201 (43.32)	121 (39.16)	88 (40.55)	3 (0.40)	4 (0.87)
Total	100.00 (n_1=1023)	100.00 (n_1=464)	100.00 (n_1=309)	100.00 (n_1=217)	100.00 (n_1=744)	100.00 (n_1=458)

The distribution of the monetary ties according to the total amount (in rupees in a whole year) is in Table 11(given below). In case of Jharkhand villages, such ties are more than 60%, whereas, in West Bengal, these are less than 20% in the village Maladanga and less than 35% in the village Raspur. In this context, we mention here that the monetary help is not necessarily a loan, because it is in most cases without interest and the borrower generally repays at his convenience. The group discussions among the people of the respective villages give possible reasons behind such difference of monetary help between two provinces. Cash income generated from ou-migration in non-agricultural jobs is more in Jharkhand villages than in West Bengal villages. Again if we look at help taken in the nature of food (mainly rice) in Table 11 (given below), such requests are more in West Bengal villages than in Jharkhand villages. Due to comparatively less land ownership of the households in Jharkhand villages, such requests become low. We have classified the requests for help by its type, as financial, material, and physical (man-power, advice etc.) in Table 10 (given below). It is obvious from this Table that non-formal help ties for money are much in the villages of Jharkhand, whereas, this is high for material exchange in case of West Bengal villages. Again, there is, in Jharkhand villages, non-formal advice or man-power help to some extent, but it is almost nil in West Bengal villages.

Table 11
Percentage distribution of ties (inside + outside vill.) of help on cash and kind in each village

Cash in Rs. (C) / Kind in Kg. (K)	Villages of Jharkhand				Villages of West Bengal	
	Hd	Cd	Bg	Mc	Md	Rp
(1)	(2)	(3)	(4)	(5)	(6)	(7)
Panel A: Cash (in Rs.)						
0 (noncash)	31.48	38.15	34.30	33.64	81.85	65.94
$0< C \leq 100$	6.35	7.76	5.18	10.60	1.75	5.90
$100< C \leq 250$	5.18	8.41	7.44	11.06	4.57	7.21
$250< C \leq 500$	11.05	8.62	16.50	15.21	7.66	11.14
$500< C \leq 1000$	11.83	6.47	10.03	12.44	2.28	5.24
$C>1000$	34.12	30.60	26.54	17.05	1.88	4.59
Total	100.00 (n_1=1023)	100.00 (n_1=464)	100.00 (n_1=309)	100.00 (n_1=217)	100.00 (n_1=744)	100.00 (n_1=458)
Panel B: Kind (in Kg)						
0 (non-kind)	90.81	92.67	82.52	77.42	16.53	33.62
$0< K \leq 20$	5.96	3.66	6.47	12.44	26.75	45.41
$20< K \leq 60$	1.86	1.08	1.94	5.07	5.78	7.86
$60< K \leq 200$	1.27	1.72	0.97	4.15	22.31	7.64
$200< K \leq 500$	0.00	0.00	0.65	0.46	17.47	3.71
$K>500$	0.10	0.86	7.44	0.46	11.16	1.75
Total	100.00 (n_1=1023)	100.00 (n_1=464)	100.00 (n_1=309)	100.00 (n_1=217)	100.00 (n_1=744)	100.00 (n_1=458)

6.3 *Purpose of Help*

The Table 12 (given below) shows that the purpose of help in Jharkhand villages poses a great level of difference with that of the villages of West Bengal. Medical facilities are poorly shaped in the villages of Jharkhand. At the same time, the villagers in Jharkhand spend lavishly in order to perform their rites in a pompous way. The villages of Jharkhand very frequently require help for medical treatment and also for performing family rites and coping with urgent consumption requirements. The villages in West Bengal do the same mostly for their maintenance of "sudden" consumption needs for social reasons. Therefore, help becomes very much essential there.

Table 12
Distribution of ties (inside + outside Vill.) by purposes ("at all") of help in each village [% in bracket]

Purposes of help served	Villages of Jharkhand				Villages of West Bengal	
	Hd	Cd	Bg	Mc	Md	Rp
(1)	(2)	(3)	(4)	(5)	(6)	(7)
Survival	20.63	32.54	20.39	17.51	68.81	32.10
Medical	46.12	36.47	43.63	27.19	3.36	4.80
Family rites	36.95	15.73	33.98	42.40	8.74	17.03
Family crisis & household requirements	4.40	3.23	1.94	5.53	11.15	25.55
Production	3.18	5.60	11.32	16.13	9.14	20.31
Others	0.10	0.22	0.65	0.00	0.00	1.09
Total	100.00 (n_1=1023)	100.00 (n_1=464)	100.00 (n_1=309)	100.00 (n_1=217)	100.00 (n_1=744)	100.00 (n_1=458)

7 Conclusions

As a way of conclusion, we would discuss about different aspects on the basis of SN data what we have presented in various ways.

In Jharkhand the ties - both inside and outside, have developed mainly among the members of the same kin group, hence the same caste the same village. Ownership pattern and principal sources of livelihood have not produced any impact on the pattern of articulation of relationship. But contrary to ties being formed within the kin group in Jharkhand, the same is articulated among the neighbours and others in Birbhum. In Jharkhand, kin-group (caste) based social networks have developed owing to social compulsion in the direction of strengthening a social safeguard or insurance on the part the weaker sections of the villages against the manifold sources of vulnerabilities in life and living, particularly aggression and exploitation by land owing caste groups, in a situation where there is no effective structural remedies to ameliorate the condition of their life and living. In fact, this raises the level of social depth of social network analysis to a higher dimension.

In fact, unlike structure - function based methodology of study, SNA does not begin with any a priority given boundary. Being able to identify empirically its boundary, parameter remains quite amenable to respond to the variations in social reality.

Our SN based findings give both "no" and "yes" as answer. In Jharkhand, ties still circulate mostly by kins (caste / community), but connect closely the section in the village with their village and with their counter part in the larger society. Thus, kinship or caste groups provide instrument of communication and contacts for linking the village with the far away job market. This is one side of the coin. The other side, the internal one, is as follows. The economic situation of the village may not be changing in the near future. The more is un-industrialisation or de-industrialisation over time, as it is happening in Giridih region of Jharkhand, the sharper will be the competition / conflicts over ownership of land and the more forcibly one will drive to mobilise support. A readily available effective instrument in this regard is to depend up on their kin or caste groups as a matter of strategy. We note that earlier land productivity in the village around Giridih, on average, was very low and that, too very much subject to uncertainties of rainfall. The region was rain fed. However, now-a-days, prospect of irrigation has opened up, through it needs individual investment, since public programme of irrigation is almost absent. Because of lack of cash income, private investment in irrigation also was lacking earlier. But since migration has resulted in flow of cash income, this has made investment possible for agricultural development. The above means "yes", but with a new meaning.

In Birbhum, on the other hand, usual structural-functional categories come out as rather residuals. The concept of social group, rather, appears to be more appropriate unit for understanding social structure. In rural West Bengal, presently the prime issue is of reducing the level of unevenness in development, i.e., which villages have moved far ahead and which ones are lagging behind. Implementation of land reforms and the Panchayat Raj have been instrumental in shifting the political and social power to the down-to-the soil villagers, and consequently the villagers are coping up with the problems of their life and living by getting themselves identified as a social class or group. This has brought about a noticeable qualitative difference in socio-economic base of networks support among the villagers in Birbhum. It comes out from our study that the savings of the villagers in West Bengal, which are generated internally, are being invested insight the rural areas in both agricultural and non-agricultural sectors, ensuring thereby a sort of recycling of the internal resources in the state.

Acknowledgements. The authors are sincerely grateful to late Professor Suraj Bandyoapadhyay and late Professor A. R. Rao for receiving their many suggestions during the project work.

Appendix A

List of occupation

Occupations	Occupations
A. Agriculture and related activities	40. Tahasildar
1. Farmer	41. Railway service
2. Cultivator	42. Service
3. Share cropper	43. Service of guard on Govt. Office
4. Soil cutter	44. Service in home clinic
5. Agricultural labourer	45. Service in private farm
6. Day labourer	46. Service in CISF
7. Bagali	**F. Labour**
B. Profession	47. Brick labour

8. Priest	48. Mica sorter/labour
9. Legal practitioner	49. Labour in chuna factory
10. Homeopathic doctor	50. Mason helper
11. Teacher	51. Khalasi in truck
12. Tution	52. Rolling mill labour
C. Business	53. Labour in candle factory
13. Barber	54. Labour in cloth
14. Washarman	55. Factory labour
15. Tailor	56. Coolie
16. Laundry	57. Labourer in stone crusher
17. Petty business	58. Duli carry
18. Small business	59. Saw mill worker
19. Milk seller	60. Watter carry
20. Hawker	61. Sweeper
21. Rice milk	62. Labour/worker in shoe shop
22. Saloon	**G. Technical work**
23. Business in domestic animal	63. Tally mistry
24. Lottery business	64. Coal carry
25. Grocery shop	65.Unspecified tech. work in mica/coal
D. Craft artisan	66. Mechanical mistry
26. Blacksmith	67. Electric mistry
27. Carpenter	68. Sanitary mistry
28. Potter	**H. Other technical work**
29. Leaf plate maker	69. Rickshaw van puller
30. Basket maker	70. Truck driver
31. Biri binding	71. Minibus/car driver
E. Service	**I. Domestic help**
32. Service in CCL/BCCL	72. Cook
33. Private Security Guard	73. Servant/maid servant
34. Service in laundry	**J. Not gainfully occupied**
35. Hotel servant	74. Pension holder
36. Clerical work	75. Dependent/remittance receiver
37. Service in Bank	76. Begging
38. Private service	77. House rent
39. Peon in private school	

Appendix B

Oral histories of settlement of caste/communities in Jharkhand villages

Village: Harsingraidih

History of migration				Outline history of occupation change	Main features of land received by early settlers of a lineage
Period (approximate number of generations back)	Settlement. of lineages of caste /group	From where migrated	Reason of migration		
7	Bavan	Tisri Block and Dhanbad area.	For cultivable land.	Cultivation was their first occupation. Now, they are involved in cultivation, service, and business.	Nearly 500 acres of land was received by them from the Zamindar.
7	Chamar	Neighbouring village.	The Bavan brought them to work as their labourers.	The past occupations were colliery related works, mica work agriculture and day labour. The ongoing occupation is non - agricultural jobs.	Dwelling land was given by the Bavans and they also cultivated land after preparing by claying jungles and levelling undulated land.

7	Turi	Motileda of Bengabad Block.	The Bavan brought them for getting their sevice as labourers and Dholok vadaks (Drum beaters) in rituals.	Past occupation was agricultural labour, mica labour, Dholock vadan. They now totally depend on Rick-show pulling, and making bricks.	It was reported that they owned 50 acres of land. But the fact of getting this land is unknown.
5	Musilm	Colliery area and Azidi village of Giridih Block.	Due to riot	They used to work in the collieries. Now they are engaged in business, agriculture and non - agricultural work.	Land purchased from the Turis.
4	Rajak	Rasko of Harzribagh district	Opression and torture by Bavan community who took already their land also. But, one Bavan of this village landlord brought them here.	First occupation was in connection with the collieries. They were involved in mica processing. From the 2nd generation they started their caste occupation. Now, they are engages in services, cultivation, business etc.	They used to pay taxes in exchange of land given to them by the Bavans also purchased land from the Turis, the Bavans, and the Chamars.

4	Gowala	Giraniya of Gandey Block	On getting the information from a Muslim friend that cultivable and cattle graging land was available in this village, they came here.	Previous occupations were agriculture and mica processing. Now only agriculture is main occupation.	Land purchased from the Bavan and the Muslim of this village and the Ghatwars of the neighbouring village.
4	Kahar	Karo village of Jamtara Block	They came here as Mica labourers.	Previous occupation was mica labour. Now, they work in Rolling mills and agriculture.	Land purchased from the Bavans
3	Ghatwar	From colliery area.	They came here to avoid some family scandals for which the entire group was ridiculed.	Their previous occupation was mica -processing and serving as labour. Now, they are engaged in business and other activities.	Land purchased from the Bavans.
2	Brahmin	Biltandsituated to the far west of this village.	Villagers brought them to worship deity in the local temple.	Performing rituals and worship were the previous occupation. Now they are engaged in teaching , services and agriculture.	Land received from the King of Gadi Sri-rampur.

2	Teli	Pachamba of Giridih	For business.	Past occupation was business. Now, they depend on petty business.	Land purchased from the Bavans.
2	Kayastha	Udnabad of Giridih Block.	For service.	Past occupation was service which is still continuing.	Land purchased from the Chamars.
2	Kumhar	Loikumartuli of neighbouring village.	Scarcity of land.	Colliery related works Presently, they are pursuing in caste occupation and other non - agricultural jobs.	Land purchased from the Rajaks
2	Lohar	Chittahit of Murger in Biher	To continue the caste occupation.	Caste occupation still going on.	Land purchased from the Bavans and the Rajaks.
3	Rajput	The Rajput and the Ghatwar are two different communities. Generations back a Ghatwar got married to Rajput girl. Since then and began to himself Rajput.			

Village: Chitmadih

History of migration				Outline history of occupation change	Main features of land received by early settlers of a lineage
Period (approximate number of generations back)	Settlement lineages of caste /group	From where migrated	Reason of migration		

7	Rajput	From village Kurchutta under Bengabad Block	Increase in population and divison of cultivable land. The village, that time,was covered with deep forest under the ownership of Zamindars.	Cultivation was their first occupation. Presently a few households migrated out of the State.	Received land from the Zamindars and the same was prepared for cultivation.
7	Koiri	From the South of this village	Huge amount of uncultivable land was of no use. Therefore, this community came here down for having land for settlement.	First and the present occupation is cultivation. This community in the village at first started migration out of the State for non-agricultural jobs.	Received land from the Zamindars and the same was prepared for cultivation.
7	Barhi	Naitand, neighbouring village under Bengabad block	They were brought by the Rajputs for making agriculture related implements and accessories.	Presently they are engaged in their traditional caste occupation and agriculture. Migration out of the State for non - agricultural jobs.	Received land from the Zamindars and the same was prepared for cultivation.

7	Napit	Unknown	They were brought by the Rajputs for hair trumming and carrying out rituals. Since the villagers had a tussle with the Napits of Kurchutta village.	Their caste occupation stands unchanged.	They have got hold of insignificant land.
7	Muslim	Village Khutaribad of Bengabad Block and bihar sharift	A man of Khadra villagee brought a few families. The rest were brought by the Rajputs for meeting the agricultural need.	In the past years they were engaged in agricultural labour. Now, they render work in collieries and others private sectors. Migration out of the State for non - agricultural jobs.	Some land was received from the neighbouring Khadra village, some provided by the Rajputs and the rest is received from zamindars.
6	Gowala	Unknown	They found the grassy barren land and jungles suitable for cattle grazing.	Caste occupation and agriculture .Migration out of the State for non - agricultural jobs.	Received land from the Zamindars and the same was prepared for cultivation.

5S	Mahuri	Naitand the neighbouring village under Bengabad Block	They came here for Business purpose as it was newly formed that time	Business alongside agricultural-both in past and at present Migration out of the State for non - agricultural jobs.	Land was received from Zamindars.
3	Chamar	Khutaribad under Bengabad Block	Villagers brought them for the purpose of dumping dead animals	Previously they were engaged in caste occupation and agriculture labour. At present they work as day labour.	Received land from the Government in the time of first settlement.
2	Dusadh	From the neighbouring village of Bengabad Block	They were brought by the Rajputs to serve them as bonded labour.	Past occupation was to serve the Rajputs as bonded labour. At present they work in the form of marginal farmers and day labour.	The Rajputs gave some land to be dwelt.
38 years	Lohar	From Jamua Block	Villagers brought them as blacksmiths.	Their traditional occupation remains unchanged. Migration out of the State for non - agricultural jobs.	They have got hold of insignificant land.

Village: Bagrah

History of migration				Outline history of occupation change	Main features of land received by early settlers of a lineage
Period (approximate number of generations back)	Settlement of lineages of caste/group	From where migrated	Reason of migration		
7	Turi	Not known	Though the cause of migration is unknown, but the oral history goes by the fact that one Chowkidar of Bengabad Police station first came here.	Their previous occupation was Drum beating, cultivation, day labour, agricultural labour and also mica labour. At present they are more or less landless and dependent on non-agricultural labour for survival. Migration out of State for non-agricultural jobs.	In the past years maximum land was occupied by this community. From the oral history it is also known that some Zamindar got land for the preparation of cultivable land.

5	Koiri	Ghutia of Bengabad block	Hulo Mahato, was the first person to came here on ground of the death of his child and he thought that his family would be non - existent if he stayed there.	Previous occupation was cultivation and Mica processing. At present, they are dependent on agricultural activities. Migration out of State for non-agricultural jobs.	Purchased land from the Turis
5	Yadav	Karanpura, Khuta-band,Khurd dighara of Bengabad block	Increase in population in previous village made them come.	Caste occupation, cultivation and mica works were their previous occupation. At present they depend on agricultural activities. Migration out of State for non-agricultural jobs.	Land purchased from the Turis.

5	Muslim	Not known	Koiri's brought Muslim in this village for rendering labour.	Previous occupation was agricultural labour and Mica processing. Now, they depend on illegal coal-carrying and non-agricultural activities. Migration out of State for non-agricultural jobs.	Some land - both agricultural and dwelling gifted by Koiris. At present they are landless more or less.
4	Napit	Some village (not recalled), 8Km. away from the present village	Yadav brought them for hair trimming and ritual purposes.	Caste occupation was their earlier occupation. Now, they depend on small farming. Migration out of the State for non - agricultural jobs.	Gowalas gave some dwelling and cultivable land.

4	Kumhar	Some village (not recalled), 9-10 Km. away from the present village	Yadav brought them for the production of potteries and Tiles for making roof.	Previous occupation was pottery and at present they depend on small farming as well as pottery. Migration out of the State for non - agricultural jobs.	Dwelling land gifted by the Yadavs and cultivable land was purchased from the Turi's.

Village: Mahacho

History of migration				Outline history of occupation change	Main features of land received by early settlers of a lineage
Period (approx-imate number of gen-erations back)	Settlem-ent of lineages of caste/group	From where migrated	Reason of mi-gration		

7	Yadav	Jamua Block	The Santhals who settled the village Mahacho firstly left it for some unknown causes. The village that time was covered by Jungles. Thereafter, a Yadav group came here from Jamua Block before 7th generation and settled here. The causes of migration of the Yadavs were: (i) the oppression and torture by the Baavan community (ii) cattle - grazing.	They were milkmen and cattle - grazer. Now, their occupation is cultivation. Migrated out of state for non-agricultural job.	The jungles of the village cut down by them to prepare cultivable land.
6	Koiri	Jamua Block	Increase in population in previous village made them come.	Cultivation is their traditional occupation. Migrated out of state for non - agricultural job.	Some land gifted by the Yadavs because that time the village population was less. They cut down forest and prepared cultivable land.

6	Chamar	Neighbouring village Telonari (1km. away from the present village)	The Yadav brought them for dumping dead animals. To serve labour in agriculture.	The previous occupation were dumping of dead animals and serving as labour .Now they are engaged in small farming /agriculture labour and non agricultural labour. Migrated out of state for non agricultural job.	Some dwelling land was gifted by Yadavs. They received land from the first settlement Survey.
5	Hazra	Jamua Block (Chitradih Village)	Poverty and population increase. Previously they were engaged in theft. The Yadavs and others kept them as their protectors.	Theft and day - labour. Now they involved in farming/Agricultural labour and day labour. Migrated out of state for non agricultural works.	Cutting forest for cultivable land.
6	Muslim	Neighbouring village (1km. away from the present village)	One Son - in - law of Telonari first came here.	The first occupation was weaving. Now, they fully depend on cultivation. Migrated out of state for non agricultural works.	Some cultivable land and dwelling land were gifted by father - in -law after purchasing.

100 yrs back	Bavan	5 km. away from Mahacho (one household)	Unknown	Cultivation	Captured land from the villagers on the pretext of their ignorance and illiteracy. Land purchased from Koiris & captured land through encroachment.

Appendix C

Caste /communities distribution of six villages

Ritual status Category	Villages[5]	Caste /community					
		Hd	Cd	Bg	Mc	Md	Rp
(1)	(2)	(3)	(4)	(5)	(6)	(7)	(8)
Hindu High Castes(H1)[6]	Bavan	39	-	-	-	-	-
	Rajput	2	37	-	-	-	1
	Brahmin	12	-	-	-	19	-
	Kayastha	1	-	-	-	2	-
	Mahuri	-	14	-	-	-	-
	Total	54	51	-	-	21	1
Hindu Middle Castes (H2)	Barhi	-	12	-	-	-	-
	Ghatwar	4	-	-	-	-	-
	Kahar	9	-	-	-	-	-
	Koiri	-	45	40	8	-	-
	Kumhar	1	-	4	-	-	-
	Lohar	17	2	-	-	-	13
	Napit	-	2	6	-	1	-
	Gowala	8	7	17	20	26	13
	Baishnab	-		-	-	2	3
	Sadgop	-		-	-	38	15
	Kamar	-	-	-	-	5	-
	Total	39	68	67	28	72	44

[5] Hd = Harsinghradih, Cd = Chitmadih, Bg = Bagrah, Mc = Mahacho, Md = Maladanga, Rp = Raspur.

[6] H1 = Hindu High, H2 = Hindu Middle, H3 = Hindu Low, ST = Scheduled Tribe, M = Muslim.

Hindu Low including Scheduled Castes (H3)	Teli	4	-	-	-	-	-
	Chamar	95	2	-	25	-	-
	Dusadh	-	1	-	26	-	-
	Pasi	-	-	-	-	1	-
	Turi	43	-	23	-	-	-
	Rajak	44	-	-	-	-	-
	Bhuia	-	-	-	-	-	10
	Hanri	-	-	-	-	18	7
	Sunri	-	-	-	-	1	39
	Bagdi	-	-	-	-	98	-
	Muchi	-	-	-	-	43	-
	Bauri	-	-	-	-	5	-
	Total	186	3	23	51	166	56
Scheduled Tribe (ST)	Santhal	-	-	-	-	54	16
	Mal	-	-		-	6	-
	Orang	-	-	-	-	-	44
	Kora	-	-	-	-	-	4
	Total	-	-	-	-	60	64
Community	Muslim (M)	34	53	14	10	22	9
Village total		313	175	104	89	341	174

References

1. Bandyooadhyay, S., Rao, A.R., Sinha Bikas, K.: Models for Social Networks with Statistical Applications, ch. 1: 1-29, ch. 6:135-181. Sage Publications, London (2010)
2. Bandyopadhyay, S., von Eschen, D.: The Condtions of Rural Progress in India, ch. 4.2.C:1-91, pp. 1–91. Report submitted to the Canadian International Development Agency, Ottawa (1980)
3. Bandyopadhyay, S., Sengupta, D., Dasgupta, A., Choudhuri, A.K., Jana, R.: Villages surrounding Giridih: A case of structural adjustments or social transformation. A technical report submitted to Sociological Research Unit, Indian Statistical Institute, Kolkata. Tech. Report No.: 3/1999/SRU (1999)
4. Bandyopadhyay, S.: A report on Evaluation of Rain-fed Farming Projrct - Phase I & Phase II. Funded by ODA, UK and undertaken by Sociological Research Unit, Indian Statistical Institute, Kolkata. Project Report No.: E-7/1995/SRU (1993-1995)
5. Bandyopadhyay, S.: Introduction to Social Network Analysis. In: Introduction to Social Network Analysis. Proceedings of Workshop on Network Analysis with Applications, funded by Department of Science and Technology, Government of India, and organized by Theoretical Statistics-Mathematics Unit, ch. 1: 1-22, Indian Statistical Institute, Kolkata (2003)
6. Choudhuri, A.K., Bandyopadhyay, S., Jana, R.: A combined report on two projects (i) Pattern of Social Relations: An Exploratory Social Network Analysis in two villages in Jharkhand, and (ii) Study of social Network in Some Villages in Giridih area in Jharkhand and Md. Bazar Area of West Bengal, Sociological Research Unit, Indian Statistical Institute, Giridih, Jharkhand. Project Report No.: I-5/2006/SRU (2006)

7. Dasgupta, A.: In-depth study on the levels of development of SCs and STs, Funded by Government of India and undertaken by Sociological Research Unit, Indian Statistical Institute, Kolkata. Project Report No.: E-4/1998/SRU (1998)
8. Fox, R.G.: Kin, Clan, Raja, and Rule. University of California Press Berkelay, USA (1971)
9. Jana, R., Bandyopadhyay, S., Choudhuri, A.K.: Reciprocity among farmers in farming system research: Application of social network analysis. Journal of Human Ecology 41(1), 45–51 (2013)
10. Jana, R.: Statistical and graph theoretic analysis of social networks of rural farmers in different agricultural activities. Ph.D. Thesis, Unpublished. Burdwan University, Burdwan, West Bengal (2005)
11. Jana, R., Choudhuri, A.K.: Studying various aspects of social networks with socio-economic changes in a rural area: A case study from West Bengal. Guru Nanak Journal of Sociology (2013) (accepted for publication in 2013 issue of the Journal)
12. Knoke, D., Kuklinski, J.H.: Network Analysis, ch. 3: 26-30. Sage Publications, USA (1988)
13. Rao, A.R., Bandyopadhyay, S., Sinha Bikas, K., Bagchi, A., Choudhuri, A.K., Jana, R., Sen, D.: Changing Social Relations. Paper presented in SURDAC Seminar, organized by Indian Statistical Institute, Kolkata (1998)
14. Rao, A.R., Bandyopadhyay, S.: Measures of reciprocity in a social network. Sankhy. Series A 49, 141–188 (1987)
15. Wasserman, S., Faust, K.: Social Network Analysis: Methods and Applications, ch. 2: 33-35. Cambridge University Press, Cambridge (1999)
16. Weisberg, H.F.: Central Tendency and Variability. In: Lewis-Beck, M.S. (ed.) Basic Statistics, vol. 1, pp. 70–76. Sage Publications, London (1993)

Evaluating the Propagation Strength of Malicious Metaphor in Social Network: Flow Through Inspiring Influence of Members

Manash Sarkar, Soumya Banerjee, and Aboul Ella Hassanien

Abstract. Interaction across social networking sites leads to different kinds of ideas, concepts and choices and sharing or some nourishing effects, which might influence others to believe or trust. Seldom may this cause some malicious effect for members and their peers only. As social network is the domain for sharing opinion and comments, subsequently it can also propagate malicious signature as well. Security and privacy is essential component to protect user profile from this kind of malicious program, which basically evolves from any close acquaintances, that also belongs to same vector plane. The degree of malicious attack of a social network depends on the number of flow links from one user to another with forward operations. It is true that the probability of malicious attack evolves from friend's community is of greater attack prone magnitude than the degree of attack from unknown members. This paper focuses the different verticals of such possibilities of attack under social network processes and also tries to investigate the rudimentary precautionary measure pertaining security algorithm behind it.

Keywords: Degree of Attack, Social Network, Attack Graph, Conditional Probability, Gaussian distribution.

1 Introduction

Social network is an online community where people can share opinion, in the form of discussion, fun, and entertainment, within a social forum, across the globe and there is a large number of these online communities these days which are functional

Manash Sarkar · Soumya Banerjee
Department of Computer Science, Birla Institute of Technology, Mesra-835215, India
e-mail: manashsarkar53@gmail.com, soumyabanerjee@bitmesra.ac.in

Aboul Ella Hassanien
Faculty of Computers and Information, Cairo University, Cairo-12613, Egypt
e-mail: aboitcairo@gmail.com

M. Panda, S. Dehuri, and G.-N. Wang (eds.), *Social Networking*
Intelligent Systems Reference Library 65,
DOI: 10.1007/978-3-319-05164-2_8,

and popular. They are popular for being proactive and easy to use. In these social networks, a rudimentary contrast between friendships is formed by introduction through a common friend in the network or because they work in the same company or had been in the same school [1]. The concept of homophily is revealed by social network studies which by indicate that *"social networks are homogeneous with regard to many socio-demographic, behaviroal and interpersonal characteristics"* [2, 3]. This common behavior under social network, which could flow from one user to another user, is one of the weak and critical point for unauthorized access of the network. If any member is affected by malicious pattern unknowingly, other members could also be affected by themselves due to this common behavior. As security is one of the most crucial aspects for social network, this could be a major concern for all social network providers.

Any existing social network, like Facebook or Twitter, exhibits some common functional attributes as *'posting photos'*, *'sending messages'*, *'likes'*, *'dislikes'*, etc. which lead to the concept of affiliation towards a community. Some of these activities could be infected and harmful not only to the user profile but to end user and other participants as well. And since most of the users are not aware in depth about the security and privacy issues of their profiles they become easy target for the attackers to attack their profiles. In this paper, we try to establish, analytically, that the probability for an attack coming from friends list is higher than from others members. Considering Facebook as a an example, any member of Facebook can send any message, wall post, send link to any member (friends, unfriends) under same the paradigm. If the send items are malicious in nature, then it will immediately start infecting the members profile when he/she clicked on it [4, 5]. The malicious signature will spread more, if the members start forwarding it to their friends and this process becomes more dramatic of it gets recursively executed. We use *Gaussian distribution* to demonstrate that if the number of flows increases then the degree of attack will decrease and also try to find the degree of attenuation of attack at certain point of time. The rest of the paper is organized as follows: Section 2 briefly depicts fundamental structure of social network from the perspective of graph theory. Section 3 gives the basis of mathematical treatment to be incorporated to establish the proposal followed by the result of present study in section 4.1 from different parametric perspectives. Finally section 5 presents conclusion with further scope of research in this regard.

2 Fundamental of Social Network: Graph Theory Perspective

One of the examples of the bipartite graph is the *Affiliation network*. This network is in association with homophily graph [6, 7, 8]. Affiliation represents the way a group of people interacts in a same community. Graphically affiliation network can be represented by denoting the group of people on the left of the graph and the communities on the right of the graph.

Bipartite Graph. A graph is said to be bipartite (bigraph) if its vertices can be divided into two disjoint sets in such a way that every edge connects a vertex in one set to a vertex in the other set. A complete bipartite graph is a special kind of bipartite graph where every vertex of the first set is associated to every vertex of the second set. Figure (1a) shows a graphical example of bipartite graph.

2.1 *Co-Evolution of Social and Affiliation Network*

Co evolution proposes of social generative models confine the statistical properties of real-world networks related only to node-to-node link formation. The structure of the social network changes depending on the day to day activities of the members. This phenomenon is caused by adding new friends and becoming members of the different communities. Hence, these changes represent a kind of co-evolution if the two persons share the same community. Figure (1b) shows an example of the social affiliation network with both friendship and community linkage.

Embeddedness. Embeddedness is the basic explanation of the economic event in social network. Embeddedness of an edge in a network can be measured by the number of common vertices at the two end points have. Figure (1c) shows an example of the Embeddedness. A $\rightarrow$ B edge has an Embeddedness of two, since A and B has two common neighbors E and F.

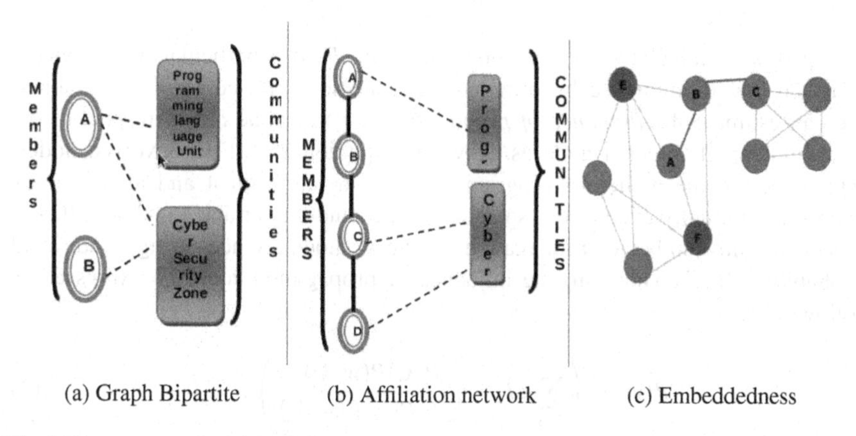

(a) Graph Bipartite (b) Affiliation network (c) Embeddedness

Fig. 1 The concept of social affiliation with some examples

Betweenness. The betweenness of a vertex in a graph is the segment of shortest paths between all pairs of vertices passing through that vertex. Centrality of a vertex within a graph can be measured by betweenness.

3 Exploring Mathematical Treatments

In a social network, malicious signature propagates from one node to another node by the click and forwarding on it. And as such, people could be influenced by their friends who have already been affected by this signature [9]. Any user can create and propagate malicious programs t not only to his/her friends but also to any other members in a the same social network. When the program propagates from one node to other nodes (members), calculating the degree of attack at a particular node requires the computation of the number of flows (how many nodes it traverses) from that node to others as distributed over the edge [10]. Considering a social network as a graph and a member in the social network as a node, mathematically, this can be expressed as follows:

m $\rightarrow$ Number of users present in a social network.
n $\rightarrow$ Number of users receive the malicious program.
k_i $\rightarrow$ Number of users clicks for propagating the malicious program.
t_i $\rightarrow$ Number of distance flow by the malicious program.
r_i $\rightarrow$ Rate of influence by previous adjacent node.
And all the members will not intend to click on the post, hence, $k_i \leq n$

If the node propagates the program, then it will be send to maximum $n-1$ nodes. Then the frequency of attack F_i propagate as follows:

$$F_i = \begin{cases} 1 \text{ propagate,} \\ 0 \text{ otherwise} \end{cases}$$

$$F_i = r_i t_i \quad (1)$$

Equation 1 calculates the frequency of attack F_i at a particular node i where r_i is the rate of influence by previous adjacent node. Based on this frequency, we can estimate the *influence of propagation* d(w) at node i with respect to the total number of infected node as shown in equation 2. All the involved nodes, which receive the malicious program, not sure to click on it and all this click events are mutually exclusive associated with a random variables w that indicates the maximum number of propagation by the ith node. Hence, using conditional probability (Bay's Theorem) the influence of propagation could be expressed as follows [11]:

$$d(w) = \left(\sum_{i-1}^{n} k_i F_i + \frac{P(k_i)P(w/k_i)}{\Sigma_{i=1}^{m} P(k_i)P(w/k_i)} \right) + \lambda_i \quad (2)$$

Where the error due to network delay is given by λ_i

The degree of attack $d(t)$ of a particular node after receiving the malicious metaphor could be calculated as follows:

$$d(t) = \frac{1}{t_i} [k_i.Cos(d(w))] + \frac{k_i}{\Sigma_{i=1}^{n} t_i} \quad (3)$$

3.1 Proposed Algorithm

The following proposed algorithm finds the degree of attack of a particular node influenced by the previous adjacent node of a social network community

Algorithm 1. Finds the degree of attack of **social networks**

Begin

Initialize m number of user present in the social network community.
Initialize n number of user receive the malicious program.
  k_i denotes the number of users click for propagation the malicious program.
  t_i denotes number of distance flow by the malicious program.
  r_i denotes rate of influence by previous adjacent node.
Consider all the members are not intended to click on the post
while $k_i \leq n-1$ **do**
  if k_i *is true* **then**
    $F_i = r_i t_i$
  else
    F_i is zero
  end
  Calculate influence of propagation d(w) by applying equation 2
  Calculate the degree of attack f(t) by applying equation 3
end

3.2 Experimental Results and Discussion

To establish the proposed mathematical model a data set has been envisaged for the proposed simulation. We consider the six values for flow of malicious program through the social network. Malicious program would propagate different number of users present in the social network. User can forward the malicious program by click on it and post to their friends profile. Among the entire user some would intend to click the program for next propagation. The probability of the click is shown in Table 1.

Table 1 Value of click by user with total number of user to be send and the probability of click

Propagation of user click (k_i)	Total Number of users receive the message (n)	Probability of click $P(k_i)$
6	14	0.42
8	23	0.34
12	18	0.66
10	19	0.52
10	35	0.28
11	21	0.52

Table 1 contains the three attribute values namely, ' Propagation of user click (k_i)', 'Total Number of users receive the message', and 'the probability of click $P(k_i)$'. The probability of click would be calculated by using the value of k_i and (n).

By using the value of $P(k_i)$, in equation 2 we get the value of degree of influence of propagation $d(w)$ in Table 2.

Table 2 Value of click by user and degree of influence of a item

Propagation of user click (k_i)	Degree of influence of propagation d(w)
6	0.706
8	1.719
12	3.851
10	4.275
10	5.103
11	6.86

The empirical result demonstrates that if the number of forward is increased then the degree of influence of a particular item will be increased. If more numbers of users receive it then the degree of popularity is increased also. It also shows that the degree of influence of propagation is dependent on the number of flow of a particular item. If the flow of message interaction increases, then degree of influence also increases and hence the value of d(w) is also high. After calculating the degree of influence of propagation d(w) we plot the value against propagation of user click (k_i) and get Fig. 2.

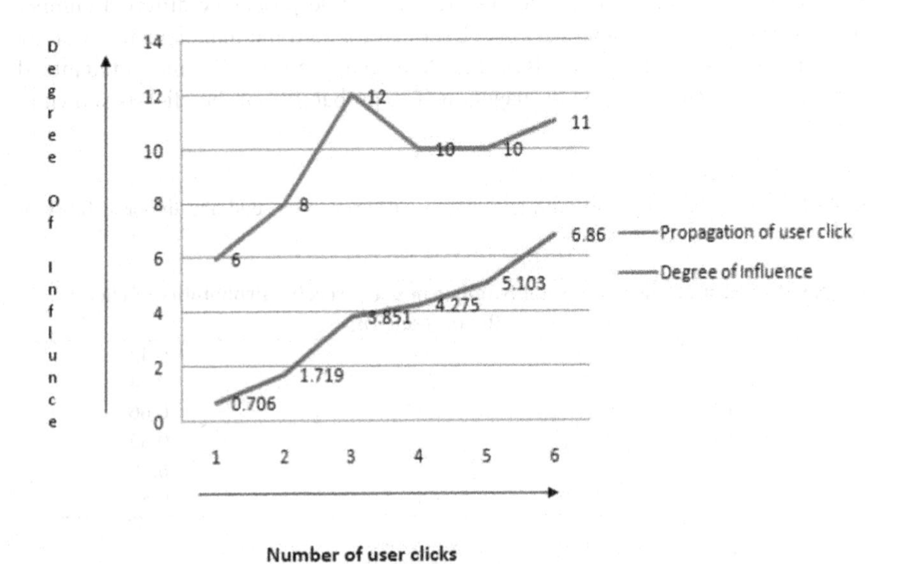

Fig. 2 Propagation of user click with degree of influence

Table 3 Number of flows and degree of influence of propagation

Number of Flow (t_i)	Degree of influence of propagation d(w)
1	0.706
2	1.719
3	3.851
4	4.275
5	5.103
6	6.86

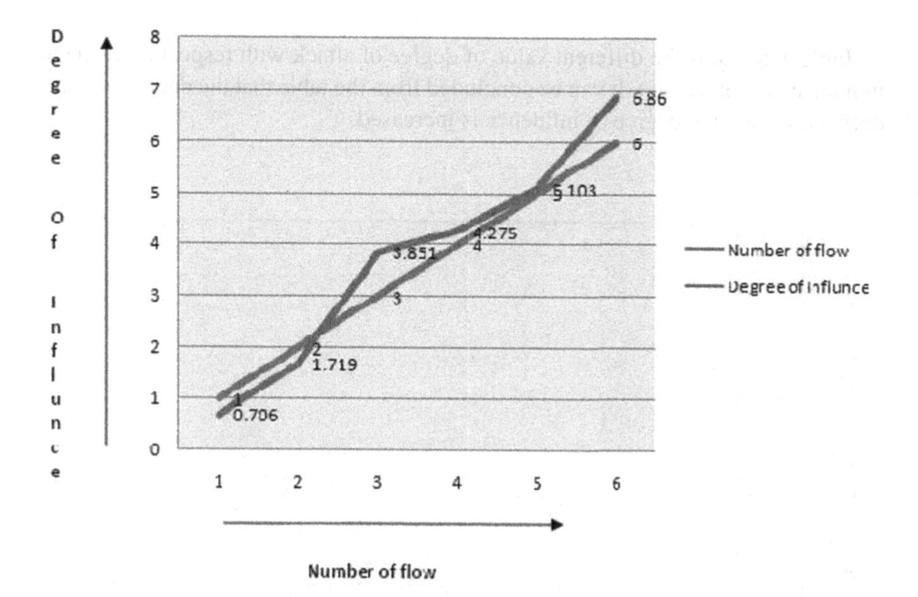

Fig. 3 Propagation of user click with degree of influence

Fig. 2 shows how the degree of influence varies with respect to the number of user clicks, it also shows that if the value of user click is increased then the degree of influence is also increased.

Table 3 shows that the degree of influence of propagation is dependent on the number of flow of a particular item. It has been observed that, if flow of message interaction increases, then degree of influence also increases. We plot the value of Table 3 and derive Figure 4 as shown in bellow.

Figure 3 shows the relation between total number of flows and the degree of influence of propagation of a particular item. If any item flows through more users, then the degree of influence will be high. We get the degree of attack by using degree of influence using equation 4.

Table 4 Degree of influence of propagation and degree of attack

Degree of influence of propagation d(w)	Degree of Attack d(t)
0.706	11.99
0.919	6.658
3.851	5.99
4.27	3.493
5.103	2.65
6.86	2.34

Table 4 contains the different value of degree of attack with respect to degree of influence of propagation. It can be concluded from the table that the degree of attack decreases when the degree of influence is increased.

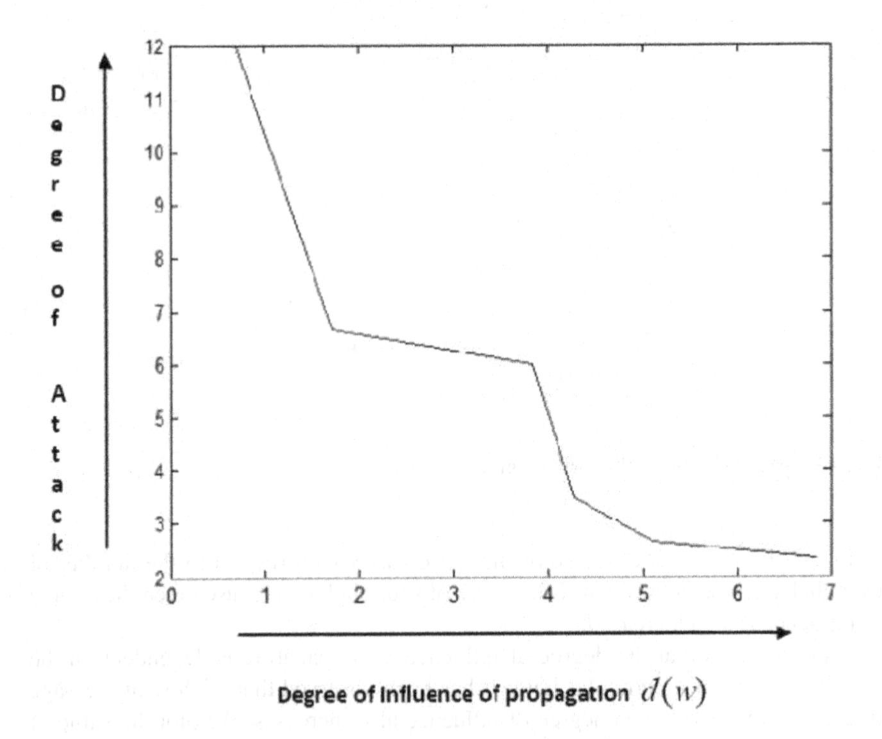

Fig. 4 Degree of attack decrease with increase of degree of influence

Figure 4 depicts that degree of attack for a particular propagation of a malicious domain is decreased when the degree of influence is increased. In social network when the influence is very high for any malicious item, then the degree of attack gradually decreases.

Table 5 Number of flow and Degree of attack

Number of Flow (t_i)	Degree of Attack d(t)
1	11.99
2	6.658
3	5.99
4	3.493
5	2.65
6	2.34

Table 5 depicts the degree of attack with respect to the number of flows of that malicious item through members in the social network. We plot the value of Table 5 and derive Figure 5 as a result of that.

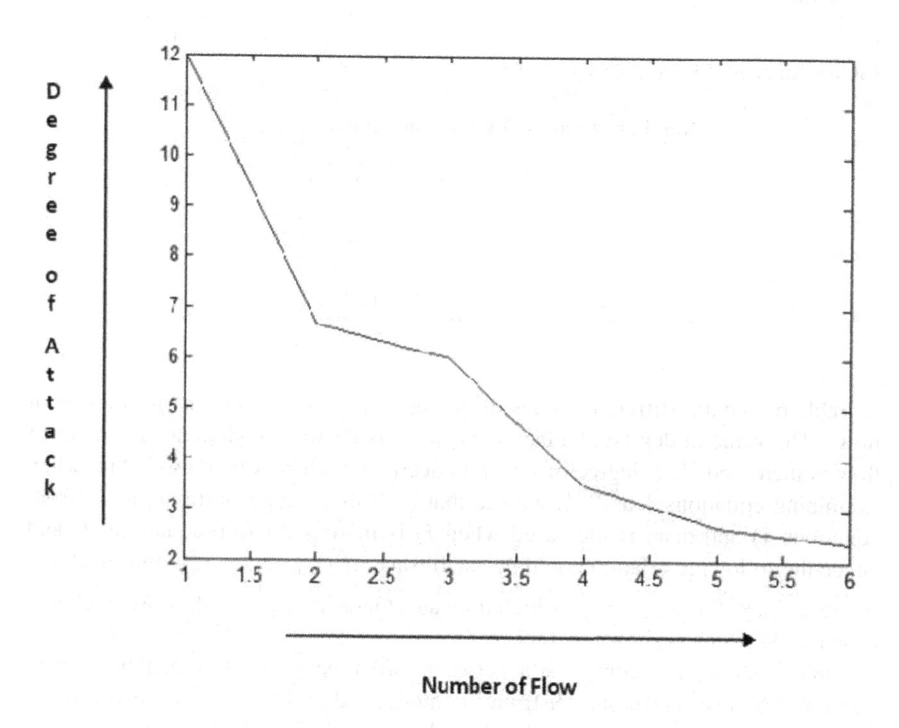

Fig. 5 Degree of attack is decreased with the increase of the number of flows

Figure 5 shows that the degree of attack can gradually decrease with the increases in the number of flows. If any malicious item flows through more users then degree of attack is diminished due to its high popularity.

3.3 Comparison with Statistical Model

For justification of the, we compare the results with Gaussian distribution model. The Gaussian distribution is a continuous function with random variable. In the proposed model the number of flows is a continuous function of number of user clicks and the degree of influence is a continuous random variable which depends on the click events. Hence the function $d(t)_{gaussian}$ is to demonstrate the attenuation of the degree of attack due to propagation effect. Using Gaussian distribution, the degree of attack $d(t)_{gaussian}$ could be formulated as follows:

$$d(t)_{gaussian} = \frac{1}{\sqrt{2\Pi}} e^{-x/2} \text{ Where } x = \frac{d(w) - \mu}{\sigma} \tag{4}$$

Where μ is the mean and σ is the standard deviation of the random variable d(w). Using the values of d(w) from Table 3, the values of $d(t)_{gaussian}$ could be calculated using equation 4 as shown in Table 6.

Table 6 Number of Flow and Degree of Attack

Number of Flow (t_i)	**Degree of Attack** $d(t)_{gaussian}$
1	4.933
2	4.817
3	4.531
4	4.486
5	4.397
6	4.183

Table 6 contains different values of degree of attack with respect to number of flows. The value of degree of attack $d(t)_{gaussian}$ is decreased when d(w) number of flow is increased. The degree of attack is decreased when is increased. And when examining equations 1 and 2, we see that F_i is directly proportional to t_i (from equations 1) and d(w) is increased when F_i is increased (from equations 2) and hence the following relation could be established: $d(t)_{gaussian} \propto \frac{1}{d(w)}$ and $d(w) \propto t_i$ so we can say $d(t)_{gaussian} \propto \frac{1}{t_i}$. which indicates that $d(t)_{gaussian}$ is decreased when t_i is increased which is shown in Table 6.

Table 7 shows the comparison between two types of results that have been calculated by using Gaussian distribution model and proposed model respectively. To measure the performance we plot the value of Table 7 to obtain Figure 6.

Figure 6 shows the comparison between the results that have been derived by using two different techniques. It is shown from the figure that the proposed model is able to demonstrate that the degree of attack differs sharply with respect to different numbers of flows. In the case of Gaussian distribution model the degree of attack differs very small than the proposed model with respect to the same number of flow. It is clear that the experimental value is quite sharp than the value calculated by Gaussian distribution method.

Table 7 Degree of Attack from Different Experiment with Respect to Number of Flow

Number of Flow (t_i)	Degree of Attack From Gaussian distribution $d(t)_{gaussian}$	Degree of Attack from experiment d(t)
1	4.933	11.99
2	4.817	6.658
3	4.531	5.99
4	4.486	3.493
5	4.397	2.65
6	4.183	2.34

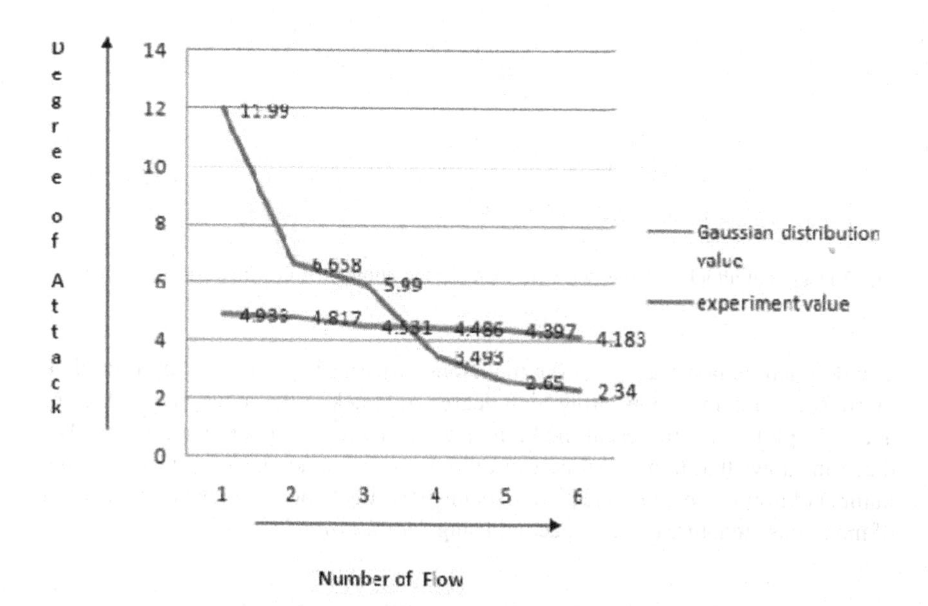

Fig. 6 Degree of attack is decreased with the increase the number of flows

4 Graphical Representation

Graphically, as depicted in Figure 7, it is clearly shows that the malicious signature evolves through more users then the degree of attack is steadily shrunk. Same malicious signature is detrimental in different degree based on the number of flow in a social network.

Fig. 7 denotes the graphical representation of a social network where every node is considered as a user. Every node in the graph represents the user in social network. According to the algorithm number of node present in graph denoted by m, is equal to number of user present in a social network. Numbers of nodes which are affected by malicious signature are denoted by n and always it is less than equal to m.

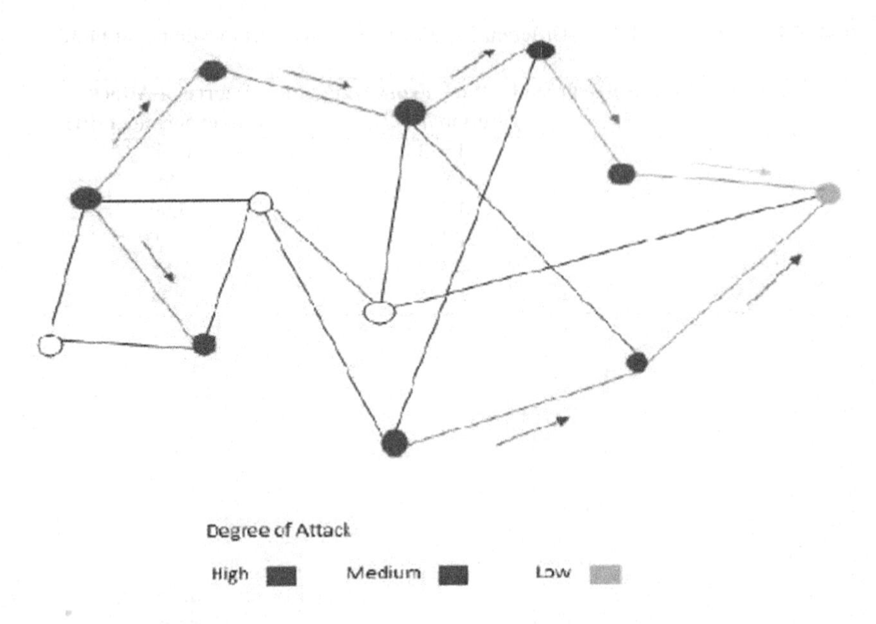

Fig. 7 Degree of attack is decreased with increase the number of flow in a social network

The diagram demonstrates that the malicious program is propagated by user click on it. The red color representing high degree of attack at that node, the pink color and light pink color represent medium and low degree of attack respectively. The diagram shows that in a social network the degree of attack is decreased when the number of flows is increased [12]. In a social network the node which is actual origin of malicious signature can be trace out using *Attack graph* [13, 14].

5 Conclusion and Further Scope of Research

In this paper, we try to explore certain mathematical anomalies on the degree of malicious program in a social network due to members of social network forwarding messages, photos and other items to their nearest friends. These forwarded items become popular due to the number of times they are forwarded directly to the intended friends as well as indirect and transitive ones. In this paper, we concluded that if the forwarded item is malicious and also influenced by the members only, then the degree of attack will be decreased because total number of forwarding is quite high. If a malicious item is forwarded transitively for a long time, then the degree of attack is gradually decreased. And on the other hand the degree of attack can be maximum when malicious item comes from next adjacent node, in other words, the malicious threat coming from the friends list is more dangerous compared to others unknown members. This is possible because the members have no time to know the behavior of that shared item. The idea here is to investigate how a member in a social network can be affected seriously by malicious program through his/her friends recommendation. In this paper emphasizes is only on the notion that the degree of attack can gradually decrease, when it is forwarded from one domain to another domain and on the calculation of the influence of propagation at a particular domain. The degree of attack model could be extended to include finding out the actual source or origin of malicious program in a social network using *Attack graph scenario* as future research.

References

1. McPherson, M., Smith-Lovin, L., Cook, J.M.: Birds of a feather: Homophily in social networks. Annual Review of Sociology 27(1), 415–444 (2001), `http://www.annualreviews.org/doi/abs/10.1146/annurev.soc.27.1.415` doi:10.1146/annurev.soc.27.1.415
2. Golub, B., Jackson, M.O.: How homophily affects the speed of learning and best-response dynamics. The Quarterly Journal of Economics 127(3), 1287–1338 (2012)
3. Bisgin, H., Agarwal, N., Xu, X.: Investigating homophily in online social networks. In: 2010 IEEE/WIC/ACM International Conference on Web Intelligence and Intelligent Agent Technology (WI-IAT), vol. 1, pp. 533–536. IEEE (2010)
4. Ceglowski, M., Coburn, A., Cuadrado, J.: Semantic search of unstructured data using contextual network graphs. National Institute for Technology and Liberal Education 10 (2003)
5. Suchal, J.: On finding power method in spreading activation search. In: SOFSEM 2008, vol. II - Student Research Forum, pp. 124–130 (2007)
6. Jackson, M.O.: Average distance, diameter, and clustering in social networks with homophily. In: Papadimitriou, C., Zhang, S. (eds.) WINE 2008. LNCS, vol. 5385, pp. 4–11. Springer, Heidelberg (2008)
7. Sekine, K., Imai, H., Tani, S.: Computing the tutte polynomial of a graph of moderate size. In: Staples, J., Katoh, N., Eades, P., Moffat, A. (eds.) ISAAC 1995. LNCS, vol. 1004, pp. 224–233. Springer, Heidelberg (1995)
8. Breiman, L., Friedman, J.H., Olshen, R.A., Stone, C.J.: Classification and regression trees. Wadsworth & Brooks, Monterey (1984)
9. Security, I., Committee, S.A., et al.: Ssac. sac 025-ssac, advisory on fast flux hosting and dns (2008)
10. Konte, M., Feamster, N., Jung, J.: Dynamics of online scam hosting infrastructure. In: Moon, S.B., Teixeira, R., Uhlig, S. (eds.) PAM 2009. LNCS, vol. 5448, pp. 219–228. Springer, Heidelberg (2009)
11. Veerarajan, T.: Probability, Statistics and Random Processes, 3rd edn. Tata McGraw-Hill Education (2008)
12. Bayer, U., Kruegel, C., Kirda, E.: Ttanalyze: A tool for analyzing malware. In: 15th European Institute for Computer Antivirus Research (EICAR 2006) Annual Conference (2006)
13. Choi, H., Lee, H., Lee, H., Kim, H.: Botnet detection by monitoring group activities in dns traffic. In: 7th IEEE International Conference on Computer and Information Technology, CIT 2007, pp. 715–720. IEEE (2007)
14. Friedman, J., Hastie, T., Tibshirani, R.: Regularization paths for generalized linear models via coordinate descent. Journal of Statistical Software 33(1), 1 (2010)

Social Network Analysis: A Methodology for Studying Terrorism

Aparna Basu

Abstract. This chapter aims to bring to the reader an overview of the work done since the 9/11 terrorist attack, in the field of Social Network Analysis as a tool for understanding the underlying pattern /dynamics of terrorism and terrorist networks. SNA is particularly suitable for analyzing terrorist networks as it takes relationships into account rather than merely attributes, which are difficult to obtain for covert networks. Using graph theoretic methods and measures and open source data it has been possible to map terrorist networks and examine roles of different actors, as well as identify groups and structures within the network. The methodology is illustrated by reviewing two case studies: the 9/11 terrorist network study by Krebs, that used data from a single terrorist attack, and a study by Basu that used data from about 200 terrorist incidents in India to create a network of terrorist organizations for predictive purposes.

Keywords: Social Network Analysis, SNA, terrorist networks, co-occurrence, graph theory, multidimensional scaling, structural equivalence

1 Introduction

When studying terrorism, it is necessary to have a look at what different people mean when they use the word "terrorism". What are the common attributes that scholars ascribe to terrorism? The term has undergone changes in the last 200 years of its existence. The difficulty in defining terrorism hinges upon the fact that one man's terrorist could easily be another man's freedom fighter [1]. Suffice it to say that 'terrorism' is a form of political violence, where extreme anxiety states are precipitated in the population at large by targeting and mounting violent, even fatal attacks on large

Aparna Basu
CSIR-National Institute of Science Technology and Development Studies
and Dr. K.S. Krishnan Marg, Pusa Gate, New Delhi 110012, India
e-mail: aparnabasu.dr@gmail.com

M. Panda, S. Dehuri, and G.-N. Wang (eds.), *Social Networking*
Intelligent Systems Reference Library 65,
DOI: 10.1007/978-3-319-05164-2_9,

numbers of innocent and uninvolved victims. Terrorist attacks occur within a state, using terror at any time and place. The security threat posed by them differs from earlier security threats such as war, where battles were fought with military power. Because of the subversive nature of terrorism, the key to combating terrorism lies in understanding how terrorist organizations operate, using both available information, as well as better methods of analyzing that information [2].

Among analytical tools, the methodology of Social Network Analysis (SNA) has proved to be particularly useful for studying terrorist networks. SNA was first used by anthropologists and sociologists to study social structures. It is a mathematical approach using Graph Theory that uses relational information between actors rather than their attributes. Terrorist groups are often non-hierarchical network structures where the binary relations between the actors or individual terrorists can be represented as a graph, with actors being represented by nodes and binary ties represented by lines or edges in a graph. Graph theoretic measures that depend on the interactions between nodes can then be used to map out the role of individuals within the network. They answer questions such as who is most connected or otherwise important in a network. In the case of terrorist or other covert networks it can be difficult to gather data on the attributes. On the other hand, relational information may be obtained, for example through family and kinship ties, school/college ties, participation in the same events, telephone and e-mail contact, etc. Unlike hierarchical structures, the network structure is a flexible organizational form. Individuals can directly connect to each other especially through the tools of modern telecommunications and the Internet. Loosely structured networks can move quickly and be adaptive without being constrained by hierarchical functioning [2].

Our approach in this chapter is pedagogical in nature. We will examine aspects of Social Network Analysis as applied to terrorism, taking empirical data, and in particular open source data based studies into account. We briefly review the nature of the two phases in the development of SNA including the more recent trends in computational terrorism research. (For a detailed coverage of the literature on social networks and terrorism, see van der Hulst, 2011 [3]). Some data sources that can be used for conducting terrorism research are described, followed by a brief review of SNA. Next, two case studies are reviewed, to illustrate the method of SNA in the study of terrorism. The first is the seminal network study by Krebs of the attacks by Al Qaeda on the twin Trade Towers in New York in September, 2001 (9/11) [4]. This study covers a major terrorist event and is based on investigations following it. The next case study by Basu [5] uses data from a large number of terrorist incidents in India, in the period 2000-2002, to obtain a network of terrorist organizations, drawing inferences about the structure of the network using multivariate statistical tools.

2 Two Phases in the Development of Social Network Analysis

SNA was developed in two distinct phases. The first phase was undertaken by anthropologists, sociologists and graph theorists, while the second phase was undertaken by physicists and computer scientists. The first phase was based on the premise that individual nodes (which could be people, organizations, events, etc.) are connected by relationships that form networks [2]. There could be multiple types of relationships, depending on the nature of the tie, each being depicted by a different network. The network is treated in graph theoretic terms, and graph theoretic indices are used to differentiate between the roles and positions of individuals in the network. This chapter will deal primarily with the approach taken in the first phase. (For an introduction to Social Network Analysis and Graph Theory, see Scott [6]; Wasserman and Faust [7]; Freeman [8]; Harary[9].)

For completeness we include some results in network analysis obtained in the second phase, which deals more closely with the statistical properties of large networks. The significant findings were the 'six degrees of separation' by Milgram, the 'strength of weak ties' by Granovetter, and the 'small world hypothesis by Watts and Strogatz. Milgram conducted an experiment in 1967 to understand how people are connected to others. Random people were asked to forward a package to any of their acquaintances who they thought might be able to reach a target individual. Milgram found that most people were connected by not more than six acquaintances [10]. Mark Granovetter wrote on "the strength of weak ties" in his 1973 paper where he argued that "weak ties", i.e., one's relationships with acquaintances, are more important than "strong ties", one's relationships with family and close friends, because of the novel and non-redundant aspect of the information transferred through them [11]. Another important concept that entered this research domain was the 'small world hypothesis' by D.J. Watts. Small world networks of Watts and Strogatz [12, 13] have a high degree of clustering but short average path length that enables one to go from one node to another in a small number of steps. Many networks in the natural and man-made world are highly clustered, but with a few ties that quickly span the network. (For a more mathematical analysis of networks refer to Brandes and Ehrlebach [14]).

The two phases in the development of SNA have grown independently, the later phase developed by physicists rarely citing the earlier work done by social scientists. This is clearly seen in the diagram which gives citations between the two phases (Fig. 1, from Freeman[15]).

3 Social Network Analysis and Terrorism

The role of networks in terrorism was recognized well before the attacks of September 11, 2001-the 9/11 attacks- where passenger planes were used to attack the twin Trade Towers in New York in 2001 in an incident that shocked the whole world. In a book titled Networks and Netwars, published in 2001 before the 9/11terrorist attacks, John Arquilla and David Ronfeldt stated that the new face of war was "netwar, a lower-intensity battle by terrorists, criminals, and extremists with a networked organizational structure" (Arquilla & Ronfeldt [16] as cited in [2]). Prior to that, there had been applications of 'link analysis' to cases in criminal investigations of fraud and conspiracy [17], but it was primarily the network study of Valdis Krebs of the 9/11 hijackers [4] in late 2001 that drew the attention of academics, the

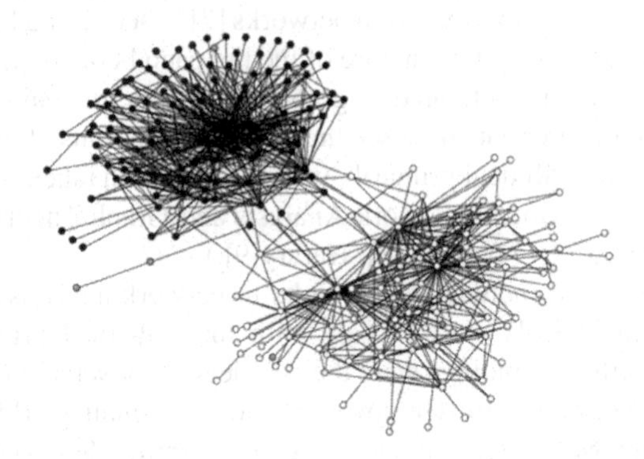

Fig. 1 Graph showing the relative lack of citation links between literature in the two phases in the development of Social Network Analysis (open circles - social science, phase 1 and filled circles-physics and computer science, phase 2). Source: Freeman, 2008[15]

government, and the media in the USA to the fact that social network analysis could be important to understanding terrorism. Krebs was called to Washington for discussions, and authors of popular network books, such as Linked, Antonio-Laszlo Barabasi [18] were interviewed extensively on television and radio programs on how the knowledge of social networks could be used to fight terrorism. There was a spurt of research interest in the area, as was seen from the queries on the listserv of the social network organization International Network for Social Network Analysis (INSNA). The affiliated journal Connections devoted an entire issue to social network analysis and terrorism in late 2001. This issue carried several important papers; the paper by Valdis Krebs where he mapped the 9/11 terrorist network by collecting publicly available data on the hijackers, and used network software [19] to depict them graphically [4], a paper by Carley and others where social network analysis and multi-agent modeling were used to destabilize terrorist networks [20], and a paper by Rothenberg analyzing the structure of the al Qaeda terrorist network based on newspaper and radio reports [21], [2]. This marked the beginning of a trend of using 'open source' data and social network analysis in the study of terrorism.

In 2004, Marc Sageman published Understanding Terror Networks where, by using public sources, he was able to collect biographies of over 100 Islamic terrorists affiliated to the global Salafi jihad-the Islamic movement led by al Qaeda, and establish that there were four major terrorist clusters spread across several countries around the globe [22]. Several studies similar to the first study by Krebs [4] using open source data followed major terrorist strikes. These include the bombing in Madrid in 2004 [23], the Bali bombings in 2002 [24], and the 2008 Mumbai attack [25]. Other studies were on the Israeli Jewish Underground terrorist group [26], and the network of terrorist organizations in India [5]. Two of these studies will be examined in greater detail under case studies.

4 Data Collection and Data Sources

Data collection in SNA involves identifying appropriate actors and gathering relational data between them. In simple networks, such as friendship networks and organizational networks, this is relatively simple and achieved through interviews, while in more complex cases structured questionnaires are used. Networks could be both, directed or undirected, and weights can be attached to indicate strength of a link. Ego networks are networks of associates centred on a particular individual. Once these associates are mapped, one can go further to map associates of associates in what is known as 'snowball sampling'. Within the same set of actors there could be different relationships, such as kinship ties, working or training together, etc., where each type of relationship could be represented by its own network.

Clearly, these methods cannot be readily applied to terrorist organizations or other covert networks. It is a difficult exercise to gain reliable information on terrorist networks. Terrorist organizations do not provide information on their members, and intelligence data is not generally available to researchers. Some researchers obtain data from public records such as court records. As stated earlier, Valdis Krebs was one of the first to collect data using newspaper reports on the hijackers after the 9/11 attack [4], from which he was able to create a representation of the al Qaeda network. The information was based on credit card records, vehicle registration records, telephone logs and information on prior contacts between the hijackers. More recently there has been an attempt to collect and organize data in structured databases which are publicly available.

4.1 Structured Data Sources

Academic interest and new grants in terrorism led to the development of several large structured data portals after the 9/11 attack. The following is an indicative but not exhaustive list of publicly available data sources. The terrorism web portal at the University of Arizona's Artificial Intelligence Centre [27] stems from the Dark Web project, which aims to study and understand international terrorism using a computational, data-centric approach. Their aim is to collect all web content generated by international terrorist groups, including website, forums, chat rooms, etc. The laboratory has collected data from jihadist forums over a number of years and provides access to 28 fora and over 10,000,000 messages through the Dark Web Forum Portal. The portal also provides statistical analysis and social network visualization functions. The approaches and methods developed contribute to the field of Security Informatics.

Another terrorism related effort in the USA is START [28], the National Consortium for the Study of Terrorism and Responses to Terrorism, in Maryland. Launched in 2005, it is a university-based research center for the scientific study of the causes and human consequences of terrorism in the United States and around the world. The research investigations cover questions such as under what conditions do individuals turn to terrorism, the nature of the recruitment process, attack patterns, the evolution of terrorist behaviour, future trends, etc. Further, how can societies enhance their resilience in the face of potential impacts of future attacks? START has developed educational programs on terrorism at the graduate and undergraduate levels. It is supported by the U.S. Department of Homeland Security's Science and Technology Directorate and also receives funding and support from Federal agencies, private foundations, and universities. START maintains the Global Terrorism Database [29] and Terrorist Organization Profiles [30].

The Global Terrorism Database (GTD) is an open-source database including information on terrorist events around the world from 1970-2011. It includes data on international terrorist incidents, and gives the date and location of the incident, weapons used, nature of the target, number of casualties, and the identity of the culprit.

The Terrorist Organization Profiles (TOPs) data collection provides background information on more than 850 organizations that have been known to engage in terrorist activity around the world over the last 40 years. Information is included on a variety of characteristics such as bases of operation, strength, ideology, and goals. These data were collected for the Terrorism Knowledge Base (TKB), managed by the Memorial Institute for the Prevention of Terrorism (MIPT) until March 2008. As stated on their website, this data has not been reviewed nor verified by START, but is maintained as a service to the research community.

Terrorism & Preparedness Data Resource Center (TPDRC) The Terrorism & Preparedness Data Resource Center (TPDRC) at the University of Michigan's Inter-university Consortium for Political and Social Research (ICPSR), archives and distributes data from a variety of sources. It facilitates access to research and administrative data from across the world, for academics and researchers on terrorism. The Centre is jointly managed by researchers from the University of Maryland's START, Michigan State University's School of Criminal Justice, and the University of Michigan's National Archive of Criminal Justice Data [31].

4.2 Unstructured Relational Data Sources

Apart from structured databases mentioned above, relational information on terrorists and terrorist organizations can be obtained from a variety of sources, such as,

Credit files, bank accounts and related transactions
Telephone calling records
Electronic mail, instant messaging, chat rooms, and web site visits
Court records
Business, payroll and tax records
Real estate and rental records
Vehicle sale and registration records
(From DIA [32], as cited in Krebs [4]).

Relational records can be used to build up a suspect's ego network. Using snowball sampling one can further enlarge the network. In general, the work using SNA to analyze terrorist networks is more descriptive, using visualization and graph-theoretic measures to draw conclusions, but not invoking other complex simulation modelling tools.

5 Simulation, Modelling and Computational Terrorism

Complex modelling tools are used in another research stream on terrorism, based on simulation and predictive analysis. The Computational Analysis of Social and Organizational Systems (CASOS) group at Carnegie Mellon University led by Carley was one of the earliest to have a number of projects in networks and terrorism. Their studies dealt with a variety of terrorism-related issues using a computational tool known as DyNet, and looked at ways to destabilize terrorist networks. Other works relate to network text-analysis, turning raw text related to the middle-eastern covert networks into a pictorial network representation using the software Automap [33], and the effectiveness of wiretapping programs in mapping networks of rapidly evolving covert organizations [34]. Another software, NETEST uses Bayesian inference models for estimation [35]. At University of Arizona Dark Web Terrorism Research Center, complex models have been used by Chen and his team, where they used social network tools to study extremist-group web forums, and were able to construct social network maps and organization structures [27, 36]. In another study, Carpenter developed algorithms for social network analysis of terrorist networks using optimization [37].

The use of social network analysis has taken a new direction with the growth of a number of new online Internet sites based on social network principles. MySpace, Facebook and Twitter are websites that allow users to connect with friends, and friends of friends, to share messages. These sites map out each user's network of friends and acquaintances. The effect of terrorist incidents on new media such as twitter has been mapped by Gupta and Kumaraguru for the Mumbai attack in 2008 [38]. A number of computational studies of terrorism have been undertaken by Subrahmanian and others, which do not all use SNA. A stochastic model used by them takes longitudinal data on terrorist groups to automatically learn the group's behaviour and predict future behaviour [39]. Game theoretic approaches *et al.* computational techniques have been used to study the behaviour of terrorist groups [40]. Several articles organized around the theme of computational approaches to counter-terrorism have been published as a Handbook in 2013 [41, 42]. New methods emerging in the social science and computer science literature may also be applicable in the area of terrorism research in future. This is an area where a multi disciplinary collaborative effort can give maximal results. Studies in terrorism created since the 1970s can provide a proper context within which simulation and predictive analyses can be interpreted.

6 Essentials of Social Networks and Centrality Measures

In this section we give some mathematical preliminaries that relate to the methodology of SNA and the steps involved in the analysis. Once the subjects under study or actors are known, and the data on relationships between them are extracted, an Adjacency Matrix can be created which shows the linkages between the actors. The elements of the Adjacency Matrix a_{ij} are numerical values attached to the relationship between pairs of actors or nodes. For example, in a friendship network, strong bonds may be represented by higher values than bonds with acquaintances. A visual representation can be made in the form of a network graph with the nodes representing actors and the edges representing the relationships between them. Typically thicker lines represent strong links and thin lines, weaker links. The network is unaffected by changes in the actual positions of nodes on the visual representation, provided all the edges remain the same. It follows that the length of edges in a graph are inconsequential. The shortest path between any two nodes, measured by the number of nodes or edges that have to be traversed in order to reach the target node, is called a *geodesic*.

The actors in a network are not all equivalent. Some may be highly connected and others relatively isolated. Some may connect two otherwise unconnected parts of a network, taking on a bridging role. It is generally accepted that these roles set apart the actor from other members of the network. Roles can be quantified through the calculation of Centrality Measures, based on the Adjacency Matrix. Simple measures are Degree Centrality which measures how well connected an actor is, the Betweenness Centrality where the actor acts as a bridge between parts of the network, and Closeness Centrality which measures in how many steps an actor can reach all the other actors in the network. We give below the definitions of the graph theoretic centrality measures used in SNA and in the case studies that follow. The relative centrality measures normalize the measures with respect to the size of the network, and are used in order to make comparisons of centrality across graphs of different sizes.

Degree Centrality

The Relative Degree Centrality of a point p_k in a graph of n nodes is given by,

$$C'_D = \frac{1}{n-1} \sum_{i=1}^{n} a(p_i, p_k) \tag{1}$$

where $a(p_i, p_k) = 1$ if the points p_i and p_k are connected and 0 otherwise.

Betweenness Centrality

The Relative Betweenness Centrality C'_B of a point p_k in a graph is given by,

$$C'_B = \frac{2C_B(p_k)}{n^2 - 3n + 2} \tag{2}$$

where C_B is the Betweenness Centrality,

$$C_B(p_k) = \sum_{i=1}^{n} i < j \sum_{j=1}^{n} \frac{g_{ij}(p_k)}{g_{ij}} \tag{3}$$

and where g_{ij} is the number of geodesics between p_i and p_j, and $g_{ij}(p_k)$ is the number of those geodesics on which p_k will lie.

Closeness Centrality

The Relative Closeness Centrality is given by,

$$C'_C(p_k) = \frac{n-1}{\sum_{i=1}^{n} d(p_i, p_k)} \tag{4}$$

where $d(p_i, p_k)$ is the distance between the points measured as the number of steps on the geodesic connecting them.

Other slightly more complex definitions of centrality have also been used but for the present pedagogic purposes these definitions due to Freeman [43] should suffice. Another frequently used centrality measure is the Bonacich Eigenvector Centrality [44], which takes into account the degree of the nodes being connected to in the calculation of a point's centrality. In other words it gives a weight to the persons you are connected to depending on their connections, in order to compute your centrality.

There are several software programs that use social network analysis and compute centrality measures and other statistical parameters relating to the network, such as density etc. Networks are visually displayed. Frequently used softwares are Pajek and UCINET [45, 46]. (For a recent review of software used in social network analysis, see Huisman [47].)

7 Case Studies

In order to illustrate the method of SNA, we have examined two cases where social network analysis has been applied to terrorism, but the approach used in each case was different. The first case covers the study by Krebs [4] of the 9/11 hijacker network. It has been selected as it was the first analysis that used SNA, drawing data from newspaper reports immediately following a major terrorist event. It underlined the fact that data available in the open domain could be used to analyze terrorist networks. We have gone into some detail about how the data was gathered and the networks mapped. Subsequent studies that enquired into properties of the network are also covered. The second case study by Basu [5] and Saxena [48] differs from the first case in that it takes terrorist organizations as nodes and uses data from a large number of terrorist events in India to draw a network of terrorist organizations. This approach is useful when details of individual terrorists are not known but the organization that is the perpetrator of an attack may be known. It is at a higher level of aggregation, where linkages between different organizations and their role in a larger terror network can be inferred. The analysis also suggests counter terrorism measures.

7.1 Case I: The 9/11 Attack on the Trade Towers in New York

Very soon after the September 11, 2001 attack by hijacked passenger planes on the Trade Towers in New York, Valdis Krebs attempted a network analysis of the hijackers [4]. The data sources were publicly released information reported in major newspapers; the New York Times, the Wall Street Journal, the Washington Post, and the Los Angeles Times. Information from the investigation became public within a short interval of time after the attack. There were 19 hijackers mainly from Arab countries, and the news reports gave which planes they were on, and which nation's passports they had used to get into the country. In his now well-known paper in *Connections*, Krebs noted, "Once the names of the 19 hijackers were public, discovery about their background and ties seemed to accelerate. From two to six weeks after the event, it appeared that a new relationship or node was added to the network on a daily basis. In addition to tracking the newspapers..., I started to search for the terrorists' names using the Google search engine. Although I would find information about each of the 19 hijackers, rarely would I find information from the search engine that was not reported by the major newspapers I was tracking. Finding information that was not duplicated in one of the prominent newspapers made me suspicious. Several stories appeared, ... (which) were proven false within one week. This made me even more cautious about which sources I used to add a link or a node to the network." [4]. Kreb's remark shows that analysis of terror attacks does not lie exclusively in the domain of intelligence, but that the judicious use of news reports in the open domain can lead to sufficient information to map a terrorist network.

A few months after the attack, the website of the Sydney Herald carried a matrix from which various connections between the hijackers could be extracted [49]. Krebs notes that he started his mapping project on seeing this matrix which was the first attempt he had seen of a visual organization of the information that emerged from the investigations [4]. Initially, he mapped the trusted prior contacts of the hijackers, which he defined as ties formed through living and learning together [50]. Using social network methods that he routinely used to map organizational structure as part of his consulting profession, he created the network of hijackers. As more information came from newspapers [51] and from court records of trials [52], Krebs updated his network. However, it was clear even at that stage that the information was far from being complete and could even be inaccurate.

In his maps, Krebs used links of three strengths (and corresponding thicknesses). The tie strength was determined by the amount of time spent together by a pair of terrorists. Those living together or attending the same school or the same classes or training had the strongest ties. Those travelling together and participating in meetings together had ties of moderate strength and medium thickness, and those who were recorded as having had a financial transaction together, or an occasional meeting, and no other ties, were labelled by him as dormant ties. These ties were shown with the thinnest links in the network [4]. The nodes representing the hijackers were colour coded by the fight they were on, while persons not on the flights were in gray. The software used by Krebs for visualizing the terror network was *Inflow* [19].

The first network map, showing only the hijackers, was long, with very few connections between the hijackers (Fig. 2, reproduced from Krebs [4]). Each node appeared to be connected to a few others but not to the rest. Most of the hijackers were beyond the 'horizon of observability' from each other. The expression refers

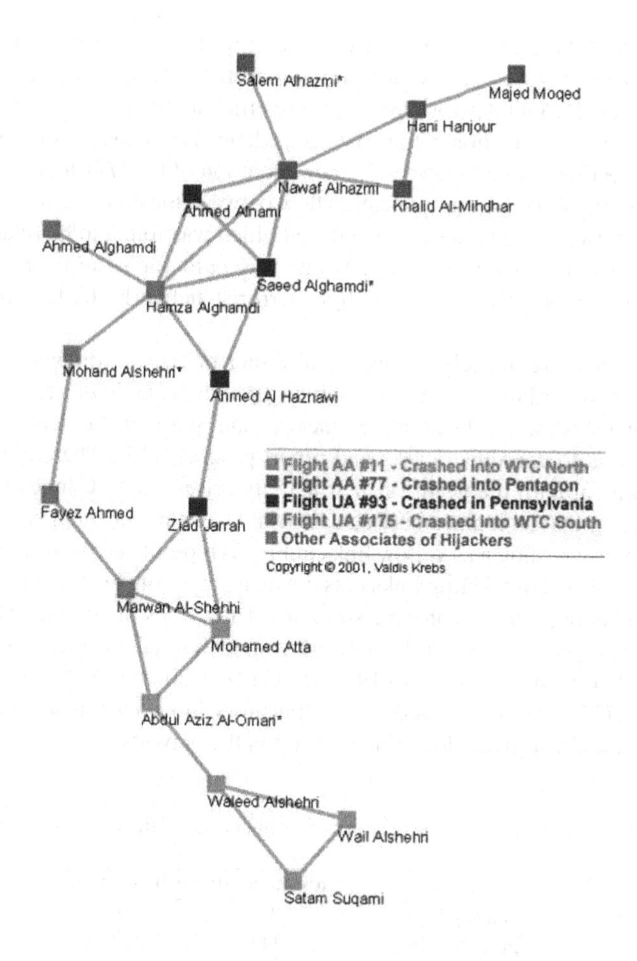

Fig. 2 Network map showing the links between the 19 hijackers of the planes that attacked the Twin Towers in New York on Sep. 11, 2001(Source: Krebs, 2001)

to the fact that they were more than two steps removed from most of the others on the map [53]. This network of 19 hijackers is also referred to as the action segment, while their associates are referred to as the complementary segment.

Structure, Security and Efficiency in the 9/11 Network. What was remarkable in the structure presented by the 9/11 network (Fig. 2) was its sparse connections. Although the hijackers were part of a single plot, there were very few observable ties between them. Krebs described the shape as snake-like. In a covert network, keeping individuals distant from each other minimizes damage to the network if one of them is captured. This was apparently part of the strategy adopted by Al Qaeda. In a video tape found in Usama bin Laden's house in Afghanistan when the Americans attacked, bin Laden mentions: "Those who were trained to fly didn't know the others. One group of people did not know the other group", ([54] as cited in [4]). In other words, a sparse or thinly connected network ensures secrecy and security. A densely knit network may be more efficient, but is also more vulnerable to attack. In order to function, a terrorist network needs to adopt an optimal strategy in terms of security and efficiency. The configuration of the 9/11 hijackers network had high security but low communication flow between the nodes. Using the average path length between the nodes as a measure (which was found to be relatively high at 4.75) Krebs concluded that the network was built for security. He suggested that in terms of organizational structure, terrorist networks trade efficiency for secrecy [4].

If the terrorists are sparsely connected, how then do they accomplish the tasks set out for them? According to Krebs this is done through the judicious use of transitory links or connections, i.e. those that connect distant parts of the network for short periods of time to coordinate tasks and report progress [13]. Thereafter, the links become dormant until the need for their activity arises again. Connections arising from a meeting that took place between some al Qaeda members in Las Vegas, and other meetings, are shown as yellow links in Fig. 3 (reproduced from Krebs [4]). We see that the addition of meeting links cuts down average path lengths in the network, giving the network a more cohesive structure. The values of Average Path Length and Density are recomputed for the network with additional meeting links (Table 1). It is found that the density rises to 19% from 16%, while the Average Path Length falls from 4.75 to 2.79 by the addition of transitory links, increasing efficiency but at the same time marginally lowering security in the network.

Table 1 Statistical parameters of the network of 19 hijackers (with and without meeting links)

	no shortcuts	with shortcuts
Group Size	19	19
Potential Ties	342	342
Actual Ties	54	66
Density	16%	19%
Average Path Length	4.75	2.79

Source: Krebs 2001; 2002

Complementary Participants. The trade-off between security and efficiency in covert networks can also be achieved through the addition of complementary participants to the network. In this case they would essentially be co-conspirators who did not board the planes but who served as "conduits for money and also provided needed skills and knowledge" [55]. When complementary participants were introduced into the network, it was expected that the larger network would have greater geodesic distances and average path lengths. Instead, it was found that the average path length decreased by 38%, from 4.75 to 2.94. Shorter average path length translates to greater efficiency in the network. This was therefore consistent with Krebs' interpretation that the efficiency of the 19 hijackers increased with the

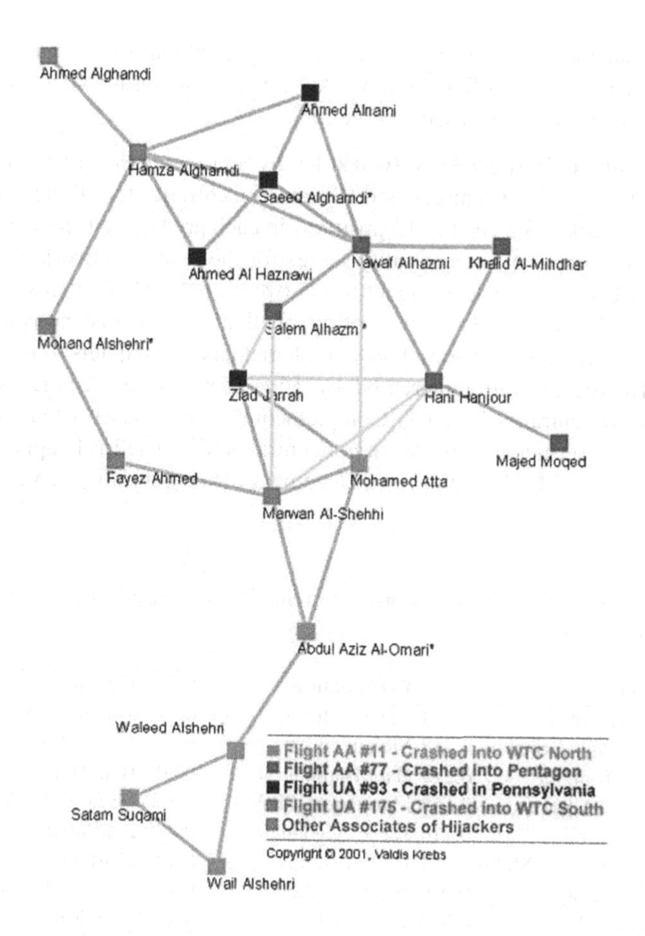

Fig. 3 Network of the 19 hijackers with the addition of meeting links (Source: Krebs, 2001[4])

addition of ties (or participants) beyond the action segment. The terrorist network's action segment (Fig. 2) was strong in security and weak in communication flow. The addition of complementary participants brought "shortcuts" to the network and improved the flow of communication. It is assumed that it also placed the overall network at a greater risk of detection. In Fig. 4 we see the 19 hijackers and their associates (not all of whom may have been involved in the terror plot.)

If one compared the features of this network to those of other criminal networks such as drug trafficking networks, it was found that the latter were more clustered than terrorist networks. Overall path lengths were smaller and the addition of other non-traffickers, such as lawyers, accountants, etc., increased the path length. In other words, while in drug trafficking networks the *security* of the network was increased by the addition of complementary actors, in terrorist networks the *efficiency* of the network was increased by the addition of complementary actors, while simultaneously increasing its *vulnerability*. Structurally, criminal enterprise networks, such as drug trafficking networks, are built outward from a core, whereas terrorist networks apparently lack a core [55].

Centrality Indices in the 9/11 Network. Krebs determined the Freeman Centrality scores, the Degree, Betweenness, and Closeness Centrality for all the actors in the expanded network [43]. The top 10 positions in each are reproduced in Table 2.

Krebs' assessment of the centrality scores for the overall network, and changes in centrality from the action segment to the overall network were also reproduced by Morselli [55] (Table 3). It was observed that the terrorist network became more centralized after the addition of complementary participants to the hijackers' segment. Degree centrality increased by 56%, Betweenness centrality by 13% and Closeness centrality by 84%. Complementary participants acted as shortcuts within the hijacker network but were not central actors. Their role appears to have been indirect, namely through the enhancement of the centrality scores of certain hijackers.

Table 2 Top 10 positions in the Freeman Centrality Indices for the 9/11 terrorist network (Source: Krebs, 2001)

Degrees	Betweenness	Closeness
0.417 Mohamed Atta	0.334 Nawaf Alhazmi	0.571 Mohamed Atta
0.389 Marwan AI-Shehhi	0.318 Mohamed Atta	0.537 Nawaf Alhazmi
0.278 Hani Hanjour	0.227 Hani Hanjour	0.507 Hani Hanjour
0.278 Nawaf Alhazmi	0.158 Marwan AI-Shchhi	0.500 Marwan AI-Shehhi
0.278 Ziad Jarrah	0.116 Saeed AIghamdi*	0.480 ZiadJarrah
0.222 Ramzi Bin al-Shibh	0.081 Hamza Alghamdi	0.429 Mustafa al-Hisawi
0.194 Said Bahaji	0.080 WaleedAlshehri	0.429 Salem Alhazmi*
0.167 Hamza Alghamdi	0.076 ZiadJarrah	0.424 Lotfi Raissi
0.167 Saeed Alghamdi*	0.064 Mustafa al-Hisawi	0.424 Saeed AIghamdi*
0.139 Lotfi Raissi	0.049 Abdul AzizAl-Omari*	0.419 Abdul Aziz AI-Omari*
0.128 MEAN	0.046 MEAN	0.393 MEAN

The Key Player Concept. Key players are those actors that play a significant role in the network by virtue of their placement in the network and their connections [56]. Centrality figures can estimate which are the significant positions in terms of connectivity. Do the centrality scores reveal who were the leaders or key players in the 9/11 network? From Table 2, it can be seen that Mohammad Atta had the highest rank on Degree centrality and Closeness centrality, and the second highest rank on Betweeness centrality. Krebs notes that after just a month of investigation, it was common knowledge that Mohamed Atta was the ring leader of the conspiracy. This was also verified from the bin Laden video tape [54]. However, practitioners of network analysis advocated caution in drawing this conclusion from the metrics and network map (Fig. 4) which clearly showed that he had the most connections. According to them, these metrics did not necessarily confirm his leadership status.

Table 3 Freeman's Centrality Indices for the 9/11 Network for the action and complementary networks (Source; Morselli [55])

Degree, betweenness, and closeness centrality for Krebs' terrorist network						
Participant's name	Action network			Action + complimentary network		
	Degree centrality (rank)	Betweenness centrality (rank)	Closeness centrality (rank)	Degree centrality (rank)	Betweenness centrality (rank)	Closeness centrality (rank)
(1) M. Moqed	.056 (5)	0 (12)	.214 (14)	.028 (11)	0 (24)	.340 (22)
(2) H. Hanjour	.167 (3)	.111 (11)	.269 (9)	.278 (3)	.227 (3)	.507 (3)
(3) K. Al-Midhar	.111 (4)	0 (12)	.265 (10)	.111 (8)	.011 (18)	.396 (13)
(4) N. Alhamzi	.333 (1)	.386 (2)	.340 (6)	.278 (3)	.334 (1)	.537 (2)
(5) S. Alhamzi	.056 (5)	0 (12)	.257 (12)	.083 (9)	.007 (20)	.429 (6)
(6) A. Alnami	.167 (3)	0 (12)	.316 (7)	.083 (9)	0 (24)	.391 (14)
(7) S. Alghamdi	.222 (2)	.135 (8)	.367 (3)	.167 (6)	.116 (5)	.424 (7)
(8) A. Alghamdi	.056 (5)	0 (12)	.286 (11)	.056 (10)	.004 (22)	.319 (25)
(9) H. Alghamdi	.333 (1)	.395 (1)	.391 (1)	.167 (6)	.081 (6)	.414 (9)
(10) A. Alhaznawi	.167 (3)	.347 (3)	.375 (2)	.083 (9)	.030 (13)	.404 (11)
(11) M. Alshehri	.111 (4)	.131 (9)	.360 (4)	.056 (10)	.012 (17)	.343 (21)
(12) F. Ahmed	.111 (4)	.111 (11)	.340 (6)	.083 (9)	.031 (12)	.387 (15)
(13) Z. Jarrah	.167 (3)	.327 (4)	.353 (5)	.278 (3)	.076 (8)	.480 (5)
(14) M. Al-Shehhi	.222 (2)	.247 (6)	.360 (4)	.389 (2)	.158 (4)	.500 (4)
(15) M. Atta	.167 (3)	.119 (10)	.316 (7)	.417 (1)	.318 (2)	.571 (1)
(16) A. Aziz Alomari	.167 (3)	.294 (5)	.277 (8)	.083 (9)	.049 (10)	.419 (8)
(17) W. Alshehri	.167 (3)	.209 (7)	.231 (13)	.111 (8)	.080 (7)	.340 (22)
(18) W. Alshahri	.111 (4)	0 (12)	.191 (15)	.056 (10)	0 (24)	.271 (27)
(19) S. Al-Suqami	.111 (4)	0 (12)	.191 (15)	.111 (8)	.033 (11)	.298 (26)
(20) R. Bin Al-Shibh	-	-	-	.222 (4)	.010 (19)	.414 (9)
(21) S. Bahaji	-	-	-	.194 (5)	.004 (22)	.409 (10)
(22) L. Raissi	-	-	-	.139 (7)	.015 (16)	.424 (7)
(23) Z. Essabar	-	-	-	.139 (7)	0 (24)	.400 (12)
(24) A. Budiman	-	-	-	.111 (8)	0 (24)	.396 (13)
(25) M. El-Motassadeq	-	-	-	.111 (8)	0 (24)	.391 (14)
(26) M. Alhisawi	-	-	-	.111 (8)	.064 (9)	.429 (6)
(27) N. Al-Marabh	-	-	-	.111 (8)	.026 (14)	.330 (23)
(28) R. Abdullah	-	-	-	.111 (8)	.002 (23)	.350 (20)
(29) A. Shaikh	-	-	-	.083 (9)	0 (24)	.360 (18)
(30) M. Darkazanli	-	-	-	.083 (9)	0 (24)	.387 (15)
(31) O. Awadallah	-	-	-	.083 (9)	0 (24)	.360 (18)
(32) R. Hijazi	-	-	-	.083 (9)	.016 (15)	.327 (24)
(33) B. Alhazmi	-	-	-	.056 (10)	0 (24)	.343 (21)
(34) F. Alsalmi	-	-	-	.056 (10)	0 (24)	.343 (21)
(35) Z. Moussaoui	-	-	-	.056 (10)	0 (24)	.371 (16)
(36) A. Khalil Alani	-	-	-	.028 (11)	0 (24)	.367 (17)
(37) M. Abdi	-	-	-	.028 (11)	0 (24)	.353 (19)
Mean	.158	.148	.298	.128	.046	.393
Centralization	.196	.261	.202	.306	.296	.372

Source: Krebs (2001).

Fig. 4 Network of the 19 hijackers and complementary members in the 9/11 attack of the Twin Towers in New York by Al Quaeda (Source Krebs, 2001[4])

There would be missing nodes and ties in the network. Centrality measures could be sensitive to minor changes in nodes and links. Discovery of a new conspirator along with new ties, or the uncovering of a tie amongst existing nodes could influence Freeman centralities. (The question of sensitivity of centrality measures to missing data was addressed later by Borgatti, et al., who showed that they are fairly robust to small changes [57].)

Although not apparent in the initial action network (Fig. 4), H. Hanjour, Z. Jarrah, M. Al-Shehhi and M. Atta, the 4 pilots of the hijacked planes, emerged as key nodes once the complementary segment was added (Fig. 4). N. Alhamzi retained his high status within the centrality ranks after the network expanded, while three hijackers, S. Alghamdi, H. Alghamdi and A. Alhaznawi became less prominent.

Loss of Terrorists from a Network. What would happen to a terrorist network if one of the terrorists was captured or otherwise removed? The 9/11 network has been used by Latora in 2004 to assess the decrease in efficiency of information transmission in the network as a result of the elimination of individual nodes [58]. The efficiency of transmission E is defined in terms of distances between nodes (essentially the harmonic mean of the distances). The change in efficiency on removal of a node, Del E, is obtained for each node in the network. Table 4 shows the results of the estimation.

Table 4 Effect of the elimination of individual nodes in the 9/11 network (Source: Latora, 2004)

	Removed node	**E(G-node i)**	**Del E/E**	k__
1	Mohamed Atta	0.4291	0.150	16
2	Salem Alhazmi	0.4484	0.112	8
3	Hani Hanjour	0.4554	0.098	10
4	Mamoun Darkazanli	0.4586	0.091	4
5	Marwan Al-Shehhi	0.4587	0.091	14
6	Nawaf Alhazmi	0.4611	0.086	9
7	Hamza Alghamdi	0.4646	0.080	7
8	Satam Suqami	0.4656	0.077	8
9	Abdul Aziz Al-Omari	0.4667	0.075	9
10	Fayez Banihammad	0.4710	0.067	7

The efficiency of the original network was 0.5047.The removed node is listed in the first column, the efficiency of the graph once the node is removed is listed in the second column, while the relative drop of efficiency is in the third column. The last column, as an alternative measure of the importance of a node, lists the degree of the removed node. The ten most important nodes of the network have been ordered according to the reduction of the efficiency they cause. We notice that Mohammad Atta shows up again as an important node, as there is highest loss of efficiency on his removal from the network.

Predicting a Terrorist Network. Till now, networks discussed have been mapped and inferences drawn after a terrorist event. An important question remains about the potential of SNA to uncover terrorist plots before the actual attack. This question has been addressed by Krebs in the 9/11 case [59]. He showed that two of the hijackers were in fact known to be linked to Al Qaeda as early as January 2000, and seen to be attending an Al Qaeda meeting in Malaysia. Another person attending the same meeting was Khallad, who was a suspect in the attack on the U.S. ship USS Cole in Aden in October 2000. According to Krebs, by mapping nodes two links away from the suspects, almost the entire hijacker network could have been mapped prior to the attack.

One of the limitations of these studies remains the general inaccessibility of intelligence data and consequent reliance on open source data. In covert networks, individuals are expected to keep a low profile before an attack. If data is missing, and this may not be uncommon, it may affect the network results. As mentioned earlier, the degree of robustness of network analysis to missing data has been discussed by Borgatti, et al. [57].

7.2 Case II: Mapping a Network of Terrorist Organizations in India

India is at the geographical centre of a belt of terrorist, insurgent and separatist violence that includes Pakistan and Afghanistan on the west and Burma and Bangladesh on the east [60]. Internally, it is faced with terrorist violence in Jammu and Kashmir, separatist forces in the North East, and Maoist and Naxal forces in several states. Consequently, it is one of the countries with the highest incidences of terrorism. There has been an increasing trend in the number of terrorist incidents over the last decade as seen in Fig. 5, with a large number of terrorist organizations, both from within India, and across the borders, fuelling the violence [61]. Terrorist activities are no longer localized. Links between militant organizations can be surprisingly long and can span continents. For example, one of the terrorists freed by India in the Kandahar hijacking of a commercial jetliner in 1999 was suspected to have financed a hijacker in the Sep. 11th (9/11) bombing of the World Trade Towers in USA [62].

Studying an individual terrorist attack does not give a picture of the global connections in a network. Information from a large number of terrorist attacks can provide a key to the organizations involved and their respective connections. The objective in this section is to see how statistical analysis can be used to uncover links that would not be apparent from a single incident, using the studies by Basu [5] and Saxena, et al. [48]. Detailed information on individual terrorists is often difficult to obtain in a terrorist attack, unless they are captured, but organizations involved are usually mentioned. The level of association between organizations may be inferred by how frequently they are mentioned together in reports of terrorist violence. Organizations with similar linkage patterns may be identified, and the roles of organizations inferred using centrality measures.

In the studies by Basu and Saxena, data on terrorist attacks in India were taken from a database on terrorism (Terror Tracker or T2), built at the Institute of Defence Studies and Analysis (IDSA) in New Delhi [63]. The database was used at IDSA for profiling militant organisations which had been active in the Indian state of Jammu & Kashmir, bordering Pakistan [64]. Open source information from structured sources (e.g., Institute for Counter-Terrorism, Israel; Institute for Terrorism and Political Violence, USA), as well as from news reports were used to build the database. Reports on each incident were de-duplicated and categorised as Category

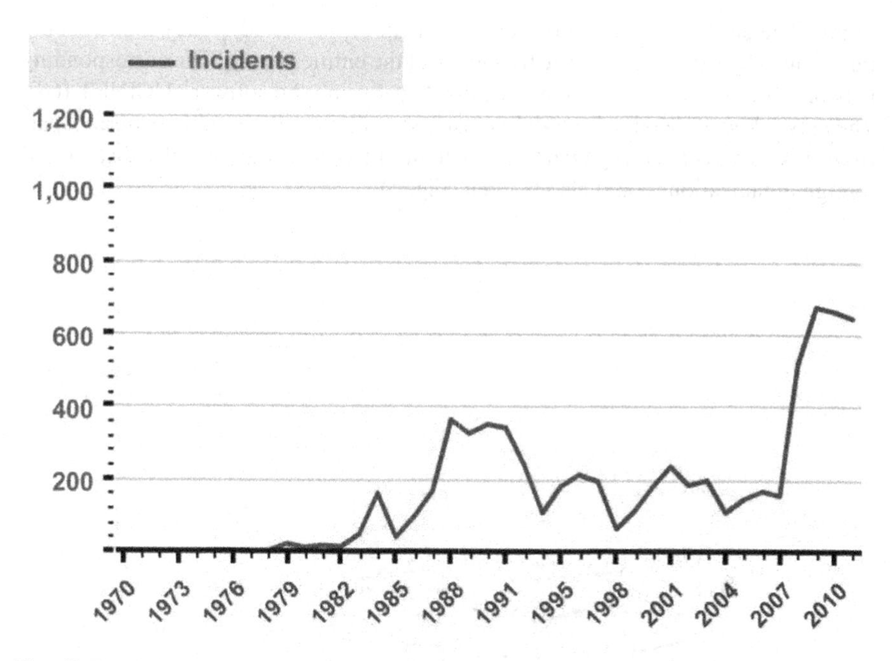

Fig. 5 Trend in the number of terrorist incidents in India over four decades: Source START [61]

1: Violent attacks, Category 2: Police reprisals, etc. More than 200 incidents in the first category were taken in the period 2000-2003 for the study. The basic unit of analysis used here is the organization.

Co-occurrence Frequency as an Indicator of Link Strength. Links between organizations are not real links but links by *association*, which are measured by *co-occurence*. An incident where more than one organization is named in a report of terrorist violence indicates a *co-occurrence*, and has been taken to suggest a probable link between the organizations. The frequency of co-occurrences indicates the strength of a link. The nature of the link remains unspecified. Co-occurrence analysis is useful in a wide variety of fields including scientometrics and bibliometrics to reveal associations or structure and order within a field of knowledge [65, 66]. Using co-occurrence data for incidents in Kashmir, Saxena et al. [48] examined links of terrorist organisations operating in Jammu & Kashmir using the graphical software Visone [67]. According to Saxena et al., the network graphs correctly displayed the linkages between militant organisations operating in Jammu and Kashmir. However, they commented on the low co-occurrence of ISI in all the graphs. ISI is the Inter Services Intelligence wing of Pakistan, and long suspected of backing militant organisations operating in Jammu and Kashmir.

In a subsequent study covering 200 terrorist incidents across India, Basu [5], and again using textual links as a proxy for real links, 61 organizations were found whose names co-occurred with each other. The co-occurrence frequency was used to populate an Adjacency Matrix, each element a_{ij} of the adjacency matrix being represented by the co-occurrence frequency of the entities i and j. The corresponding network of terrorist organizations was displayed using the software UCINET [68]. The network was found to be tightly clustered, without a discernible pattern. Apart from linkages between organizations in India, links to geographically distant and foreign organizations could also be seen (Fig. 6).

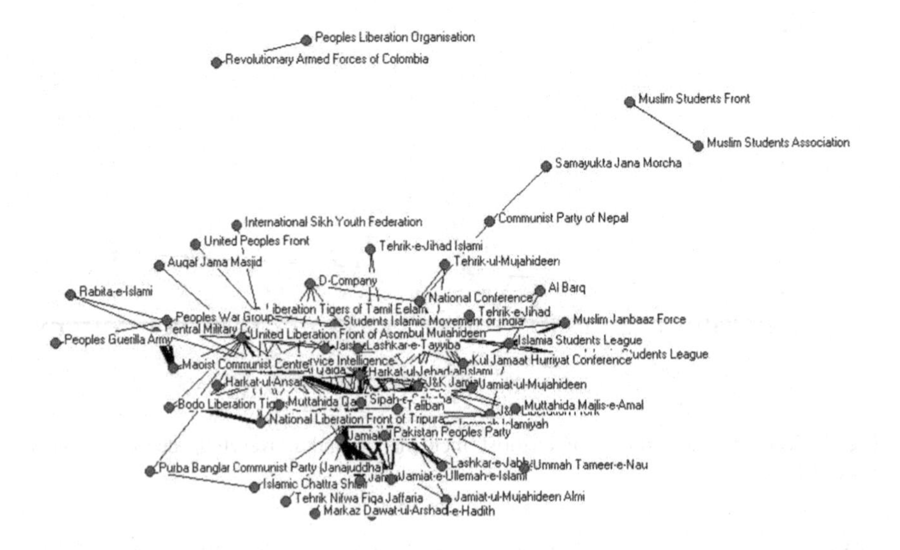

Fig. 6 Network of 61 organizations named together in reports of terrorist violence in India in the period 2000-2002: Source: Basu, 2005 [5]

Thresholding. Some of the links displayed in Fig. 6 could be spurious. The data could be noisy, in that every co-occurrence might not represent a real or meaningful link. There could be casual allusions to some organizations. Some textual errors could also be introduced through name variants, or conversely, the same name (or acronym) being used for different organizations. For statistical validity, very infrequent occurrences suggesting spurious or chance linkages were filtered out using a principle of 'thresholding'. This form of data truncation is used in SNA to reduce clutter and help in arriving at clearer results. After eliminating links of frequency less than five in two years, 35 organizations remained in the network. The first few of these organizations with the largest number of links, totalling 85% of all links, are shown in Table 5. The percentage of links of any organization indicates the degree of connectivity of that organization. (The table demonstrates the applicability of the 80-20 Rule to terrorism, which states that 20 percent of the entities account for 80 percent of the activity).

Ideological and Regional Groups Using Structural Equivalence. To bring out structure within the network Basu [5] used structural equivalence. Structural equivalence is used to identify similarities and dissimilarities between nodes in a network

Table 5 Percentage of links of terrorist organizations in India, reflecting degree of connectivity

Rank	Organisation	Percent links
1	Jamiat Ulema-e-Hind	11.26%
2	J&K Jamiat-e-Islami	8.85%
3	Taliban	8.09%
4	Sipah-e-Sahaba	6.98%
5	Jamiat-e-Islami	6.84%
6	Jamiat Ulema-e-Islam	4.57%
7	Jamiat-e-Ullemah-e-Islami	4.57%
8	Central Military Commission	4.06%
9	Lashkar-e-Tayyiba	3.85%
10	Maoist Communist Centre	3.73%
11	Inter Service Intelligence	2.76%
12	Al Badr	2.64%
13	Al Qaida	2.60%
14	People's War Group	2.34%
15	All Tripura Tiger Force	2.13%
16	United Liberation Front of Asom	2.04%
17	Islamia Students League	2.00%
18	Jaish-e-Mohammad	1.86%
19	National Liberation Front of Tripura	1.83%
20	Hizbul Mujahideen	1.51%

in terms of their linkages to the rest of the nodes [69, 70, 71]. Thus nodes having a similar pattern of linkages can be grouped together into equivalence clusters. This brings out latent structure in the network which may not be apparent from a visual examination, say of Fig. 6. For ease of visual interpretation organizations with similar link structure are placed closer together on the map. The network of 35 linked nodes that remained after eliminating weak ties was mapped using Structural Equivalence and the Multidimensional Scaling (MDS) [72, 73, 74] routine in UCINET. The routine used the co-occurrence based Adjacency Matrix to create a Proximity Matrix to reflect similarity in the link structure of the nodes. Multidimensional scaling was used to convert from the higher dimensional data to a two-dimensional map which could be visually interpreted (Fig. 7). Nodes with similar node structure are expected to be closer together on this map.

The map in Fig. 7 shows several groups or clusters, as depicted by the colours of the nodes. The grey nodes on the right are Islamist organizations, which could be from J&K in India, or across the border in Pakistan. The international organizations, Al Qaida and Taliban also appear in grey, as well as the Dubai based D-company. Some organizations like the National Conference are political organizations, but are linked to the terrorist organizations through the reports on violent attacks, where one of their members could be a victim.

On the left is a group of four militant organizations which are all located in the North Eastern region of India. The United Liberation Front of Asom (ULFA),

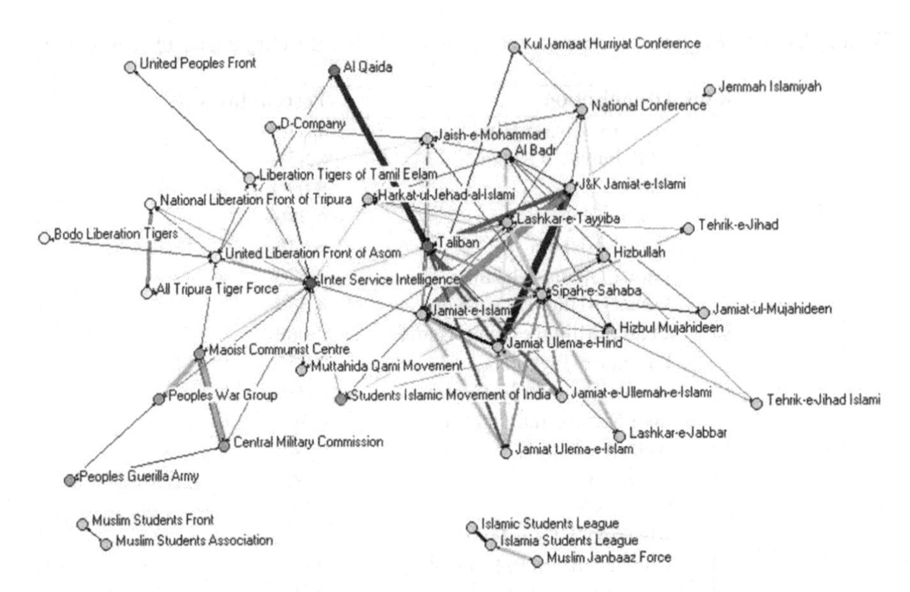

Fig. 7 Group structure in the map of organizations named in terrorist incidents in India (2000-2002) Source: Basu 2005[5]

Bodo Liberation Front, All Tripura Tiger Force and the National Liberation Front of Tripura are all organizations from this region, constituting a regional group. Below them is another group of four organizations, the Maoist Communist centre, People's Guerilla Army, People's War Group, and Central Military Commission. These are left-wing groups, which operate in Nepal and several states including some south and south-eastern states of India. This is an ideological grouping.

Two nodes that do not form part of the clusters appear on the top left, and are the Liberation Tigers of Tamil Eelam (LTTE) which operated in the state of Tamilnadu and Sri Lanka, and the United Peoples Front. The ISI, the Inter Services Intelligence of Pakistan, at the centre of the map, and SIMI, the students' movement from central India do not form part of any cluster. The thickness and colour of the links indicate the intensity of textual links connecting any two nodes.

Role and Position in the Network. The role of a node in the network is determined by the nature of connectivity to other nodes. In this network it was found that the Betweenness Centrality had the largest discriminating power (Table 2). The node with the highest Betweeness Centrality is the ISI of Pakistan. This underscores the important role that the ISI has in India's terror network, in spite of the fact that it is not highly connected (see Saxena [48]). Other organizations like Jaish-e-Mohammad and Lashkar-e-Taiba are among the top ten on the list on Betweeness Centrality (Table 6).

The network shown below has nodes sized by betweeness values for ease of interpretation (Fig. 8). Nodes are positioned using structural equivalence and multi-dimensional scaling as before. The Inter Services Intelligence (ISI) with the highest

Table 6 Betweenness Index of organizations based on textual analysis of reports on terrorist incidents in India: Source: Basu 2005 [5]

Organization	Betweenness
Inter Service Intelligence, Pakistan	12.60
Jamiat-e-Islami	5.25
Taliban	4.34
Sipah-e-Sahaba	4.14
Jaish-e-Mohammad	3.00
J&K Jamiat-e-Islami	2.96
United Liberation Front of Asom	2.69
Liberation Tigers of Tamil Eelam	2.35
Harkat-ul-Jehad-al-Islami	1.74
Lashkar-e-Tayyiba	1.64
Hizbul Mujahideen	1.35
Hizbullah	1.35
Al Badr	1.09
Students Islamic Movement of India	0.91
Peoples War Group	0.90
Central Military Commission	0.90
Jamiat Ulema-e-Hind	0.83
National Conference	0.76
Al Qaida	0.66

betweeness value in the network is seen at the centre of the figure. It dominates the network and connects to all the major groups.

The analysis points to the key role of ISI in India's terror network. It lies at the centre of a diverse set of militant groups. In the earlier study discussed here by Saxena et al. [48], it was remarked that the visibility of ISI was unexpectedly low. The analysis by Basu indicates that even though the frequency of co-occurrence is low, the high betweenness centrality of ISI sets it apart as having an important *'brokerage'* or bridging role in the terror network. India has always maintained that the ISI has been funding terror attacks in India indirectly through other organizations [75]. Organizations like Lashkar-e-Taiba are known to have been responsible for the terrorist attack on Mumbai in 2011, with suspected ISI backing [76]. While some caution in interpretation may be needed as centrality indices are sensitive to changes in nodes and links, other studies suggest that such changes may not be so significant [57].

The statistical analysis of a very large number of links, as discussed above, brings out latent patterns while suppressing noise. Complex inter-linkages are readily perceived though the visualization techniques and computation of graph theoretic indices helps identify important players with diverse roles. Knowledge of linkages between organizations formed on the basis of past data can also inform counter-terrorism, as pre-emptive strategies and targets may be formed based on the indications of the influence pattern of individual organizations.

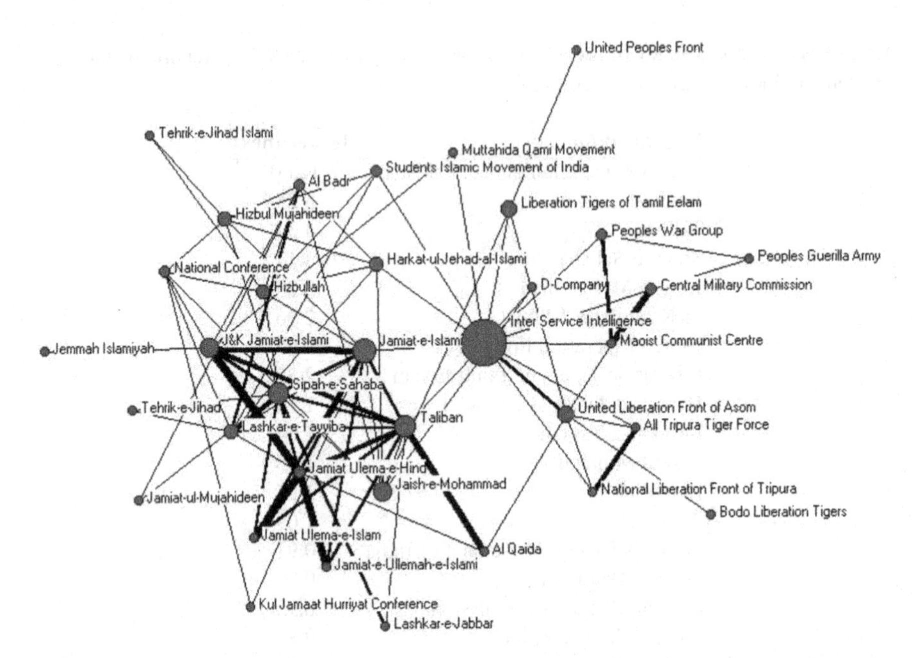

Fig. 8 Structural Equivalence and MDS map of organizations named in terrorist incidents in India, with nodes sized by Betweeness Centrality. (Source Basu, 2005[5])

8 Conclusions

Terrorism today is one of the major concerns of most nations. In terrorist attacks, while the perpetrator may be an individual (or a few individuals), he almost invariably has the backing of an organization. These in turn often have linkages with other organizations with similar if not common ideology, and not infrequently, common funding sources. It is this feature - these network-like connections - that makes them so much more resilient, and for the same reason so much more necessary to know and understand, so that nations can take pre-emptive steps against terror on a real time basis. In this context Social Network Analysis, as a research methodology, has already found a niche space in terrorism studies, enabling as it does decision making based on interpretation of the structure of networks, and the role of individuals and organizations within it. The value of social network theory versus other political science and sociological approaches is its focus on the network structure rather than the characteristics of the individual. The network structure of an organization (in this case a terrorist organization) affects its ability to access new ideas, resources, recruit new individuals, etc., and is therefore the key to its stability and continued existence.

Acknowledgements. The author acknowledges a grant from the Council of Scientific and Industrial Research under their Emeritus Scientist Scheme in the period this article was written. Discussions with Dipankar Basu are gratefully acknowledged.

References

1. Defining terrorism. Transnational Terrorism, Security and the Rule of law, European Commission Working Paper 3 (2008), http://www.transnationalterrorism.eu/tekst/publications/WP3%20Del%204.pdf
2. Ressler, S.: Social network analysis as an approach to combat terrorism: Past, present, and future research. Homeland Security Affairs 2(2), 1–10 (2006)
3. van der Hulst, R.C.: Terrorist networks: the threat of connectivity, The SAGE Handbook of Social Network Analysis, p. 256 (2011)
4. Krebs, V.E.: Mapping networks of terrorist cells. Connections 24(3), 43–52 (2002)
5. Basu, A.: Social network analysis of terrorist organizations in india. In: North American Association for Computational Social and Organizational Science (NAACSOS) Conference, pp. 26–28 (2005)
6. Scott, J.: Social network analysis: A Handbook. Sage Publications (1994)
7. Wasserman, S.: Social network analysis: Methods and applications. Cambridge University Press (1994)
8. Freeman, L.C.: The development of social network analysis. Empirical Press Vancouver (2004)
9. Harary, F.: Graph theory. Addison-Wesley (1969)
10. Milgram, S.: The small world problem. Psychology Today 2(1), 60–67 (1967)
11. Granovetter, M.S.: The strength of weak ties. American Journal of Sociology 78(6), 1360–1380 (1973)
12. Watts, D.J., Strogatz, S.H.: Collective dynamics of small-world networks. Nature 393(6684), 440–442 (1998)
13. Watts, D.J.: Networks, dynamics, and the small-world phenomenon 1. American Journal of Sociology 105(2), 493–527 (1999)
14. Brandes, U., Erlebach, T. (eds.): Network Analysis. LNCS, vol. 3418. Springer, Heidelberg (2005)
15. Freeman, L.C.: Going the wrong way on a one-way street: Centrality in physics and biology. Journal of Social Structure 9(2) (2008)
16. Arquilla, J., Ronfeldt, D.: Networks and netwars: The future of terror, crime, and militancy. Rand Corporation (2001)
17. Sparrow, M.K.: The application of network analysis to criminal intelligence: An assessment of the prospects. Social Networks 13(3), 251–274 (1991)
18. Barabási, A.L., Frangos, J.: Linked: The New Science of Networks. Basic Books (2002)
19. Krebs, V.E.: Network metrics. Flow 3.0 Users' Manual (2001)
20. Carley, K.M., Reminga, J., Kamneva, N.: Destabilizing Networks. Connections 24(3), 79–92 (2001)
21. Rothenberg, R.: From whole cloth: Making up the terrorist network. Connections 24(3), 36–42 (2001)
22. Sageman, M.: Understanding terror networks. Univ. of Pennsylvania Press (2004)
23. Rodríguez, J.A.: The th Terrorist Network: In its Weakness Lies Its Strength. In: XXV International Sunbelt Conference, Los Angeles (2005)
24. Koschade, S.: A social network analysis of Jemmah Islamiyah: The applications to counterterrorism and intelligence. Studies in Conflict & Terrorism 29(6), 559–575 (2006)
25. Azad, S., Gupta, A.: A quantitative assessment on 26/11 mumbai attack using social network analysis. Journal of Terrorism Research 2(2), .4-1-4 (2011)

26. Perliger, A., Pedahzur, A.: Social network analysis in the study of terrorism and political violence. Political Science and Politics 44(1), 45 (2011)
27. Dark web and geopolitical web research, AI Laboratory, University of Arizona, http://ai.arizona.edu/research/terror
28. National consortium for the study of terrorism and responses to terrorism (START), http://www.start.umd.edu/start/about/overview/
29. Global terrorism database, http://www.start.umd.edu/gtd/
30. Terrorist organization profiles, http://www.start.umd.edu/start/data_collections/tops/
31. Terrorism and preparedness data resource center, http://www.start.umd.edu/start/data_collections/tpdrc/
32. Defense Intelligence Agency (USA). Criminal Network Analysis Training Course (2000)
33. Diesner, J., Carley, K.M.: Using network text analysis to detect the organizational structure of covert networks. In: Proceedings of the North American Association for Computational Social and Organizational Science (NAACSOS) Conference (2004)
34. Tsvetovat, M., Carley, K.M.: On effectiveness of wiretap programs in mapping social networks. Computational and Mathematical Organization Theory 13(1), 63–87 (2006)
35. NETEST: Estimating a Terrorist Network's Structure. In: 11th European Intelligence and Security Informatics Conference (EISIC), Athens (2011)
36. Abbasi, A., Chen, H.: Applying authorship analysis to extremist-group web forum messages. IEEE Intelligent Systems 20(5), 67–75 (2005)
37. Carpenter, T., Karakostas, G., Shallcross, D.: Practical issues and algorithms for analyzing terrorist networks. In: Proceedings of the Western Simulation Multi Conference (2002)
38. Gupta, A., Kumaraguru, P.: Twitter explodes with activity in mumbai blasts! a lifeline or an unmonitored daemon in the lurking? Mumbai (2012)
39. Sliva, A., Subrahmanian, V., Martinez, V., Simari, G.I.: The soma terror organization portal (STOP): Social network and analytic tools for the real-time analysis of terror groups. In: Social Computing, Behavioral Modeling, and Prediction, pp. 9–18. Springer, Heidelberg (2008)
40. Dickerson, J.P., Mannes, A., Subrahmanian, V.: Dealing with Lashkar-e-taiba: A multi-player game-theoretic perspective. In: European Intelligence and Security Informatics Conference (EISIC), pp. 354–359. IEEE (2011)
41. Subrahmanian, V. (ed.): Handbook of Computational Approaches to Counterterrorism. Springer (2013)
42. Mannes, A.: Qualitative analysis & computational techniques for the counter-terror analyst. In: Handbook of Computational Approaches to Counterterrorism, pp. 83–97. Springer (2013)
43. Freeman, L.C.: Centrality in social networks: Conceptual clarification. Social Networks 1(3), 215–239 (1979)
44. Bonacich, P.: Power and centrality: A family of measures. American Journal of Sociology 92(5), 1170–1182 (1987)
45. Networks / Pajek: Program for large network analysis, http://vlado.fmf.uni-lj.si/pub/networks/pajek/
46. Borgatti, S.P., Everett, M.G., Freeman, L.C.: Ucinet for windows: Software for social network analysis. Analytic Technologies, Harvard (2002)
47. Huisman, M., van Djuin, M.A.J.: A readers guide to SNA Softwares. In: Sage Handbook on Social Network Analysis, pp. 578–600 (2011)

48. Saxena, S., Santhanam, K., Basu, A.: Application of social network analysis (SNA) to terrorist networks in Jammu & Kashmir. Strategic Analysis 28(1), 84–101 (2004)
49. The hijackers.... and how they were connected, Sydney Morning Herald (Septembr 22, 2001), http://www.smh.com.au
50. Erickson, B.H.: Secret societies and social structure. Social Forces 60(1), 188–210 (1981)
51. The Plot: A Web of Connections, Washington Post (September 24, 2001), http://www.washingtonpost.com/wp-srv/nation/graphics/attack/investigation_24.html
52. Indictment of Zacarias Moussaoui. U.S. Department of Justice (December 11, 2001), http://www.usdoj.gov/ag/moussaouiindictment.htm
53. Friedkin, N.E.: Horizons of observability and limits of informal control in organizations. Social Forces 62(1), 54–77 (1983)
54. Transcript of bin Laden Video Tape, United States Department of Defense (December 13, 2001), http://www.defense.gov/news/Dec2001/d20011213ubl.pdf
55. Morselli, C., Giguère, C., Petit, K.: The efficiency/security trade-off in criminal networks. Social Networks 29(1), 143–153 (2007)
56. Borgatti, S.P.: Identifying sets of key players in a social network. Computational & Mathematical Organization Theory 12(1), 21–34 (2006)
57. Borgatti, S.P., Carley, K.M., Krackhardt, D.: On the robustness of centrality measures under conditions of imperfect data. Social Networks 28(2), 124–136 (2006)
58. Latora, V., Marchiori, M.: How the science of complex networks can help developing strategies against terrorism. Chaos, Solitons & Fractals 20(1), 69–75 (2004)
59. Krebs, V.: Uncloaking terrorist networks. First Monday 7(4) (2002)
60. Chellaney, B.: Fighting terrorism in southern asia: The lessons of history. International Security 26(3) (2001)
61. START: Global terrorism database, http://www.start.umd.edu/gtd/search/Results.aspx?country=92
62. CNN: Suspected hijack bankroller freed by india in 1999 (2001), http://www.cnn.com/2001/US/10/05/inv.terror.investigation/index.html
63. Saxena, S., Santhanam, K.: Design approach to creating a terrorism database with open source information. IDSA Internal Report (2001)
64. Santhanam, K.: Jihadis in Jammu and Kashmir: A Portrait Gallery. Sage (2003)
65. Leydesdorff, L., Vaughan, L.: Co-occurrence matrices and their applications in information science: Extending ACA to the web environment. Journal of the American Society for Information Science and Technology 57(12), 1616–1628 (2006)
66. Pitel, G., Millet, C., Grefenstette, G.: Deriving a priori co-occurrence probability estimates for object recognition from social networks and text processing. In: Bebis, G., Boyle, R., Parvin, B., Koracin, D., Paragios, N., Tanveer, S.-M., Ju, T., Liu, Z., Coquillart, S., Cruz-Neira, C., Müller, T., Malzbender, T. (eds.) ISVC 2007, Part II. LNCS, vol. 4842, pp. 509–518. Springer, Heidelberg (2007)
67. Brandes, U., Wagner, D.: Analysis and visualization of social networks. In: Jansen, K., Margraf, M., Mastrolli, M., Rolim, J.D.P. (eds.) WEA 2003. LNCS, vol. 2647, pp. 321–340. Springer, Heidelberg (2003)
68. Borgatti, S.P., Everett, M.G., Freeman, L.C.: Ucinet for windows: Software for social network analysis. Harvard Analytic Technologies (2002)
69. Lorrain, F., White, H.C.: Structural equivalence of individuals in social networks. The Journal of Mathematical Sociology 1(1), 49–80 (1971)

70. Faust, K., Wasserman, S.: Blockmodels: Interpretation and evaluation. Social Networks 14(1), 5–61 (1992)
71. Moody, J., White, D.R.: Structural cohesion and embeddedness: A hierarchical concept of social groups. American Sociological Review, 103–127 (2000)
72. Kruskal, J., Wish, M.: Multidimensional scaling. Sage (1978)
73. Borg, I., Groenen, P.J.F.: Modern Multidimensional Scaling: Theory and Applications. Springer (2005)
74. Breiger, R.L., Boorman, S.A., Arabie, P.: An algorithm for clustering relational data with applications to social network analysis and comparison with multidimensional scaling. Journal of Mathematical Psychology 12(3), 328–383 (1975)
75. Bhalla, A.: India submits proof that Pakistan funds terror activities, builds international pressure for action, India Today (2013), http://indiatoday.intoday.in/story/india-builds-pressure-against-pakistan-isi-let-terrorism-counterfeit-currency/1/284080.html
76. Pakistan Intelligence services aided Mumbai terror attacks. The Guradian (October 18, 2010), www.guradian.co.uk/world/2010/Oct/18/pakistan-isi-mumbai-terror-attacks

Privacy and Anonymization in Social Networks

B.K. Tripathy, M.S. Sishodia, Sumeet Jain, and Anirban Mitra

Abstract. As the Internet continues to grow, the proliferation of online social networks raises many privacy concerns. The users of these OSNs are divulging endless details about their lives online. This personal information can be used by attackers to perpetrate significant privacy breaches and carry out attacks such as identity theft and credit card fraud. The privacy concerns arise from not just the users posting their personal information online, but also from OSNs publishing this information for analysis. Driven by Web 2.0 applications, more and more social network has been made publicly available. Preserving the privacy of individuals in this published data is an important concern. Although privacy preservation in data publishing has been studied extensively and several important models such as k- anonymity and l-diversity as well as many efficient algorithms have been proposed, most of the existing studies deal with relational data only. Those methods cannot be applied to social network data straightforwardly. Anonymization of social network data is a much more challenging task than anonymizing relational data. Firstly, in relational databases, attacks come from identifying individuals from quasi-identifiers. But in social networks, information such as neighbourhood graphs can be used to identify individuals. Secondly, tuples can be anonymized in relational data without affecting other tuples. But in social networks, adding edges or vertices affects the neighbourhoods of other vertices in the graph as well. In this chapter, we give a brief overview of the privacy concerns in online social networks and provide a detailed description of our algorithm, GASNA, a greedy algorithm for social network anonymization. This algorithm provides structural anonymity and sensitive attribute protection by achieving k-anonymity and l-diversity in social network data.

B.K. Tripathy · M.S. Sishodia · Sumeet Jain
SCSE, VIT University, Vellore, TN, India
e-mail: tripathybk@vit.ac.in,
{mayanksshishodia,sumeet.jain44}@gmail.com

Anirban Mitra
Dept. of CSE, MITS, Rayagada, Odisha, India
e-mail: mitra.anirban@gmail.com

M. Panda, S. Dehuri, and G.-N. Wang (eds.), *Social Networking*
Intelligent Systems Reference Library 65,
DOI: 10.1007/978-3-319-05164-2_10,

We also discuss the challenges faced by the existing algorithms/models for social network data privacy and suggest techniques to counter these challenges. The issues discussed are the high cost of achieving k-anonymity when the value of k is fixed and the need for a better anonymity model which suits the current scenario of social networks. We also propose a new model called partial anonymity which can help reduce the number of edges added for anonymization when the value d of d-neighbourhood is greater than 1.

1 Introduction

"You have zero privacy anyway. Get over it", Sun Microsystems' then-CEO Scott McNealy is widely reported to have declared in 1999. At the time, it was a shocking thing for someone in McNealy's position to say. But now, just 14 years later, it is clear that he was right. The Internet has over 2.4 billion users and is the largest source of information worldwide. Online social networks (OSN) have become ubiquitous. Face book, world's largest online social network, has 1.11 billion users as of March, 2013 while Google+ has 500 million as of December 2012 and Twitter has 200 million as of February 2013. The users of these OSN are divulging endless details about their lives online. Face book founder, Mark Zuckerberg said in 2010, "in the last five or six years, blogging has taken off in a huge way and all these different services have people sharing all this information. People have really gotten comfortable not only sharing more information of different kinds, but more openly and with more people". With so much information being shared online, user privacy in OSN has become an important concern.

The privacy threats in OSN can be divided into two categories. The first includes the privacy threats as a result of user activity and privacy settings on these websites. People share personal information like their date of birth, home address, family ties, pictures and activities. Such information can be used by attackers to perform attacks such as credit card fraud or identity theft. When we post how we are looking forward to that family vacation next week at Hawaii, everyone will know that our house will be empty for a whole week. How can we trust that our TV which we just purchased will not be stolen? Similarly, if someone posts about layoffs in a company, it may indicate to the shareholders that the company is doing poorly. They will sell their shares which will reduce the company value. Users should be careful about what information they post and should only add those people to their contact list whom they know and trust for sure. This will make them less vulnerable to malicious attacks. The social networking websites must also explain clearly the information privacy policies and methods to its users. The second category of privacy threats in OSN deals with user privacy in published social network data. Social network analysis has become a popular field of research due to the popularity of social networking websites such as Face book, Google+, Twitter, etc. An adversary may intrude privacy of some victims using the published social network data and some background knowledge. When social network data is published, privacy can't be protected by merely replacing the identifying attributes such as name and ID number of users by meaningless unique identifiers.

1.1 Chapter Structure

In the forthcoming sections, we discuss about social network privacy and mention some steps users can take to prevent their privacy. We then discuss about privacy concerns that arise when social network data is published and give emphasis on background knowledge attacks. In section 9, we give a detailed explanation of our algorithm GASNA ([25,28]) and compare it with previous literature in section 11. We finally conclude in section 12 and present some future issues regarding social network data anonymization.

2 Social Networks

A social network consists of a finite set or sets of actors (social entities, which can be discrete individual, corporate or collective social units) and the relation (a collection of the defining feature that establishes a linkage between a pair of actors) or relations defined on them [29]. The presence of these relations is a critical and defining feature of a social network. For example, the entities may be individuals and the connections between them being relationships, friendships, or flow of information.

3 Privacy in Social Networks

Social networking sites have seen a boom in their popularity starting from the late 2000s. Today, if you have an internet connection and a real life social network, most likely you have an online social network as well. In contrast with directed communication with an individual via emails or phone calls, OSN have a typical user share his information with friends, family members and any other random person who happens to land on his profile. Face book, which started as a communication website for students in a few universities, now has almost half of the world's Internet users visit it daily to get updates on the social lives of others. Initially, users worried about their privacy, would seek to join private networks limited to friends and family. Face book, however, forced users to make some of their information such as name, profile picture, gender, current city, networks, friend list and pages visible to everyone on the Internet in 2009 [9]. Face book founder Mark Zuckerberg told users"we viewed that as a really important thing, to always keep a beginner's mind and what we would do if we were starting the company now and we decided that these would be the social norms now and we just went for it". This changed the perception of privacy not just on Face book, but across the entire web. People now are comfortable sharing with everyone, the information they perceived private a decade ago.

Users can reduce privacy threats by only adding those people as their contacts whom they know very well in real life. They should also properly go through each and every privacy setting offered by the website. While some information about a user such as name and picture is publicly available to everyone, other information is visible only to some contacts as determined by the user's privacy settings. However, when the relationships between people deteriorate, social networking websites avoid all responsibility for the actions of their users. Face book's policies state"You are solely responsible for your interactions with other users of the Application. We will not be responsible for any damage or harm resulting from your interactions with other users of the Application. We reserve the right, but no obligation, to monitor interactions between you and other users." Users should expect nothing from social networking sites apart from anonymization of their information in the published data.

Users should also be careful as to what they share online. If you don't want something to spread, don't post it online. As an example, consider three Face book friends - Mike, Tom and Jessica. Mike works in a company where Jessica is his boss. Tom is their mutual friend. Mike wanted to go to a picnic one day, but Jessica wouldn't have granted him a leave for that. So Mike made up a story about a family member being hospitalized. Now Mike goes to the picnic and uploads a status on Facebook"The best picnic ever!". He obviously does not want Jessica to know about it, so he uses Facebook privacy settings to hide this status from Jessica. However, Tom, who also attended that picnic, shares Mike's status. Now Jessica can read Mike's status which was shared by Tom. This is something that Mike never wanted and is a breach of his privacy. To address this privacy issue, Facebook introduced a feature that status updates which are hidden from any user cannot be shared. This however, is not the case for pictures. If Mike uploaded a picture from the picnic, Tom can share it and Jessica can see the picture shared by Tom, even though the original picture by Mike was hidden from her. Mike now has an angry boss.

3.1 How Users Lose Control of Their Privacy

Initially, when the users sign up for the social networking websites, they have control on what information they want to share and whom they want to connect with. However, they do not have comprehensive and accurate idea of the information they have explicitly and implicitly disclosed. Moreover, they accept the default privacy settings, generally public, since setting online privacy is time consuming. Eventually, they lose control and face privacy risk. Privacy risk is measured by Privacy Score [15]. It takes into account the information that a user has shared and who can view that information. The two basic premises of privacy are Sensitivity and Visibility. The more sensitive the information revealed by a user, the higher his privacy risk. For example, mother's maiden name is more sensitive than his contact number. Regarding visibility, the wider the information about a user spreads, the higher his privacy risk. For example, his home address known by everyone poses a higher risk than if it were known only by friends. Other privacy risk occurs by inferring private information about a user from his profile, friendships, group memberships, etc. Private information can be inferred using methods like Majority voting [35] and Classification [35]. The basic premise is "birds of a feather flock together". It is known that users with common attributes often form dense communities. For example, consider a group of people who form a clique. Knowing that most of the people in that group work in a company X, there is a high probability that a person who is a member of that group, but has not disclosed his company, also works at X. To summarize, the main privacy challenges regarding personal privacy in OSNs come from the system complexity, leaks through inference and unskilled users.

4 Privacy in Published Social Network Data

Privacy in OSN becomes a serious concern, especially when social network data is published.

4.1 Social Network Representation

Social network analysts use two methods to represent social networks: graphs and matrices. The first method, graphs, consists of points (or nodes) to represent actors and lines (or edges) to represent ties or relations. These are called sociograms and are a very useful way of representing information about social networks. However, it becomes hard to see patterns when there are many actors and/or many kinds of relations. The other method used to represent social networks is matrices. Matrices allow the application of mathematical and computer tools to summarize and find patterns. The most common form of matrix for social network analysis is the adjacency matrix. The graph with 'n' actors is represented as an adjacency matrix of size $n \times n$. A relationship between i^{th} and j^{th} node is represented by the value in the cell i, j [12, 29].

4.1.1 Modelling a Social Network

A social network is modeled as a simple graph $G = (V, E, L, \zeta)$, where V is a set of edges, where V is a set of vertices, $E \subseteq V \times V$ is a set of edges, L is a set of labels and a labeling function $\zeta : V \rightarrow L$ assigns each vertex a label. For a graph G, V(G), E(G), L_G, ζ_G are the set of vertices, the set of edges, the set of labels, and the labeling function in G, respectively.

4.2 Social Network Analysis

Social network analysis is concerned with finding patterns in the connections between entities. It has been widely applied to organizational networks to classify the influence or popularity of individuals and to detect collusion and fraud. Social network analysis can also be applied to study disease transmission in communities, the functioning of computer networks and emergent behaviour of physical and biological systems. Some other example analyses are Motif analysis, community structure, resiliency / robustness, Homophily, correlation / causation.

4.3 Anonymization

Many naive users may do not know that the information they provide online is stored in massive data repositories and may be used for various purposes. Researchers have pointed out the privacy implications of massive data gathering, and a lot of effort has been made to protect the data from unauthorized disclosure.([20,24,26,32]). However, most of this work has been on micro data (data stored as one relational table, where each row represents an individual entity) ([1-8,11,15-19,21,22]) and models such as k-anonymity[27] and l-diversity[18] have been developed for it. But these models cannot be simply applied to social network data. Anonymization of social network data is a much more challenging task than that of micro data. Firstly, in relational (micro data) databases, attacks come from identifying individuals from quasi-identifiers. But in social networks, information such as neighbourhood graphs can be used to identify individuals. Secondly, tuples can be anonymized in relational data without affecting other tuples. But in social networks, adding edges or vertices affects the neighbourhoods of other vertices in the graph as well [25].

Technological advances have made social network data collection very easy. However, agencies and researches that collect such data often face with two undesirable problems. They can publish data for others to analyse, but this will create severe privacy threats [4], or they can withhold data because of those privacy concerns, but this will make further analysis impossible. Therefore, the goal is to enable the useful analysis of social network data while protecting the privacy of individuals [26]. The published data may contain some sensitive data of individuals in the social network, which must not be disclosed. For this, social network data must be anonymized before it is published. This anonymization should offer protection against potential re-identification attacks. Even then, graph structure and background knowledge combine to threaten privacy in many new ways [10].

4.4 Challenges in Anonymizing Social Networks Data

A major category of privacy attacks on relational data is to re-identify individuals by joining a published table containing sensitive information with some external tables modelling background knowledge of attackers. To protect against such re-identification attacks, the concept of k-anonymity was proposed [24, 27]. Specifically, a data set is said to follow k-anonymity ($k \geq 1$) if, on the quasi-identifier attributes (that is, the maximal set of join attributes to re- identify individual records), each record is indistinguishable from at least $(k - 1)$ other records. The larger the value of k, the better the privacy is protected.

As an example, consider a synthesized social network of a class of students created by the school as shown in (Fig. 1). Each vertex represents a student in the class. An edge links two students who are 'friends' with each other and interact a lot. Suppose this network is to be published. The publishing agency does a naïve anonymization (Fig. 2). Naïve anonymization is a transformation of the network in which identifiers are replaced with random numbers. In the published network, the highest degree of a vertex is 11, the second highest is 7. It is known that the

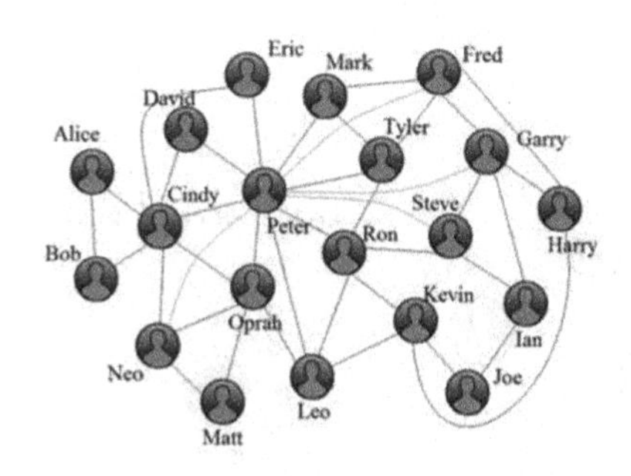

Fig. 1 Sample Social Network

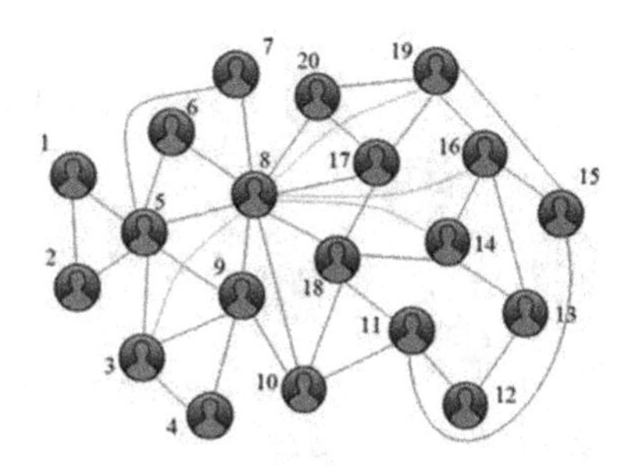

Fig. 2 Naïve Anonymization

Table 1 Sensitive Attributes in the Social Network

ID	Disease
Peter	HIV
Alice	Cancer
Bob	None
Matt	Diarrhoea
Joe	Cancer
David	None
Fred	HIV
Garry	HIV
Oprah	HIV
Ron	HIV
...	...

Table 2 Sensitive Attributes in Naíve-Anonymized Network

ID	Disease
8	HIV
1	Cancer
2	None
4	Diarrhoea
12	Cancer
6	None
19	HIV
16	HIV
9	HIV
18	HIV
...	...

class representative, Peter, has maximum connections within the class. Using this knowledge, an adversary can identify this vertex 8 of degree 11 to be Peter and know that he suffers from HIV. This intrudes Peter's privacy immediately. Now, suppose that the adversary wants to find information about a student Fred and knows that he has 5 connections in the network. Using this knowledge, he tries to find him in the network. There are 4 vertices in the network which have degree 5: 19, 16, 9 and 18 as shown in Fig. 3. Fred can be anyone among these. Thus, Fred cannot be identified in the social network with a probability greater than 1/4. If every node in the social network cannot be identified with a probability greater than 1/k, the network is said to follow k-anonymity [27]. The attacks used here are called matching attacks. In such type of attacks, the adversary matches external information to a naïvely anonymized network.

However, Machanavajjhala et al. [18] showed that a k-anonymous table may suffer from severe privacy problems due to the lack of diversity in the sensitive attributes. In particular they showed that, the degree of privacy protection is determined by the number and distribution of distinct sensitive values associated with each equivalence class and not on the size of equivalence classes on quasi-identifier attributes which contain tuples that are identical on those attributes. To overcome the drawbacks in k-anonymity, they propose the notion of l-diversity [18]. Xiao and Tao [30,31] prove that l-diversity always guarantees stronger privacy preservation

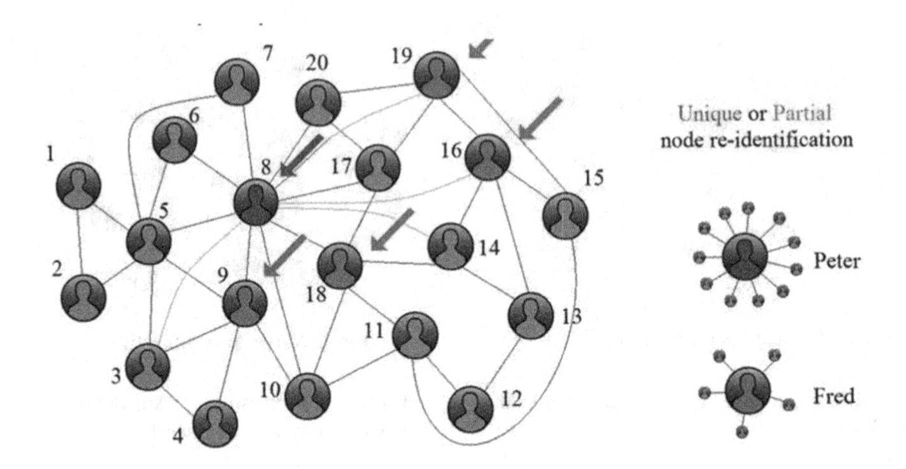

Fig. 3 A Matching Attack

than k-anonymity. Continuing with the previous example, the adversary wants to know the disease Fred is suffering from. Disease is the sensitive attribute in this network. Going through the sensitive attribute values of vertices with 5 degrees, the adversary finds out that all of them are HIV. Hence, one can make logical conclusion that Fred is HIV positive. This way, his privacy is breached even though his vertex had a similar structure to 3 other vertices. This is the limitation of k-anonymity. It is fixed by another concept called l-diversity [18]. l-diversity for social network graphs means that the structurally similar vertices have atleast l distinct values of sensitive attribute. If Gary, Ron and Oprah suffered from Cancer, Tuberculosis and Diarrhoea respectively, then all the adversary could have found out was that Fred suffers from one disease among HIV, Cancer, Tuberculosis and Diarrhoea. His privacy would have then been preserved.

Several important models and many efficient algorithms have been proposed for privacy preservation in relational data. Those methods cannot be applied to social network data straightforwardly. Anonymizing social network data is much more challenging than anonymizing relational data [36,37]. First, it is much more challenging to model background knowledge of adversaries and attacks about social network data than that about relational data. On relational data, attacks mainly come from identifying individuals from the quasi-identifier. However, in a social network, information such as vertex and edge labels, neighbourhood graphs and induced sub-graphs can be combined and used to identify individuals. Second, it is harder to measure the information loss occurred due to anonymization of social network data than that of relational data. In relational data, the information loss in an anonymized table can be measured using the sum of information loss in individual tuples. We can calculate the distance between a tuple in the original table and its corresponding anonymized tuple to measure the information loss at the tuple level. However, a social network consists of a set of vertices and a set of edges. It is hard to compare two social networks by comparing the vertices and edges individually. Two social networks having the same number of vertices and the same number of edges may have very different network-wise properties such as connectivity, between-ness, and diameter. Lastly, it is much more challenging to devise anonymization methods for social network data. Divide-and-conquer methods are extensively applied to anonymization of relational data since tuples here are separable in anonymization. In other words, anonymizing a group of tuples does not affect other tuples in the table. However, anonymizing a social network is much harder since changing labels of vertices and edges may affect the neighbourhoods of other vertices, and removing or adding vertices and edges may affect other vertices and edges as well as the properties of the network.

5 Private Information in Published Social Network Data

In privacy preservation on relational data, the attributes in a table are divided into two groups: non-sensitive attributes and sensitive attributes. The values in sensitive attributes are considered to be private for individuals. However, in social network data much more pieces of information can be considered as privacy of individuals [29]. These are:

Vertex Existence. In social network data, whether a target individual appears in the network or not can be considered as the privacy of the individual. If a target individual can be determined appearing in the network, an attacker knows the sensitive attribute value of the target.

Vertex Properties. In social network data, some properties of a vertex such as degree can be considered as privacy of the individual.

Sensitive Vertex Labels. In social network data, vertices may carry labels, which can be divided into two categories: non-sensitive vertex labels and sensitive vertex labels Similar to the case of relational data, the values of sensitive vertex labels are considered to be privacy of individuals.

Link Relationship. In social network data, an edge between two vertices indicates that there is a relationship between the two corresponding individuals. The link relationship among vertices can be considered as privacy of individuals.

Link Weight. Some social networks may be weighted. The weights of edges can reflect affinity between two vertices or record the communication cost between two individuals.

Sensitive Edge Labels. In social network data, edges may carry several labels as well. The edge labels may be divided into non-sensitive edge labels and sensitive edge labels. The values of sensitive edge labels are considered as privacy for the corresponding two individuals.

Graph Metrics. In social network analysis, many graph metrics have been proposed to analyse graph structures, such as between-ness (that is, the degree an individual lies between other individuals in the network directly or indirectly), closeness (that is, the degree an individual is near to all other individuals in the network directly or indirectly), centrality (that is, the count of the number of relationships to other individuals in the network), path length (that is, the distances between pairs of vertices in the network), reachability (that is, the degree any member of a network can reach other members of the network), and so on.

6 Background-Knowledge Attacks

A family of attacks have been identified by Backstrom, Dwork and Kleinberg [2] such that even from a single anonymized copy of a social network hiding the identifying attributes, it is possible for an adversary to learn whether edges exist or not between some specific target pairs of vertices. The attacks are based on the uniqueness of some small random subgraphs embedded in an arbitrary network using ideas related to those found in arguments from Ramsey theory. Two categories of attacks are addressed in [2]. The first category is active attacks. Before releasing the anonymized network **G** of *(n-k)* vertices, the attackers can choose a set of b target users; randomly create a subgraph H containing *k* vertices and then attack H to the target vertices. After the anonymized network is released, if the attackers can find the subgraph **H** in the released subgraph **G**, then the attackers can follow edges from **H** to locate the *b* target vertices and their locations in **G** and determine all edges among those *b* vertices. The experiments by Backstrom [2] on a real social network with 4.4 million vertices and 88 million edges show that the creation of 7 vertices by an attacker can reveal on an average 70 target vertices and compromise the privacy of approximately 2,400 edges between them.

6.1 Background Knowledge of Adversaries

In relational data, a major type of privacy attacks is to identify individuals by joining the published table with some external tables modelling the background knowledge of the users. Specifically, the adversaries are assumed knowing the values of the quasi-identifier attributes of the target victims. In privacy preservation in publishing social networks, due to the complex structures of the graph data, the background knowledge of adversaries may be modelled in various ways [29]:

Identifying Attribute of Vertices. A vertex may be linked uniquely to an individual by a set of attributes, where the set of identifying attributes play a role similar to a quasi-identifier in the reidentification attacks on relational data. Vertex attributes are often modelled as labels in a social network. An adversary may know some attribute values of some victims. Such background knowledge may be abused for privacy attacks.

Vertex Degrees. The degree of a vertex in the network captures how many edges the corresponding individual is connected to others in the network. Such information is often easy to collect by adversaries. For example, the neighbour of a target individual may easily estimate the number of friends the victim has. An adversary equipped with the knowledge about the victim's degree can reidentify the target individual in the network by examining the vertex degrees in the network.

Link Relationship. An adversary may know that there are some specific link relationships between some target individuals. For example, in a social network about friendship among people, edges may carry labels recording the channels people use to communicate with each other such as phone, e-mail, and/or messaging. An adversary may try to use the background knowledge that a victim uses only e-mails to contact her friends in the network to link the victim to vertices in the network. Privacy attacks using link relationship as the background knowledge are studied.

Neighbourhoods. An adversary may have the background knowledge about the neighbourhood of some target individuals. For example, an adversary may know that a victim has four good friends who also know each other. Using this background knowledge, the adversary may re-identify the victim by searching the vertices in the social graph whose neighbourhoods contain a clique of size atleast four. Generally, we can consider the d-neighbour of a target vertex, that is, the vertices within a distance d to the target vertex in the network, where d is a positive integer.

Embedded Graphs. An adversary may embed some specific subgraphs into a social network before the network is released. After collecting the released network, it is possible for the adversary to re-identify the embedded subgraph if the subgraph is unique. As shown in [16], the creation of 7 vertices by an attacker can reveal an average of 70 target vertices.

Graph Metrics. As mentioned above, graphs have many metrics, such as betweenness, closeness, centrality, path length, reachability, and so on. Graph metrics can be used as background knowledge for the adversaries to breach the privacy of target individuals.

7 Related Work

Before the social networks, many privacy and anonymization models were proposed for relational data. Generalization and perturbation are two basic and popular techniques used to anonymize relational data. In context of social network data, Hey et al. [13] proposed a clustering-based method which clusters vertices and edges into groups and anonymizes a subgraph into a super-vertex. This method ensured that the details about individuals can be hidden properly. Since the released network only contained structural information of the network, the question of uncertainty in analysis arises for the same. Another clustering based approach was proposed by

Zheleva et al. [35]. The focus was to prevent the network from link re-identification attacks. To achieve this, the authors considered two types of edges; sensitive and non-sensitive edges. To protect sensitive edges against adversary attacks two approaches were proposed, either remove sensitive edges or remove some observed edges which passed some criteria or threshold. These clustering based models may reduce the size of original network significantly which compromises the utility of the network.

To improve on these disadvantages, another set of models based on graph modification were introduced. Liu and Terzi [16] considered the identity disclosure scenario where the identities of individuals associated with vertices are revealed. To model the background knowledge of an adversary, the authors considered possible re-identification attacks against individuals by an adversary using the prior knowledge of the degree of a target vertex. An adversary is assumed to know the degree of a target victim. By searching the degrees of vertices in the published network, the adversary may be able to identify the individual, even when the identities of the vertices are removed before the network data is published. Liu and Terzi [16] proposed the notion of graph k-degree anonymity, which is similar to k-anonymity in relational data. Specifically, a graph is said to be k-degree anonymous if for every vertex v in the graph, there exist at least $(k\text{-}1)$ other vertices in the graph with the same degree as v.

Another popular approach for graph anonymization through manipulation is randomization. Hay et al. [13] developed a randomized edge construction method. In this method random edges were added or deleted to the original graph to achieve anonymized graph. Zhou and Pei [36] created a model that prevents graphs from neighbourhood attacks. Neighbourhood attacks are possible when an adversary has knowledge about the neighbours of the user. Zhou and Pei [36] extended the k-anonymity model in relational data to social network data. For a social network **G**, suppose an adversary knows the neighbourhood structure of a vertex $u \in V(G)$, denoted by NeighborG(u). If NeighborG(u) has at least k isomorphic copies in G0 where G0 is an anonymization of G, then u can be re-identified in G0 with a confidence of at most 1/k. This k-anonymity was achieved through a greedy method. In the first step, the algorithm extracts the neighbourhoods of all vertices in the network. The authors developed a neighbourhood component coding technique based on minimal DFS code [36,37] to identify isomorphism in neighbourhoods. In the second step, the algorithm greedily organizes vertices into groups and anonymizes the neighbourhoods of vertices in a group to the same, until the graph satisfies k-anonymity. Later, Zhou and Pei [37] extended [36] and introduced l-diversity into social network anonymization. In this case, each vertex is associated with some non-sensitive attributes and some sensitive attributes. If an adversary can re-identify the sensitive attribute values of one target individual with a high confidence, the privacy of that individual is breached. An l-diverse graph makes sure that the adversary cannot infer the sensitive attribute values with a confidence over 1/l. Wei and Lu [34] proposed another technique to prevent background knowledge based attacks. In order to prevent the adversary from identifying a target individual in the released graph according to the degree of the target and the target's immediate neighbours, they made every vertex in each k-subgraph possess the same structure by adding and deleting some edges.

Tripathy et al. [28] proposed Three phase graph anonymization algorithm which incorporated both clustering and graph manipulation. The model prevents target individual identity from attack by an adversary. The model proposed by Tripathy et al takes into account k-anonymity for network anonymization. The authors used a fast isomorphism technique [28] to identify structurally similar nodes in the graph. In the first phase, similar nodes which satisfy k-anonymity and l-diversity are clustered together. The second phase is adjustment phase, where clusters formed are adjusted. In the third phase, l-diversity is satisfied. The three phase algorithm also works for d-neighbourhood, where $d > 2$. Based on three phase algorithm, GASNA (*Greedy algorithm for social network anonymization*) algorithm was developed by Shishodia et al. [25,28]. The details of the algorithm are given in the next section. The algorithm proposed by Shishodia et al. [25,28] is focused on minimizing the number of edges added to anonymize the graph.

All these techniques have a common disadvantage that original network is changed after manipulation and consequently, the utility of system is decreased. To tackle this, Dwork et al. presented differential privacy in the context of networks. Differential privacy is a property of an algorithm. Informally, it requires that an algorithm be insensitive to small changes in the input, such as the addition or removal a single record. The formal definition uses the concept of neighbouring databases: two databases D and D' are neighbors if they differ in at most one record, i.e., $|(D - D') \cup (D - D')| = 1$ [13]. In networks, graphs differing by a single edge are assumed to be neighbouring graphs. Using this approach, a differentially private algorithm essentially protects against edge disclosure. Differential privacy places no limiting assumptions on the input or on adversary knowledge. An adversary with knowledge of all of the edges in the graph except one cannot use the differentially private algorithm to infer the presence or absence of the unknown target edge.

8 Problem Formulation

The neighbourhood graphs of nodes in the published social network data must satisfy k-anonymity and the sensitive attribute values of these k-anonymous nodes must satisfy l-diversity.

9 GASNA: Greedy Algorithm for Social Network Anonymization [24,27]

The major advantage of three phase algorithm was anonymization for $d = 2$ neighbourhoods, but the cost to anonymize the graph was high [28]. Cost to anonymize a network is directly proportional to number of edges added to make the network anonymous. Moreover, there was no concrete algorithm provided to add/remove edges. Similar to three phases, this algorithm also has three phases. These are the clustering phase, the adjustment phase and the anonymization phase [25,28].

9.1 *Clustering Phase*

The clustering phase takes care of grouping similar nodes together into clusters. This similarity is calculated based on the cost function mentioned in previous section. The nodes having low similarity cost are clustered into one group. Similarity is calculated as follows:

9.1.1 Finding the Similarity between Neighbourhood Structure of Two Nodes

Algorithms have been proposed for determining whether two given nodes have isomorphic neighbourhoods. But these algorithms are very costly. To reduce this cost, we propose a new method to find the similarity between neighbourhoods of two nodes. This method, which calculates the number of edges that must be added to make the neighbourhood of two nodes isomorphic, can be extended for $d > 1$. Lower the cost, higher is the similarity. For making structure of a node X similar to another node Y for neighbourhood 'd', the cost of adding an edge to a node at a distance '*dis*' from X is '$d - dis + 1$'.

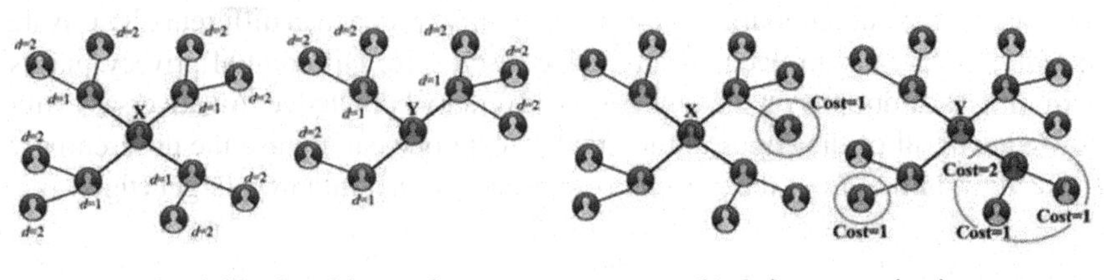

(a) d-neighborhood in graph (b) d=2; anonymization

Fig. 4

In order to make the nodes **X** and **Y** structurally similar for $d = 1$ neighborhood, one more edge must be added to **Y**. Hence the cost is 1. To make **X** and **Y** similar for *d=2*, we have to (i) add a new node of degree 3 to **Y** (cost=2+1+1), (ii) convert one of the neighbors of **X** of degree 3 to degree 4 by adding an edge (cost=1) and (iii) convert one of the neighbors of **Y** of degree 2 to 3 by adding an edge (cost=1). The total cost is 6. The size of each clustered is dependent on the values of *k* for *k*-anonymity and *l* value for *l*-diversity. Maximum size of a cluster can be *k+l-1* and the minimum size should be *k*. This constraint is kept so that each cluster has at least '*k*' similar nodes. Each element within a cluster is anonymized in the third phase. In the example given, 20 nodes of the network are input to clustering phase. After the clustering phase, nodes which are denoted through same color belong to same cluster. We obtained 5 clusters in this phase.

9.2 *Adjustment Phase*

The adjustment phase guarantees that every cluster has at least k nodes. The clusters formed in early iterations of the clustering algorithm have lower chances of having less than k nodes compared to the clusters formed towards the end of clustering phase. This is due to the fact that clusters formed in initial stage of clustering algorithm have enough nodes to satisfy k-anonymity and l-diversity, but the clusters formed in the end may not satisfy this die due to scarcity of nodes. The adjustment phase algorithm starts from the last cluster formed. The clusters having less than k nodes are merged together so that the resulting cluster has at least k nodes. If there is only one cluster left with less than k-nodes, its nodes are moved to the cluster whose nodes have maximum similarity with respect to that node. In the example given, there is no such cluster which does not satisfy k-anonymity and l-diversity, so adjustment is not required.

NOTE: There are chances that even if we add together the clusters having lesser nodes than k-each, the size of the resulting cluster may not come up to k. In such a case, nodes from the nearest cluster of size more than k can be moved to this cluster to fill it up to k elements. Same goes for l-diversity.

Table 3 Naíve Snonymized Network

Patient name	ID	Disease	Degree
Peter	8	HIV	11
Cindy	5	Cancer	7 4
Fred	19	None	5 6
Garry	16	Diarrhea	5 6
Oprah	9	HIV	5
Ron	18	HIV	5
Tyler	17	Cancer	4 1
Steve	14	Cholera	41
Kevin	11	Cancer	4
Leo	10	HIV	4
Neo	3	Diarrhea	4
Ian	13	Cancer	3 1
Harry	15	None	3
Mark	20	Diarrhea	3
Eric	7	HIV	2 1
Bob	2	None	21
Alice	1	Cancer	22
Matt	4	Cholera	22
Joe	12	HIV	22
David	6	Cancer	22

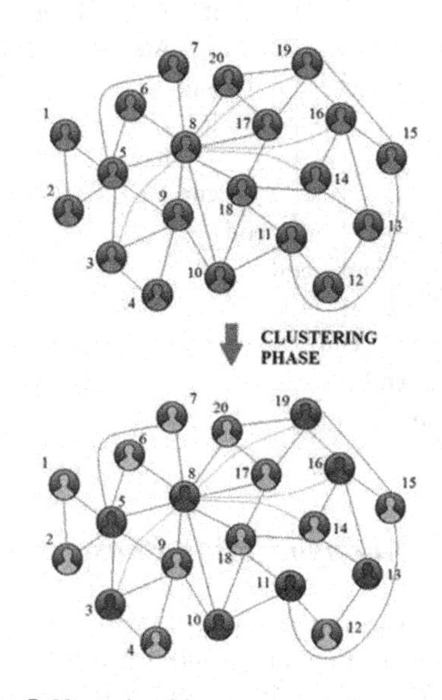

Fig. 5 Clustering Phase

```
Algorithm 2.
Data: A set T of n-nodes, the value k for k-anonymity, the value l for l-diversity and the
      value d for d-neighbourhood. Let T_i be the i^th node in the input graph.
Result: Clusters {C_1,C_2,...,C_m}; m ≥ ⌊n/(k+l−1)⌋
Order {T_i} according to degrees in descending order. Store this in array Q;
Let x=0 and cid (cluster id) = 0;
Mark visited [T_i] = 0 ∀i = 0,1,2,...,n−1;
j=0;
while (j < n) do
    if (visited[Q[j]] == 0) then
        Add Q[x] to C_cid;
        Let p = 0, y = x+1 and ca[ ] be ascending cost array.;
        while (y < n) do
            if (visited[Q[y]] == 0 and distance(Q[y], {each element in
            C_cid}) > d−1) then
                Calculate cost i.e. cost_y;
                Insert the cost into ca[];
            end
            for each element in ca[] as ca[z] do
                Let flag = 0;
                while (flag == 0) do
                    Let s(C_cid) be the set of distinct sensitive attribute values of the
                    nodes in C_cid.;
                    Let s[z] be the sensitive attribute value of ca[z].;
                    if ((|C_cid < k) or ((s_[z] ∉ s(C_cid)) and (|s(C_cid)| < l)) then
                        Add ca_[z] to C_cid.;
                    end
                    if ((|C_cid ≥ k) and (|s(C_cid ≥ l)) then
                        flag =1;
                    end
                end
            end
            j++;
        end
    end
end
```

9.3 Anonymization Phase

The nodes are anonymized with respect to other nodes in their respective clusters. This is the final phase and is involved with edge addition/deletion. This can be done in various ways. We provide below 3 such approaches for *d=1* and *d=2*.

Algorithm 3.

```
Data: Clusters {C1,C2...,Cm}; m ≥ ⌊n/(k+l−1)⌋
Result: Adjusted clusters, each of minimum size k
Let i = m − 1;
while (i ≥ 0) do
    if (|Ci| < k) then
        j = i − 1;
        while (j ≥ 0) do
            p = 0;
            while (p < |Ci|) do
                Calculate the cluster having elements with maximum similarity with
                respect to element Ci[p] and shift Ci[p] to that cluster.;
                p++;
            end
        end
    end
    i−−;
end
```

9.3.1 Edge Addition by Introducing Fake Nodes in the Cluster (AFN)

In this technique, during anonymization phase fake nodes are added to the clusters to anonymize the nodes within. The degree of fake nodes is taken as maximum of the most probable degree (The degree which maximum number of nodes in the graph possess) and the average degree of the graph. This is done to ensure that average degree of graph and the structural properties of graph are maintained. To achieve this, the edges are added to connect these fake nodes with the existing needy nodes till the degree of the former crosses this most probably degree value, following which a new fake node is created. Edges are added between the fake nodes and nodes of the clusters which require higher degree to achieve anonymity. In some cases, the degree required by all the nodes in clusters may be satisfied, but the fake nodes themselves might not be anonymized. To anonymize fake nodes we connect them to one another even if their degree exceeds the maximum degree possible. This would not affect our original network because these nodes are redundant information in themselves.

9.3.2 Edge Addition to Existing Nodes (AEN)

Another approach to satisfy degree requirements of nodes which are to be anonymized is through addition of edge among them. In this way, we are fulfilling the degree requirement of two nodes simultaneously by adding a single edge between them if they are not connected already. For instance, consider two nodes **X** and **Y** such that they belong to different clusters and are not connected already. An edge is added between them to satisfy degree requirements of both the nodes and consequently, nodes are anonymized.

Algorithm 4.

```
P = Degree possessed by maximum number of nodes.;
Let i = 0, j = 0, c = 0;
while (i < m) do
    D = max_degree(C_i);
    while (j < |C_i|) do
        Let da[] be 'degree to be added' array;
        da[c] = D - degree(C_i[j]);
        nodes[c] = C_i[j];
        c++;
        j++;
    end
    i++;
end
Order nodes[] and da[] according to their corresponding da[] values in descending order.;
Let s = Σ_{i=0}^{m} da[i];
Let a = ⌊s/P⌋ and b = da[0] i.e. max element in da[];
c = max(a,b);
Create an array fn[c] with c fake nodes containing (⌊s/P⌋ − 1) nodes of value P_d and 1 node of value P_d +
Smod(P_d);
i = 0;
j = 0;
while (j < n) do
    if (da[j] == 0) then
        j++;
    else
        if (fn[i] == 0) then
            i = (i+1)mod(c);
        else
            Connect node[j] with i^th fake node.;
            da[j]−−;
            fn[i]−−;
            i = (i+1)mod(c);
        end
    end
end
least_anonymized_degree=degree(C[m-1][0]);
i = c−1;
while (i > 0) do
    j = i−1;
    while ((P-fn[i])< lad) do
        fn[i]−−;
        connect fake_node[i] and fake_node[j];
        j = (j−1)modc;
        if (j == i) then
            Break;
        end
    end
    i−−;
end
```

If a node's degree requirement has not been satisfied and there is no other suitable node to which it can be connected, then connect that node to a node of nearest suitable cluster with odd number of elements if the degree requirement is odd. Connect it to the nearest suitable cluster with even number of elements if the degree requirement is even. This allows the rest of the nodes in the cluster to connect to each other and satisfy their mutual degree requirements. If a node's degree requirement is

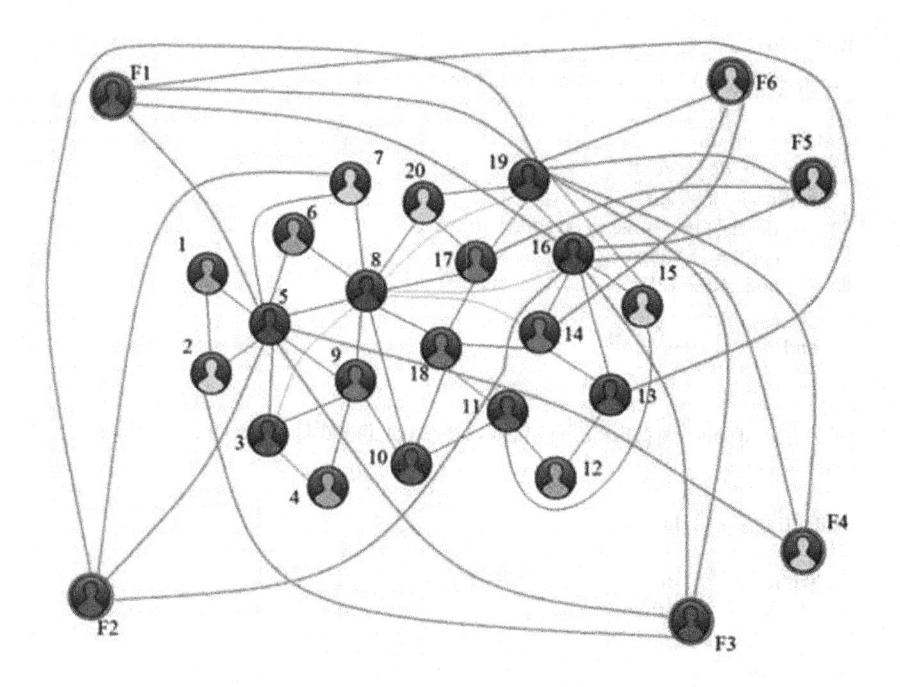

Fig. 6 Network Anonymized by Method of Addition of Fake Nodes

odd and there is no cluster with odd number of elements OR if its degree requirement is even and there is no cluster with even number of elements, the cost will increase significantly. A few fake nodes may need to be added to decrease the number of edges added otherwise.

If the degree of such a node, **X**, is odd (Let the degree be D), find the nearest cluster with at least D nodes that are not connected to **X**. Let nodes $\{Y_1, Y_2, Y_3, \ldots, Y_D\}$ be such nodes in the nearest cluster **C**. Check whether the remaining nodes of **C** can be connected to each other till **C** is left with only one node un-anonymized. If so, connect **X** to the nodes $\{Y_1, Y_2, Y_3, \ldots, Y_D\}$ and the remaining nodes of **C** to each other till only one node is left unanonymized in **C**. A fake node is created to anonymize this node and then the fake node itself is anonymized by connecting it to all nodes of the cluster of smallest size. However, if the remaining nodes of **C** cannot be connected to each other, then find another combination of nodes in **C** which can be connected to **X** or move on to the next nearest cluster. If the degree of the node left unanonymized is even and there is no cluster with even number of nodes, connect this node to an element of the nearest cluster. Connect the remaining elements of the cluster to each other. Repeat the process if this cluster also has a node left with even degree.

Algorithm 5.

```
P = Degree possessed by maximum number of nodes.;
Let i = 1; j = 0;
while (i < m) do
    D = max_degree(C_i);
    while (j< |C_i|) do
        Let da[] be degree to be added array;
        da[c] = D - degree(C_i[j]);
        nodes[c] = C_i[j];
        c++; j++;
    end
    i++;
end
Order nodes[] and da[] according to their corresponding da[] values in descending
order.;
i = 0;
while (i < n-1) do
    j = i+1;
    while (j < n) do
        if (i == j) then
            j++;
        else
            if (da[i] == 0) then
                Break;
            else
                if (da[j] == 0) then
                    j++;
                else
                    if ( i^th and j^th nodes are not connected) then
                        connect i^th and j^th nodes;
                        da[i]--;
                        da[j]--;
                        j++;
                    end
                    j++;
                end
            end
        end
    end
    i++;
end
```

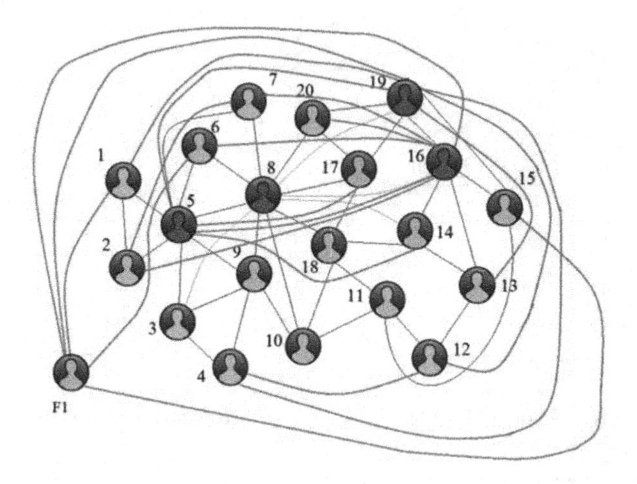

Fig. 7 Network Anonymized by Adding Edges between Existing Nodes

9.3.3 Removal of Edges

This method is used in combination with 9.3.1 and 9.3.2. If a node has higher degree than the majority of the nodes in its cluster, then this extra degree is removed by deletion of an edge from that node. This edge to be deleted is chosen as the one connecting to another node which must also lose degree. This method lowers the number of fake nodes/ fake edges to be added but can result in loss of important information when the edges are deleted. We can apply one or more of the above algorithms to make the nodes similar so that they can be anonymized. The above algorithms handle the one neighbourhood case. In the next subsection, we show how these algorithms handle the two neighbourhood case.

10 Empirical Evaluation of GASNA

In this section, we present the results obtained when GASNA was applied to social network data generated by RMAT [6]. R-MAT can generate graphs with power law vertex degree distribution and small-world characteristic, which are two of the most important properties for many real-world social networks.

R-MAT takes 4 probability parameters, namely *a,b,c* and *d*. We used the default values of 0.45, 0.15, 0.15 and 0.25 for these four parameters, respectively as was done in [34, 36]. We generated a series of synthetic data sets by varying the number of vertices from 128 to 4096 and average degree from 3 to 7. For sensitive attribute values, we generated a list of 32 city names and equally distributed them among the nodes. We have evaluated our algorithm for $d=1$ for d-neighborhood.

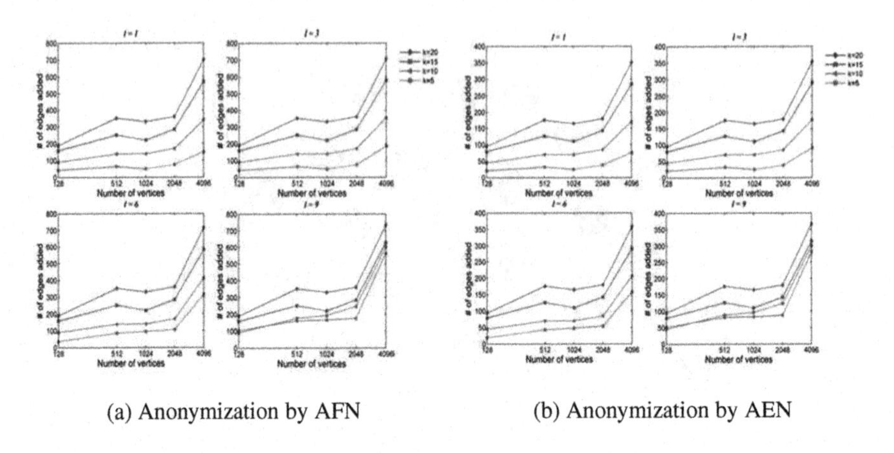

(a) Anonymization by AFN (b) Anonymization by AEN

Fig. 8 Anonymization

The complete results have been provided in [25]. Here, we provide only for the generated data with average degree of node = 7. Fig. 8a shows the results for AFN and fig. 8b for AEN.

The number of edges added in AEN is almost half than that in AFN. This is because AEN involves addition of edges between existing nodes which require additional degree(s) and are not already connected to each other. However, the number of edges is not always exactly because additional edges may be added for special cases such as anonymization of fake nodes (in AFN) and unsatisfied nodes (in AEN). This special case in AEN occurs when there is only one vertex left which requires additional degree or when there are multiple vertices, already connected to each other, requiring additional degrees [25].

One interesting result is the increase in number of edges added when the value of l for l-diversity is higher than the value k for k-anonymity. This happens because when the number of vertices in the cluster reaches k, only those vertices can further be added which have a sensitive attribute value not already present in the cluster. There may be nodes (let us represent them by a set **A**) which were structurally similar to those k nodes already in the cluster (let us represent them by a set **X**), but they cannot be added because they share a common sensitive attribute value with some node in the cluster. Nodes which are not that similar structurally are added to the cluster to make it l-diverse. This increases the cost. Moreover, the nodes **A** which should have been in this cluster now go into another cluster. Again, the same thing happens and structurally less-similar nodes are added to that cluster to make it l-diverse. The number of edges added is further increased.This effect is less prominent when $k > l$ because here, the nodes in **A** and **X** can be clustered together. After that, other nodes can be added to make the cluster l-diverse [25].
We also present the results when GASNA was applied on real world data. The real data set we used was a co-authorship network of scientists working on network theory and experiment, as compiled by M. Newman in May 2006 [23]. The version

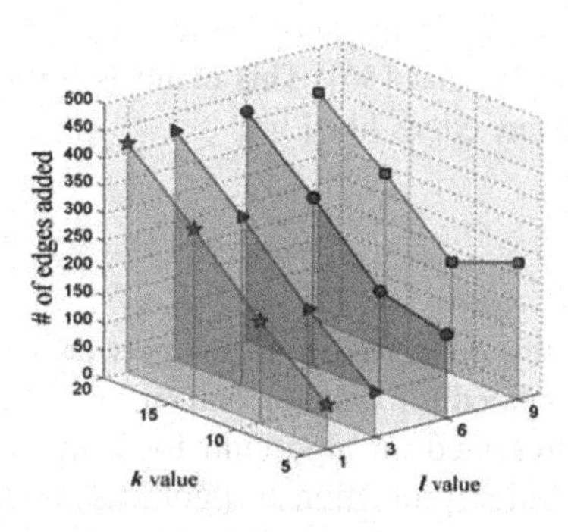

(a) Number of edges deleted for anonymization of real data by AFN

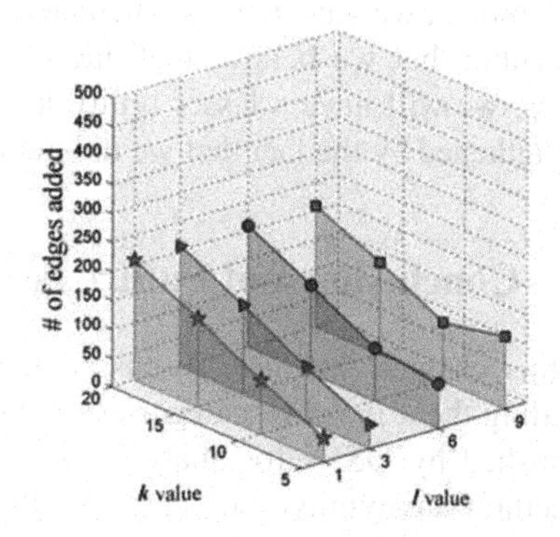

(b) Number of edges added for anonymization of real data by AEN

Fig. 9

used contained all components of the network, for a total of 1589 scientists. For *l*-diversity, we took the first letter of the first name of the scientist as the sensitive attribute value. For example, the scientist ABRAMSON, G. was given a sensitive attribute value 'G'. Making the 26 letters of the English alphabet as the domain of the sensitive attribute values is a good test for GASNA which aims to achieve both *k*-anonymity and *l*-diversity

11 Comparison with the Literature

In this section, we present the comparison of GASNA with [34] and [36].

Table 4: Since other researchers have not used fake nodes, we have compared the results of GASNA AEN with other papers in the literature. Although we did not run the program for 5000 nodes, the negative slope of lines in Fig. indicate that GASNA will require significantly less number of edge additions even for 5000 nodes. However, it must be noted that in this comparison, GASNA has been applied only for degree-based attacks.

Table 4 Comparison of Result

Avg. Deg.	k-anonymity	Zhou Pei(5000 nodes)	Wei Lu (5000 nodes)	GASNA (4096 nodes)	
				l=1	l=9
3	**k=5**	100	80	26	134
	k=20	700	500	149	155
7	**k=5**	1000	750	76	301
	k=20	2400	1900	351	367

Although we have not yet calculated the generalization cost (utility loss) for our algorithm, but we believe that due to addition of significantly lesser number of edges, we will have a lesser utility loss than [34] and [36]. This claim is further strengthened by the fact that we did not delete any edges like [34].

12 Conclusion and Future Issues

In this chapter, we have discussed the privacy concerns in social networks by dividing them into two categories: (i) user provided information and (ii) user data published by OSNs for analysis. We also presented an algorithm for achieving structural anonymity and sensitive attribute value protection in published social network data. Now, we present some future issues in social network data anonymization. We explain how a fixed value of k for k-anonymity can increase the cost of anonymization. We also present the challenges faced in achieving anonymity in modern social networks and provide insights on dealing with set valued sensitive attributes.

12.1 Varying k Value for k-anonymity According to the Degree of the Nodes Being Considered

In the current algorithms ([25,28]), k is assigned a fixed value by a user. In this approach, the cost of anonymization (number of edges added) can be high. These algorithms add nodes to a cluster till its elements satisfy the values of k and l of k-anonymity and l-diversity respectively ([25,28]). Once this cluster is k-anonymized and l-diverse, the algorithm moves onto create the next cluster. The maximum number of nodes in a cluster after the clustering phase of the algorithm given by Tripathy et al. [25,28] is $k+l-1$. Consider a case where $k=4, l=2$. Now the maximum size of cluster would be 5. Let the d value for neighbourhood be $d=1$ and the unique degree values in the graph, in ascending order, be $d_1, d_2, d_3..., d_{max}$. Suppose there are 6 nodes of degree d_{max} and out of these, five nodes are added to the cluster 1 (maximum cluster size is 5). Now, one node is left with degree d_{max}. For making this node 4-anonymous, at least 3 nodes (minimum cluster size is 4) of degree d_{max}-1 must be promoted to degree d_{max}. If the value of k was flexible, the cluster size would also have been flexible and this 6th node with degree d_{max} could have been added to the cluster 1 itself thereby reducing the cost. Figure 13, shows how the value of k will vary according to the degree of the nodes being considered for a given cluster.

12.2 Anonymity in Modern Social Networks

The current algorithms for achieving k-anonymity in social networks only consider the structural similarity of nodes (users in the network) ([25,28,34,36,37]). However, this is not sufficient to model the modern social networks where many other

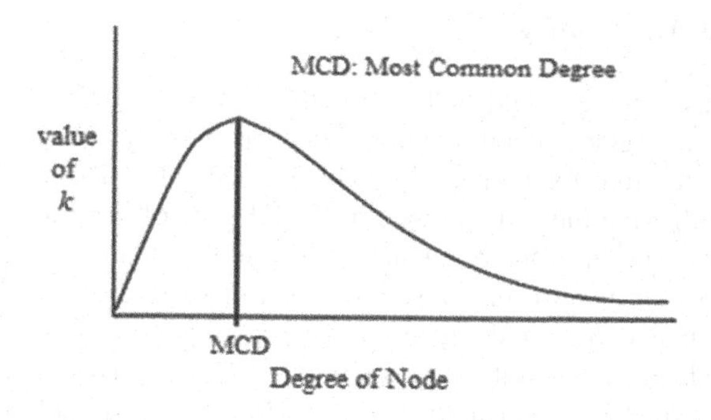

Fig. 10 Variation of k with degree of node being considered

attributes are also attached to these nodes. For example, consider the scenario of Face book, in which, the released user data, apart from friends (links/degree), may also contain other attributes such as likes, groups, family relationships, etc. These attributes link together multiple users; this makes them different from the simple quasi-identifiers in micro data. An adversary can use complete or partial knowledge of these attributes to identify a user even in a structurally anonymized network like the one shown in Table 5.

Table 5 Nodes (users) and their attributes

User	Friends	Relationships	Group	Likes
A	X,Y,Z	Son of X	VIT	L1,L2,L3
B	X,Y,Z	Wife of Y	VIT, MIT	L4
C	X,Y,Z	-	MIT	L1,L4
⋮	⋮	⋮	⋮	⋮
Z	A,B,C,F	-	Stanford	L2,L3,L4

The users *A*, *B*, *C* have same structure (friends with same *X*, *Y*, *Z*), but each of them can be identified using the other attributes such as their likes, groups and relationships with other users. Suppose the adversary needs to find a user in the database. The adversary, knowing that the user is a male and has joined the group called VIT, searches the group and finds two users *A* and *B*. *A*, being"Son of *X*", is a male while *B* is a female. Hence, the adversary can successfully identify *A*. To solve this, each node in a group must have *k*-similar nodes in that group. However, there may be nodes that aren't a member of any group. Hence, each of these nodes must also have *k*-similar nodes in the graph. This solution has to further be expanded for likes, relationships and other attributes.

12.3 Partial Anonymity

The nodes in a cluster are said to be partially anonymized if each ith node in the cluster has a neighbourhood structure which is a subset of $(i+1)^{th}$ node's neighbourhood structure for a given d-neighbourhood value. However, the nodes in the cluster will have the same degree but the degree of their neighbours may differ. Consider a cluster with '**N**' elements. The idea of partial anonymity is to make structure of ith node in the cluster similar to structure of $(i+1)^{th}$ node in it. If there is a change in i^{th} node due to making it structurally similar to $(i+1)^{th}$ node, then this change is not reflected back in $(i-1)^{th}$ node. This way, the number of edges to be added is significantly reduced. However, if this i^{th} node gains a degree, then this change is reflected back on all the previous nodes in the cluster. Hence, this concept of partial anonymity preserves k-anonymity for d=1 and partial neighbourhood structure anonymity for $d=2$ ([25,28]).

References

1. Aggarwal, G., Feder, T., Kenthapadi, K., Motwani, R., Panigrahy, R., Thomas, D., Zhu, A.: Approximation algorithms for k-anonymity. Journal of Privacy Technology (JOPT) (2005)
2. Backstrom, L., Dwork, C., Kleinberg, J.: Wherefore art thou r3579x: anonymized social networks, hidden patterns, and structural steganography. In: Proceedings of the 16th International Conference on World Wide Web, pp. 181–190. ACM (2007)
3. Bader, D.A., Madduri, K.: Gtgraph: A synthetic graph generator suite, Atlanta, GA (February 2006)
4. Bamba, B., Liu, L., Pesti, P., Wang, T.: Supporting anonymous location queries in mobile environments with privacy grid. In: Proceedings of the 17th International Conference on World Wide Web, pp. 237–246. ACM (April 2008)
5. Campan, A., Truta, T.M.: Data and structural k-anonymity in social networks. In: Bonchi, F., Ferrari, E., Jiang, W., Malin, B. (eds.) PinKDD 2008. LNCS, vol. 5456, pp. 33–54. Springer, Heidelberg (2009)
6. Chakraborty, D., Zhan, Y., Faloutsos, C.: R-MAT: A recursive model for graph mining. Computer Science Department, No. 541 (2004)
7. Chawla, S., Dwork, C., McSherry, F., Smith, A., Wee, H.: Toward privacy in public databases. In: Kilian, J. (ed.) TCC 2005. LNCS, vol. 3378, pp. 363–385. Springer, Heidelberg (2005)
8. Evfimievski, A., Gehrke, J., Srikant, R.: Limiting privacy breaches in privacy preserving data mining. In: In Proceedings of the Twenty-Second ACM SIGMOD-SIGACT-SIGART Symposium on Principles of Database Systems, pp. 211–222. ACM (June 2003)
9. Facebook Blog: New Tools to Control Your Experience, Facebook (May 12, 2013), `https://blog.facebook.com/blog.php?post=196629387130` (Web December 9, 2009)
10. Fung, B., Wang, K., Chen, R., Yu, P.S.: Privacy-preserving data publishing: A survey of recent developments. ACM Computing Surveys (CSUR) 42(4), No. 14 (2010)
11. Getoor, L., Diehl, C.P.: Link mining: a survey. ACM SIGKDD Explorations Newsletter 7(2), 3–12 (2005)
12. Gross, J.L., Yellen, J.: Graph theory and its applications. Chapman and Hall/CRC (2006)

13. Hay, M., Miklau, G., Jensen, D., Towsley, D., Weis, P.: Resisting structural re-identification in anonymized social networks. Proceedings of the VLDB Endowment 1(1), 102–114 (2008)
14. LeFevre, K., DeWitt, D.J., Ramakrishnan, R.: Mondrian multidimensional k-anonymity. In: IEEE Proceedings of the 22nd International Conference on Data Engineering, ICDE 2006, pp. 25–25 (2006)
15. HIPPA: Health insurance portability and accountàbility act (2002), http://www.hhs.gov/ocr/hipaa
16. Liu, K., Terzi, E.: Towards identity anonymization on graphs. In: Proceedings of the 2008 ACM SIGMOD International Conference on Management of Data, pp. 93–106. ACM (2008)
17. Lunacek, M., Whitley, D., Ray, I.: A crossover operator for the k-anonymity problem. In: In Proceedings of the 8th Annual Conference on Genetic and Evolutionary Computation, pp. 1713–1720. ACM (2006)
18. Machanavajjhala, A., Kifer, D., Gehrke, J., Venkitasubramaniam, M.: l-diversity: Privacy beyond k-anonymity. ACM Transactions on Knowledge Discovery from Data (TKDD) 1(1), No. 3 (2007)
19. Malin, B.A.: An evaluation of the current state of genomic data privacy protection technology and a roadmap for the future. Journal of the American Medical Informatics Association 12(1), 28–34 (2005)
20. Miklau, G., Suciu, D.: A formal analysis of Information disclosure in data exchange. In: ACM Conference on Management of Data (SIGMOD), Paris, pp. 575–586 (2004)
21. Nergiz, M.E., Clifton, C.: Thoughts on /i /k /i-anonymization. Data and Knowledge Engineering 63(3), 622–645 (2007)
22. Nergiz, M.E., Atzori, M., Clifton, C.: Hiding the presence of individuals from shared databases. In: Proceedings of the 2007 ACM SIGMOD International Conference on Management of Data, pp. 665–676. ACM (2007)
23. Newman, M.E.: Finding community structure in networks using the eigenvectors of matrices. Physical review E 74(3), No. 036104 (2006)
24. Pang, R., Paxson, V.: A high level programming environment for packet trace anonymization and transformation. In: ACM SIGSOMM, Karlsruhe, Germany (2003)
25. Shishodia, M., Jain, S., Tripathy, B.K.: communicated to "GASNA: A Greedy Algorithm for Social Network Anonymization", communicated to ASONAM (2013)
26. Stein, R.: Social Networks' sway be underestimated. Washington Post, No. 26 (May 2008)
27. Sweeney, L.: k-anonymity: A model for protecting privacy. International Journal of Uncertainty, Fuzziness and Knowledge-Based Systems 10(5), 557–570 (2002)
28. Tripathy, B.K., Mitra, A.: An algorithm to achieve k-anonymity and l-diversity anonymisation in social networks. In: 2012 Fourth International Conference on Computational Aspects of Social Networks (CASoN), Brazil, November 21-23, pp. 126–131 (2012)
29. Tripathy, B.K.: Anonymisation of Social Networks and Rough Set Approach. In: Computational Social Networks, pp. 269–309. Springer, London (2012)
30. Xiao, X., Tao, Y.: Anatomy: Simple and effective privacy preservation. In: Proceedings of the 32nd International Conference on Very Large Data Bases, pp. 139–150. VLDB Endowment (2006)
31. Xiao, X., Tao, Y.: Personalized privacy preservation. In: ACM SIGMOD International Conference on Management of Data, pp. 229–240. ACM (2006)
32. Xu, J., Wang, W., Pei, J., Wang, X., Shi, B., Fu, A.W.C.: gspan: graph-based substructure pattern mining. In: ICDM 2002, Maebashi City, China (2002)

33. Yan, X., Han, J.: gspan: Graph-based substructure pattern mining. In: Proceedings of the 2002 IEEE International Conference on Data Mining, ICDM 2003, pp. 721–724. IEEE (2002)
34. Wei, Q., Lu, Y.: Preservation of privacy in publishing social network data. In: 2008 International Symposium on Electronic Commerce and Security, pp. 421–425. IEEE (2008)
35. Zheleva, E., Getoor, L.: Preserving the privacy of sensitive relationships in graph data. In: Bonchi, F., Malin, B., Saygın, Y. (eds.) PInKDD 2007. LNCS, vol. 4890, pp. 153–171. Springer, Heidelberg (2008)
36. Zhou, B., Pei, J.: Preserving privacy in social networks against neighbourhood attacks. In: IEEE 24th International Conference on Data Engineering, ICDE 2008, pp. 506–515. IEEE (2008)
37. Zhou, B., Pei, J.: The k-anonymity and l-diversity approaches for privacy preservation in social networks against neighbourhood attacks. Knowledge and Information Systems 28(1), 47–77 (2011)

On the Use of Brokerage Approach to Discover Influencing Nodes in Terrorist Networks

Nisha Chaurasia and Akhilesh Tiwari

Abstract. Social Network Analysis is a non-conventional Data Mining technique which analyzes social networks on web. The technique is used frequently for studying network behaviors using centrality measures viz. Degree, Betweenness, Closeness and Eigenvector. Hence has also led to the concept of Terrorist Network Mining which aims at detection of the terrorist group, studying the hierarchy they follow for the communication (using SNA) and then finally destabilizing of the network activities. The chapter focuses on an approach under SNA known as Brokerage which finds brokers who serve as the leading nodes in the network. Brokerage is expected to be beneficial in case of estimating the terrorist groups where different subgroups of terrorist organization coordinate to fulfill their awful deeds. The brokerage on whole estimates the influential roles as it would be done by individually calculating the centrality measures, with much more useful information aiding to amend terrorist network analysis.

Keywords: Centrality, Data Mining, Graph Theory, Social Network Analysis.

1 Motivation

The bulk of data in the fast computing world has made Data Mining emerge as a demanding field of research with its valuable and novel approaches. In the current scenario, the wide scale availability of data on web is however an attractive thought but management and retrieval of useful information is not possible manually. Henceforth, Data Mining is seen as a most beneficial solution because of its ability to mine useful information from large amount of data available. The approaches used by Data Mining in this context ranges from Association Rule Mining, Clustering, Classification, and Outlier Analysis to a new emanating one named as Social Network Analysis.

Nisha Chaurasia · Akhilesh Tiwari
Department of CSE & IT, MITS Gwalior, India
e-mail: chaurasianisha21@gmail.com, atiwari.mits@gmail.com

M. Panda, S. Dehuri, and G.-N. Wang (eds.), *Social Networking*
Intelligent Systems Reference Library 65,
DOI: 10.1007/978-3-319-05164-2_11,

Social Network Analysis is generally considered as a data mining technique that is used to investigate the social networks present on the web. Accordingly, Social Network Analysis (SNA) has become a widely applied method in research and business for inquiring into the web of relationships on the individual, organizational and societal level [23, 24]. SNA has been extensively preferred as the analysis technique because of its powerful estimation methodology known as centrality measures. The centrality measures are the network properties which serve as the essentialities of the SNA technique. Several centrality measures are defined such as degree, betweenness, closeness, and eigenvector that can suggest the importance of node in a network [22].

The threat of terrorism from ages has made the various law enforcement agencies to work interminably in order to diminish the annoyance. Though there awfulness has presented their footmarks time to time, the severity in developing era has been noticed when they exploited internet for fulfillment of 9/11 attacks. Since then, the law enforcement agencies incorporated the concept of Social Network Analysis (SNA) to detect these illicit people (terrorists).

When the analysis done by SNA is applied for investigation of terrorist networks that are hidden among the various social networks, it is recognized as Investigative Data Mining (IDM). The IDM is considered as a combination of SNA and the Graph Theory which together perform the link and network analysis using various measures and concepts appropriately. On one hand, SNA defines the role for each user; Graph Theory on other presents the network as the graph and the corresponding network users as the nodes in the graph.

Though IDM is now intensively used for the estimation of terrorist groups on web using Data Mining as the backbone, it is also known as Terrorist Network Mining. Terrorist Network Mining can be well understood as the Data Mining process that adequately uses SNA techniques and Graph Theory for the investigation of hidden terrorist networks on web.

Apart from the various successful steps put in the direction of detecting the terrorist groups and neutralizing their network activities on web, terrorist network mining has made law enforcement agencies to learn more about behavior followed by the various terrorist groups.

The chapter presents the enhancement to the Terrorist Network Mining by introducing the concept of brokerage. Brokerage is a SNA approach that aids in determining the important nodes along with the frequency of interaction possible among the nodes in the network. In simple words, it is the analysis of the broker and their possible persistence while communication activities are performed in a network.

2 Background

The strength of the terrorists groups on web came into shed by 9/11 attacks which left the everlasting memories of their inhumanity. The attack made the intelligence and the law enforcement agencies to concern more strongly about the security as the success of the attack involved the intensive use of the internet. This is one reason for the major effort made by law enforcement agencies around the world in gathering information from the Web about terror- related activities [4]. The goal of gathering information was to prevent the future attacks possibilities.

The very first attempt in order to analyze the terrorist networks was made by Valdis Krebs in 2002 after September 2001 attacks. He used network analysis to provide an extensive analysis of the 9/11 Hijackers network, explains three problems he encountered very early on drawing on the work of Malcolm Sparrow [3]. He mentioned three problems that a social network analyst would encounter while constructing the terrorist organizations network graph. These were [3]:

1. *Incompleteness* - the inevitability of missing nodes and links that the investigators will notuncover.
2. *Fuzzy boundaries* - the difficulty in deciding who to include and who not to include.
3. *Dynamic* - these networks are not static, they are always changing.

Krebs on the basis of the knowledge he gained about the September, 2001 framed a network consisting of terrorist nodes and evaluated the importance and the contribution of each node in the attacks. The major contributors for his study were the priori study done by:

1. Malcolm Sparrow (1991), examining the application of SNA to criminal activity.
2. Wayne Baker and Robert Faulker (1993), suggests looking at previously stored or known data to find the relationship data.
3. Bonnie Erickson (1981), explaining the importance of trusted prior contacts (that came in touch before) for the effective functioning of a secret society (such as terrorist groups).

Krebs during his study of terrorist networks evaluated the links in the network on the basis of their strength. The strength of the tie depends on the time spent by users together. He categorized the tie or strength on three scales:

1. Strongest tie link reveals cluster of network players or leaders of the group.
2. The node pair with the strongest tie would largely be governing the group.
3. Moderate strength or medium thickness links reveals the nodes through which maximum transactions are done or information is forwarded.
4. Weak tie or the thinnest links reveals the nodes having a single transaction, or an occasional meeting and no other ties.

Using the SNA centrality measures, Krebs evaluated the involvement of each node or user in the attacks. This beneficial step helped law enforcement agencies to detect terrorist networks more effectively. However the centrality measures are significant for the analysis but these are very sensitive to minor changes in network connectivity [1]. After the successful analysis of the criminal network, Kreb discussed the shortcomings that were to be considered. He discussed about the difficulty faced in discovering of links that may be the stronger ties but because of their low frequency of activation they may appear to be weak ties. The second consideration was about the network detection as the less active the network, the more difficult it is to discover [1].

The next beneficial and novel effort for terrorist group detection on web was made by Y.Elovici, A.Kandel, M.Last, B.Shapira and O. Zaafrany in 2004. They as a group analyzed the behavior of terrorists on web using data mining techniques. They tried to solve the problem still faced about the dynamic switching of IP addresses and URLs by terrorist users. Hence in place of tracking terrorists on the basis of their IP addresses, they proposed a methodology by monitoring on all the ISPs traffic to detect the users accessing the terrorists" related information, keeping in mind the privacy issues. They defined three essential design goals to be fulfilled for the methodology. These are [4].

1. Training the detection algorithm should be based on the content of existing terrorist sites and known terrorist traffic on the Web.
2. Detection should be carried out in real-time. This goal can be achieved only if terrorist information interests are presented in a compact manner for efficient processing.
3. The detection sensitivity should be controlled by user-defined parameters to enable calibration of the desired detection performance.

Y.Elovici et al. methodology also includes integration of three research fields to detect as well to evaluate typical terrorists behavior. The three topics were the following:

1. *Computer Security*: It was performed using Intrusion Detection System (IDS). An IDS purpose is to detect abnormal actions if any, in the constantly monitored environment. It could an individual system, or number of computers in the network or the network itself. It tracks the activity of users on web and evaluates measures viz. accuracy, completeness, performance, efficiency, fault tolerance, timeliness, and adaptivity. Along with these, the interest measures are true positive rate (TP, the percentage of intrusive or abnormal activities, such as terror-related pages viewed, detected by the system), false positive rate (FP, the percentage of normal actions, such as the pages viewed by normal users, incorrectly detected as intrusive by the system. Accuracy which is the percentage of alarms found to represent abnormal behavior out of the total number of alarms [4].
2. *Information Retrieval*: For this the vector space model is used. The content of the web page is represented as a document in form of an n-dimensional vector. The similarity among two documents is calculated using any of the distance measuring methods such as Euclidean distance or Cosine. The cosine distance evaluated the similarity between an accessed web page and a given set of terrorists" topic of interest. The terrorists" interests are represented by several vectors where each vector relates to a different topic of interest [4].
3. *Data Mining*: This research field involved uses unsupervised clustering to cluster or partition the web pages as collection belonging to terrorists" with

same topic of interest. Cluster analysis is the process of partitioning data objects (records, documents, etc.) into meaningful groups or clusters so that objects within a cluster have similar characteristics but are dissimilar to objects in other clusters [9]. For each collection of web pages or a cluster, a centroid is determined and represented by the vector space model.

After combining these three fields as the steps, learning of Typical-Terrorist-Behavior is obtained. Typical-Terrorist-Behavior depends on the number of clusters. When the number of clusters is higher, the Typical-Terrorist-Behavior includes more topics of interest by terrorists where each topic is based on fewer pages [4]. Next step to this is to monitor the ISP traffic and detect the terrorists present on the web. For this purpose, each accessed web page by user is converted into vector named access vector using a vector generator. The access vector is then compared for similarity against all centroid vectors of the Typical -Terrorist-Behavior using cosine measure.

If the similarity among the two is higher a predefined threshold then an alarm is generated by the detector indicating user who accessed those web pages as illicit user. The sensitivity of the detection process depends on this predefined threshold value. Higher value of threshold will decrease the sensitivity of the detection process, decrease the number of alarms, increase the accuracy and decrease the number of false alarms. Lower value of threshold will increase the detection process sensitivity, increase the number of alarms and false alarms and decrease the accuracy. The optimal value of threshold, T depends on the preferences of the system user [4]. The similarity may be calculated using the inequality:

$$max\left[\frac{\sum_{i=1}^{m}(tC_{i1}.tA_i)}{\sqrt{\sum_{i=1}^{m}tC_{i1}^2\cdot\sum_{i=1}^{m}tA_i^2}},\cdots,\frac{\sum_{i=1}^{m}(tC_{in}.tA_i)}{\sqrt{\sum_{i=1}^{m}tC_{in}^2\cdot\sum_{i=1}^{m}tA_i^2}}\right] \quad (1)$$

The whole process can be viewed through the Fig. 1 and Fig. 2 [4].

Next step to this research was made by Nasrullah Memon and Abdul Rasool Qureshi in 2005 by introducing a data mining tool named Investigative Data Mining (IDM) for the terrorist network analysis. As opposed to traditional data mining aiming at extracting knowledge form data, mining for investigative analysis, called Investigative Data Mining (IDM), aims at discovering hidden instances of patterns of interest, such as patterns indicating an organized crime activity [6]. An important problem for identifying of terror/crime networks was resolved by IDM, based on available intelligence and other information. It is also capable to determine the subgroups if present in the network.

In contrast to traditional data mining where mining is performed on large databases, the IDM analyses and predicts the risks from the access behavior and the association existing among the terrorists network. With this, it has also been a helping hand in the destabilizing of the network.

Nasrullah Memon and Henrik Legind Larsen in 2006 proposed the two highly beneficial algorithms for the destabilizing of the terrorist networks. N. Memon along with his fellow mates, after the deep study about the criminal networks proved various possible solutions to enhance the strategies for the efficient detection of terrorist networks on web.

He proposed the hierarchy of two hierarchical algorithms such that destabilization is achieve in a promising manner. The algorithms basis was the SNA approach along with the measures defined previously. The first algorithm objective was to convert the network undirected graph obtained by SNA approach into a directed graph. Meanwhile the second algorithm works specifically for destabilization by constructing a tree from the graph, by calculating the dependency of each node to other nodes in the network. They also introduced a new centrality measure called dependence centrality. The dependence centrality (DC) of a node is defined as how much that node is dependent on any other node in the network [10, 12]. The nodes with less DC are predicted as the key player (leader or the gateway) nodes as they are the nodes with highest number of direct links to other nodes and do not depend on any other nodes in the network for communication.

Same year, Muhammad Akram Shaikh, Wang Jiaxin presented the effectiveness of IDM along with its working in identification of key nodes in the network. They defined IDM as "The technique which is used for determining associations and predicting criminal behavior by analyzing network structure in order to identify key nodes for the purpose of destabilizing criminal/terrorist networks" [13]. According to the study, IDM offers the capability to track the criminal network more effectively and also to analyze the interaction patterns within the network. The IDM generally uses the graph theory and the SNA techniques for the estimation of the network.

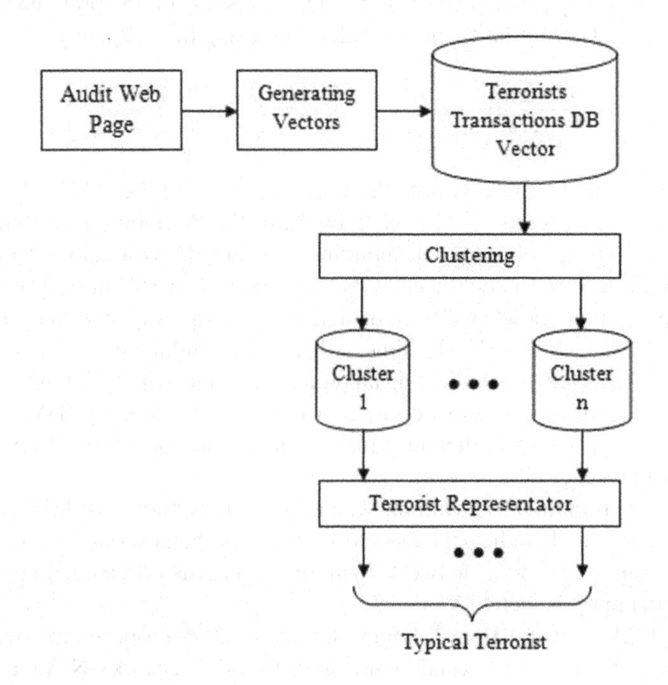

Fig. 1 Learning of the terrorist behavior

Using the SNA technique and Graph theory, IDM considers the network as an adjacency matrix of size n x n, where n represents the number of vertices in the network. An adjacency matrix, A is defined as the matrix with value 1 if two nodes are connected otherwise set with value 0, written as:

$$A_{ij} = \begin{cases} 1 \text{if} i and j are connected \\ 0 \text{if} else \end{cases} \tag{2}$$

Also the matrix is a symmetric matrix i.e.

$$A_{ij} = A_{ji} \tag{3}$$

For example, consider the network (Fig. 3) as a graph with each network user as node of the graph.

The adjacency matrix of the above shown graph is shown in Table 1.

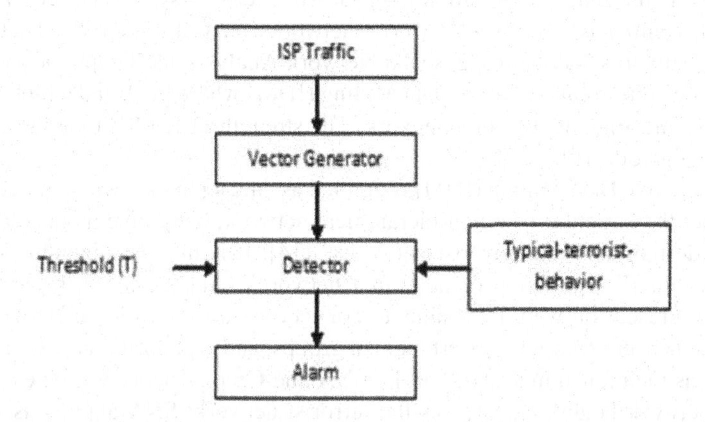

Fig. 2 Detection of terrorist networks on web

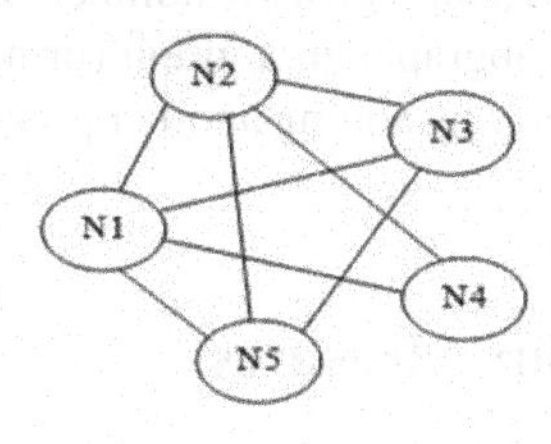

Table 1 Adjacency Matrix

	N1	N2	N3	N4	N5
N1	-	1	1	1	1
N2	1	-	1	1	1
N3	1	1	-	0	1
N4	1	1	0	-	0
N5	1	1	1	0	-

Utilizing the adjacency matrix, the SNA uses the following centrality measures to determine the importance of each node in the graph.

1. Degree defines the leader or the hub of the network.
2. Betweenness finds the extent to which a particular node lies between other nodes in a network.
3. Closeness, unlikely to betweenness, estimates the extent of farness of a node with respect to other nodes in a network.
4. Eigenvector defines the influence of a node on its neighbouring nodes.

An elaboration to these centrality measures is covered in succeeding sections

3 Terrorist Network Mining

Because terrorists operate worldwide, data associated with their activities will be mixed with the data about people who are not terrorists [11]. Hence, Terrorist Network Mining is a study of terrorist networks on web where the analysis of terrorist networks is done using Social Network Analysis and Graph theory. It is a term evolved from Investigative Data Mining (IDM) where SNA and Graph Theory help in connecting the dots in a network. The strength of IDM is to assist analysts and investigators [16].

Investigative Data Mining (IDM) is defined as "the technique which models data to predict the structure of a non-hierarchical network, determine associations and help in destabilizing the terrorist networks" [14]. It is the investigation of social networks for determination of abnormal networks hidden among the legitimate networks present on web. IDM aims to connect the dots between individuals and map and measure complex, covert, human groups, and organizations [18].

IDM as mentioned makes use of the SNA and Graph Theory for the estimation of the network being suspected as the terrorist network. SNA applies its various approaches that help in assessment of the roles played by the network users as leader or gatekeeper. Graph Theory gives a number of concepts and procedures that aims to detect maximal subgraphs in a graph (or network) that have a certain property and looses this property by adding another point and its relationships to the subgraph [14].

4 Calculating Centrality Measures

The general notion of centrality encompasses a number of different aspects of the "importance" or "visibility" of actors within a network [15]. The centrality of a node in a network is interpreted as the importance of the node [21].To understand and analyze how actually the underlying centrality measures of SNA are calculated, a dataset of 26/11, Mumbai attacks held in 2008 is taken into account. All the centralities are calculated considering the dataset and as a result revealing the role possibilities for each attacker.

The dataset consists of 13 attackers of Lashkar-e-Taiba (LeT), involved in the Mumbai attacks. Out of the 13 attackers, 10 were responsible for carrying out the attack while 3 other operated simultaneously from Pakistan. The 10 attackers divided themselves into five groups targeting 5 locations in Mumbai viz. Taj Mahal Hotel; the Oberoi Trident Hotel; Caf Leopold; Chhatrapati Shivaji Terminus; and the Nariman (Chabad) House. Table 2 lists the names of these 13 terrorists along with their location of operation during 26/11.

Table 2 26/11 Attackers Names and their Location of operation

Sl. No.	Name of Terrorist	Location
1	Abu Kaahfa	Pakistan
2	Wassi	Pakistan
3	Zarar	Pakistan
4	Hafiz Arshad	Caf Leopold
5	Abu Umer	Caf Leopold
6	Javed	Taj Mahal Hotel
7	Abu Shoaib	Taj Mahal Hotel
8	Abdul Rehman	Oberoi Trident Hotel
9	Fahadullah	Oberoi Trident Hotel
10	Baba Imran	Nariman House
11	Nasir	Nariman House
12	Ismail Khan	Chhatrapati Shivaji Terminus
13	Ajmal Amir Kasab	Chhatrapati Shivaji Terminus

The attack was planned well in advance and a Pakistani American David Headley was employed to gather critical information about Mumbai - which was later confirmed by Headley''s confession in the Chicago case [20]. He guided the 13 attackers for the attack and among the three dominating attackers; Wassi has come out as the attacker who was significantly involved in the maximum communication. The software used for the construction of 26/11 network is UCINET which represents, analyzes, visualizes, and simulates nodes (attackers) and ties (relationships) from the input data [20]. Fig. 3 shows the manner in which these terrorists were linked forming a network.

As already been discussed Terrorist Network Mining combines the SNA technique and Graph Theory to analyze the terrorist network. It may be concluded that the social network approach for the study of behavior involves two main themes: (a) the use of formal theory organized in mathematical terms, (b) followed by the systematic analysis of empirical data [25].Using the Graph Theory, the network is visualized as a graph where two communicating users (called nodes) are linked through an edge sharing some relationship. The graph is then represented in a matrix form or an adjacency matrix, A_{ij} represented as:

$$A_{ij} = \left\{ 1 if i and j are connected 0 else \right. \tag{4}$$

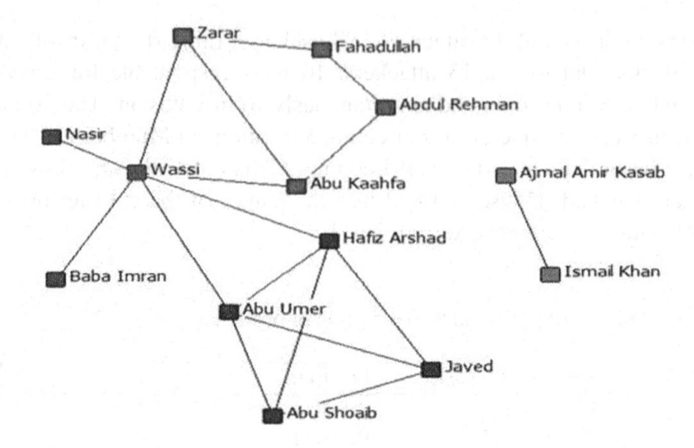

Fig. 3 26/11 terrorist network

and

$$A_{ij} = A_{ji} \tag{5}$$

The adjacency matrix for the 26/11 terrorist network with respect to figure is shown in Table 3:

Table 3 Adjacency matrix for 26/11 Attackers

A : Abu Kaahfa
B : Wassi
C : Zarar
D : Hafiz Arshad
E : Javed
F : Abu Shoaib
G : Abu Umer
H : Abdul Rehman
I : Fahadullah
J : Baba Imran
K : Nasir
L : Ismail Khan
M : Ajmal Amir Kasab

	A	**B**	**C**	**D**	**E**	**F**	**G**	**H**	**I**	**J**	**K**	**L**	**M**
A	0	1	1	0	0	0	0	0	0	0	0	0	0
B	1	0	1	1	0	0	1	0	0	1	1	0	0
C	1	1	0	0	0	0	0	0	0	0	0	0	0
D	0	1	0	0	1	1	1	0	0	0	0	0	0
E	0	0	0	1	0	0	1	0	0	0	0	0	0
F	0	0	0	1	1	0	1	0	0	0	0	0	0
G	0	1	0	1	1	1	0	0	0	0	0	0	0
H	1	0	0	0	0	0	0	0	1	0	0	0	0
I	0	0	1	0	0	0	0	1	0	0	0	0	0
J	0	1	0	0	0	0	0	0	0	0	0	0	0
K	0	1	0	0	0	0	0	0	0	0	0	0	0
L	0	0	0	0	0	0	0	0	0	0	0	0	1
M	0	0	0	0	0	0	0	0	0	0	0	1	0

The centrality matrix for each node is calculated. These can be defined as:

4.1 Degree Centrality

The degree of a node is its number of links. An individual"s having a high degree, for instance, may imply his leadership. The degree D_i of a vertex i is [6]:

$$D_i = \sum_{i=1}^{n} A_{ij} \quad (6)$$

The degree calculated on the 26/11 dataset results are listed in Table 4:

Table 4 Degree measure for 26/11 Attackers

Sl. No.	Name of Terrorist	Degree
1	Abu Kaahfa	3
2	Wassi	6
3	Zarar	3
4	Hafiz Arshad	4
5	Abu Umer	4
6	Javed	3
7	Abu Shoaib	3
8	Abdul Rehman	2
9	Fahadullah	2
10	Baba Imran	1
11	Nasir	1
12	Ismail Khan	1
13	Ajmal Amir Kasab	1

4.2 Betweenness Centrality

Betweenness is the number of geodesics (shortest paths between any two nodes) passing through it. An individual with high betweenness may be a gatekeeper in the network. It is given by [6]:

$$B_a = \sum_{a}^{n} \sum_{j}^{n} g_{ij}(a) \quad (7)$$

g_{ij}(a) indicates whether the shortest path between two other nodes i and j passes through node a.

Betweeness measure for 26/11 terrorists are listed in Table 5:

Table 5 Betweenness measure for 26/11 Attackers

Sl. No.	Name of Terrorist	Betweenness
1	Abu Kaahfa	7.500
2	Wassi	54.000
3	Zarar	7.500
4	Hafiz Arshad	12.500
5	Abu Umer	12.500
6	Javed	0.000
7	Abu Shoaib	0.000
8	Abdul Rehman	0.500
9	Fahadullah	0.500
10	Baba Imran	0.000
11	Nasir	0.000
12	Ismail Khan	0.000
13	Ajmal Amir Kasab	0.000

4.3 Closeness Centrality

Closeness centrality of a vertex is the number of other vertices divided by the sum of all distances between the vertex and all others [8]. In other words, Closeness is the sum of all geodesics between the particular node and every other node in the network [6]:

$$C_a = \sum_{i=1}^{n} l(i,a) \tag{8}$$

l (i, a) is the length of the shortest path connecting nodes i and a. The calculated closeness measure for each terrorist of 26/11 is shown in Table 6:

Table 6 Closeness measure for 26/11 Attackers

Sl. No.	Name of Terrorist	Closeness
1	Abu Kaahfa	26.667
2	Wassi	30.000
3	Zarar	26.667
4	Hafiz Arshad	27.273
5	Abu Umer	27.273
6	Javed	23.529
7	Abu Shoaib	23.077
8	Abdul Rehman	8.333
9	Fahadullah	8.333
10	Baba Imran	24.490
11	Nasir	24.490
12	Ismail Khan	8.333
13	Ajmal Amir Kasab	8.333

4.4 Eigenvector Centrality

Eigenvector centrality acknowledges that not all connections are equal. If we denote the centrality of vertex i by x_i, then we can allow for this effect by making x_i proportional to the average of the centralities of i"s network neighbors [6].

$$x_i = \frac{1}{\lambda} \sum_{i=1}^{n} A_{ij} X_j \tag{9}$$

λ is constant. Defining the vector of centralities x = $(x_1; x_2; : : :)$, we can rewrite this equation in matrix form as:

$$\lambda.x = A.x \tag{10}$$

Hence we see that x is an eigenvector of the adjacency matrix with eigen value λ. Table 7 shows Eigenvector measure calculated for 26/11 attackers:

Table 7 Eigenvector measure for 26/11 Attackers

Sl. No.	Name of Terrorist	Eigenvector
1	Abu Kaahfa	0.206
2	Wassi	0.444
3	Zarar	0.206
4	Hafiz Arshad	0.456
5	Abu Umer	0.456
6	Javed	0.358
7	Abu Shoaib	0.358
8	Abdul Rehman	0.081
9	Fahadullah	0.081
10	Baba Imran	0.125
11	Nasir	0.125
12	Ismail Khan	0.000
13	Ajmal Amir Kasab	0.000

Though these all measures were sufficiently capable of deciding the roles of nodes in the network, they all needs to be calculated individually and after the required measures are calculated, the appropriate analysis is enforced thereafter.

There is a concept of brokerage that searches for the brokers and in addition determines the dominating nodes in a graph. The brokerage on whole estimates the various roles as it would be done individually by calculating the centrality measures, with much more useful information aiding to amend terrorist network analysis.

5 Brokerage

Brokerage is an approach introduced by Ronald S. Burt to understand the manner of embedding a node in its neighborhood that would be fruitful to analyze power, influence and dependency effects. Brokerage is the activity of people who live at the intersection of social worlds, who have a vision advantage of seeing and developing good ideas, an advantage which can be seen in their compensation, recognition, and the responsibility they're entrusted with in comparison to their peers [7].

Although social networks are based on the interactions between two actors but it is noticed that the social network analysis of two nodes relationships are maximally affected by third parties. The third party may be a trusted party that allows transparent communication or an unfaithful party that may affect the communication. Such transmitting actors - often referred to as brokers play a decisive role in the connectedness of social structures and hence, in determining the existing amounts of social capital available to the members of a network [5].

Moreover brokerage is the term friendly in case of ego networks where ego is the broker in the graph and brokerage is the relation the broker share while communication. Ego networks are those networks that analysis social characteristics such as size, relationship strength, density, centrality, prestige and roles.

Brokerage in terms of a network can be understood as a node that acts as a broker or the agent facilitating two nodes either of the same group or from different groups to communicate. Brokers are the bridges and gatekeepers who are controlling information flow in the organizations [19]. The broker can itself be a node from network of either of the two nodes or apart from the two networks or both. Thus any flow of information, knowledge, or products from one node to another essentially propagates through the broker node. The brokerage is expected to be beneficial in case of estimating the terrorist groups where different subgroups of a terrorist organization coordinate to fulfill their awful deeds.

6 Roles Detected by Brokerage Approach

As discussed in previous section, brokerage is about coordinating people between whom it would be valuable, but risky, to trust [7]. The roles played by a broker depend on its location while communication is being carried out. On the basis of this, a broker node may play the following influential roles viz. coordinator, consultant, gatekeeper, representative and liaison facilitating communication.

To understand these roles, consider three users nodes n1, n2 and n3, where n1 and n3 are the two parties among which communication is made while n2 is the broker node through which brokerage is carried out.

6.1 Coordinator

A brokerage node is assumed to act as the coordinator if all the communications within a group a carried out through the broker. In other words, if the node n2 belongs to the same group within which nodes n1 and n3 lies, then the node n2 is called as the Coordinator of nodes within the same group of itself.

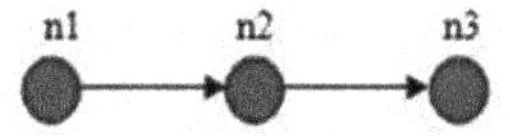

6.2 *Consultant*

If the brokerage node i.e. n2 lies outside the group to which nodes n1 and n3 belongs but is an intermediator among them, then it is assumed as the consulting brokerage node and is called Consultant node while communication.

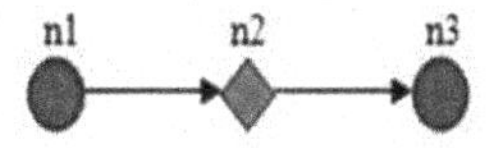

6.3 *Gatekeeper*

The broker is expected to act as Gatekeeper when broker is at the boundary of a group and controls the communication by allowing the outsiders to pass through it in order to let them access within the group. It can be understood as; if the node n2 and n3 belongs to same group then the outsider node n1 needs to pass node n2 to communicate to node n3.

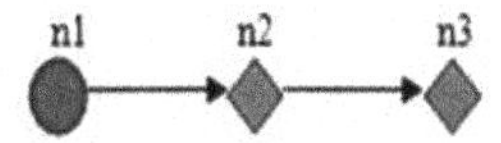

6.4 *Representative*

A broker is declared as Representative when the broker node is in the same group with a node which wishes to initiate communication and acts as the contact point or representative through which communication to outsiders can be made. Diagrammatically, if node n2 lie in same group as node n1 who wants to communicate to outsider node n3, the node n2 is said to be the representative.

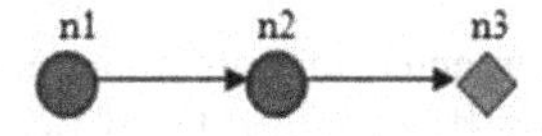

6.5 Liaison

Brokerage result

Liaison is the broker who does not belong to either of the group of the nodes among whom it facilitates the communication. Again, it is the case when node n2 i.e. the broker node acts as the third party through which node n1 from a group wants to communicate node n3 of other group.

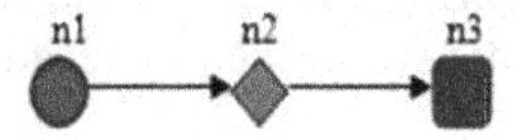

7 Experimental Analysis through Brokerage

Again considering the 26/11 dataset, the above mentioned five roles are estimated through brokerage utilizing UCINET once again. The standard method implemented in Ucinet calculates how many times every actor in a network plays these brokerage roles between groups, giving the total amount of brokerage roles as well [17]. The brokerage can be calculated under ego networks permitting the analysis of the actors playing influential roles in the network. The nodes are considered as ego (broker) individually with their extending connections is estimated. The nodes link when a particular node acts as a broker node may be visualized through the Fig. 5-17.

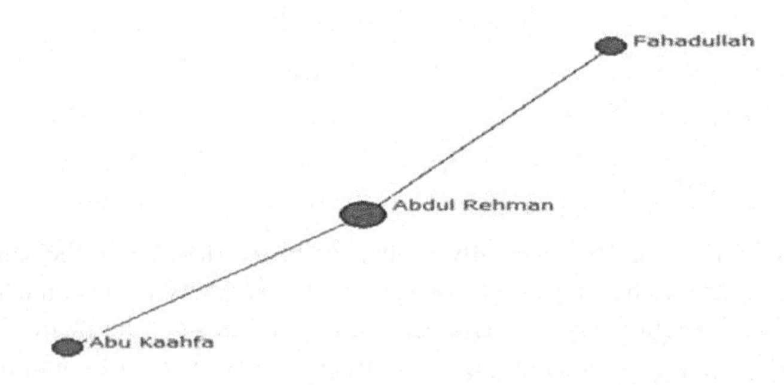

Fig. 4 Abdul Rehman acting as a broker

The standard method implemented in Ucinet calculates how many times every actor in a network plays these brokerage roles between groups, giving the total amount of brokerage roles as well [2]. When the same dataset is analyzed, following brokerage results were obtained.

The analysis results in division of the actors into 2 groups according to the influence of roles played by the actor. This may be view that among the 13 terrorists,

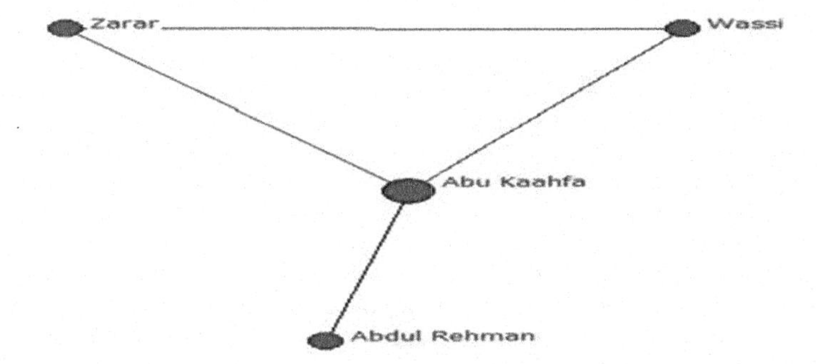

Fig. 5 Abu Kaahfa acting as a broker

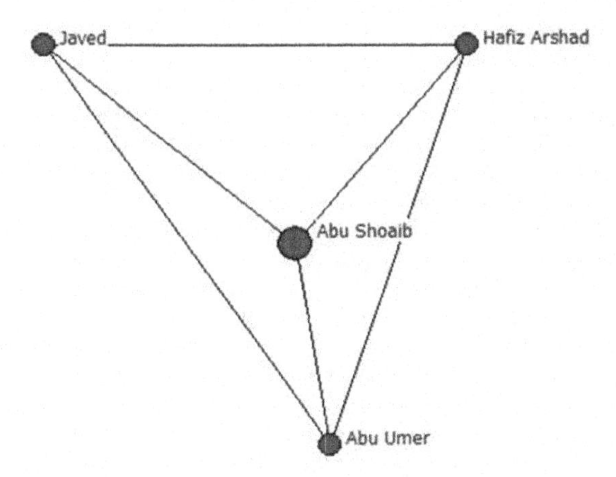

Fig. 6 Abu Shoaib acting as a broker

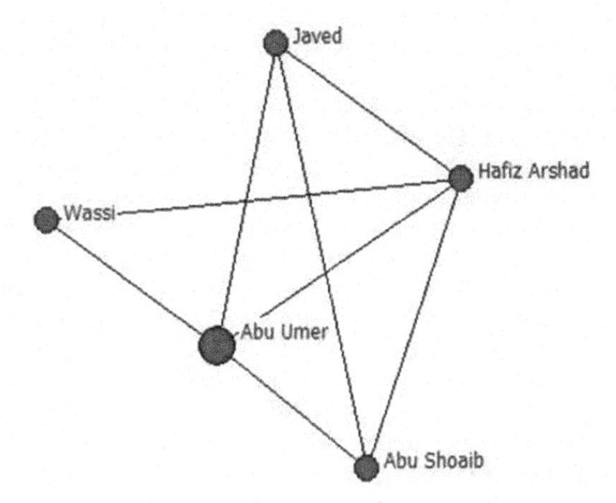

Fig. 7 Abu Umer acting as a broker

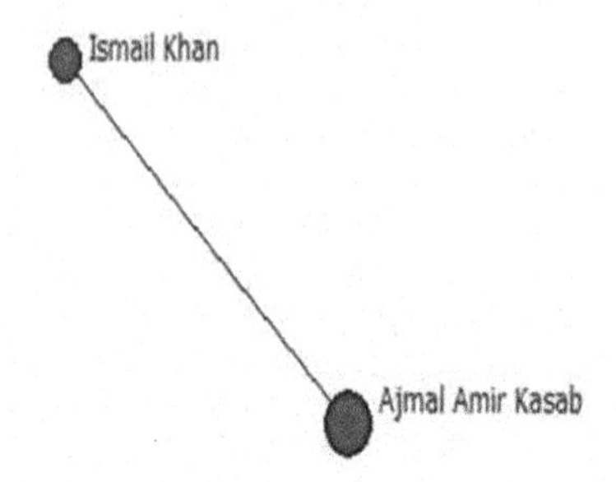

Fig. 8 Ajmal Amir Kasab acting as a broker

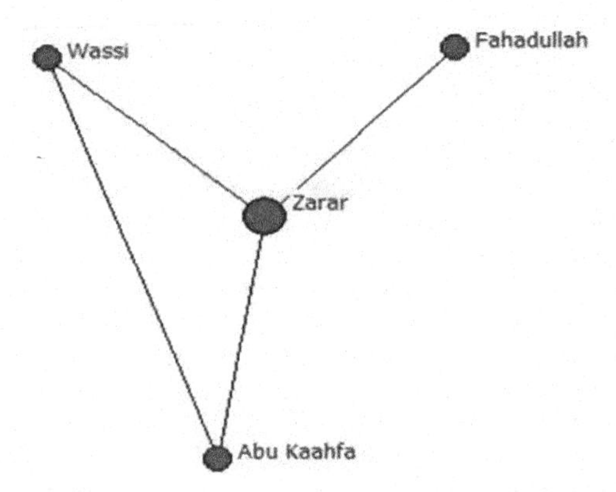

Fig. 9 Zarar acting as a broker

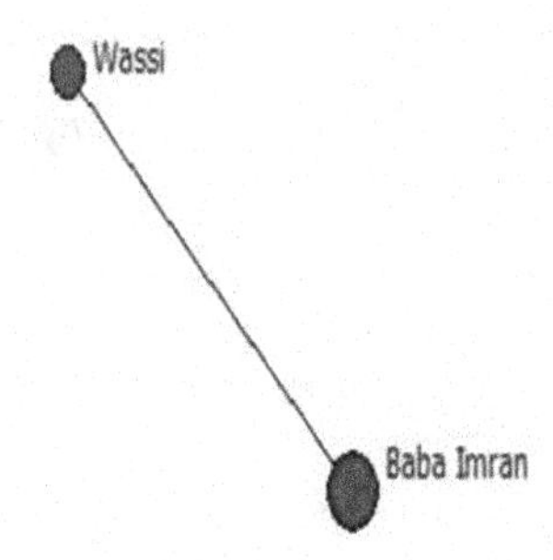

Fig. 10 Baba Imran acting as a broker

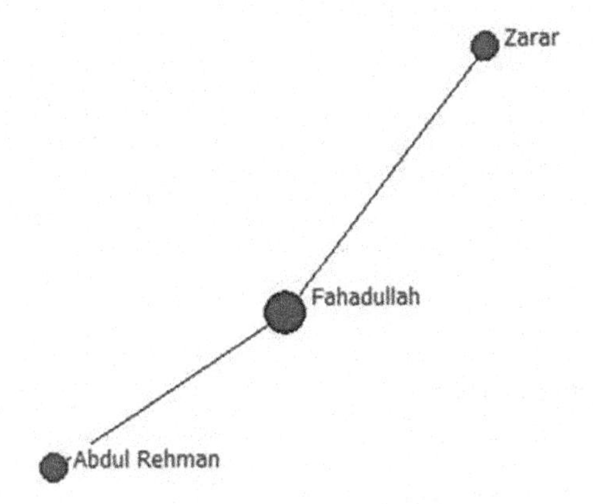

Fig. 11 Fahadullah acting as a broker

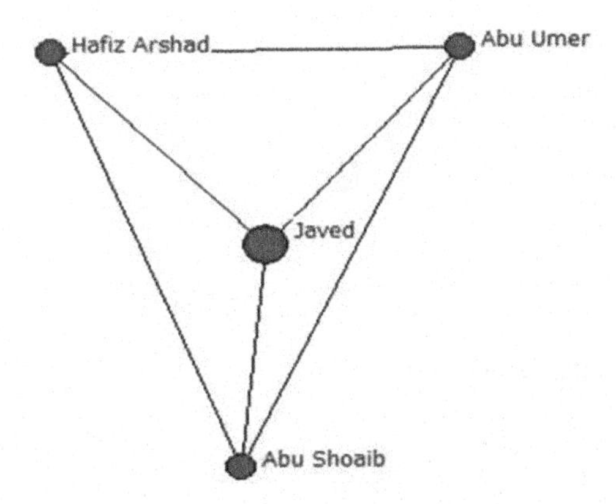

Fig. 12 Javed acting as a broker

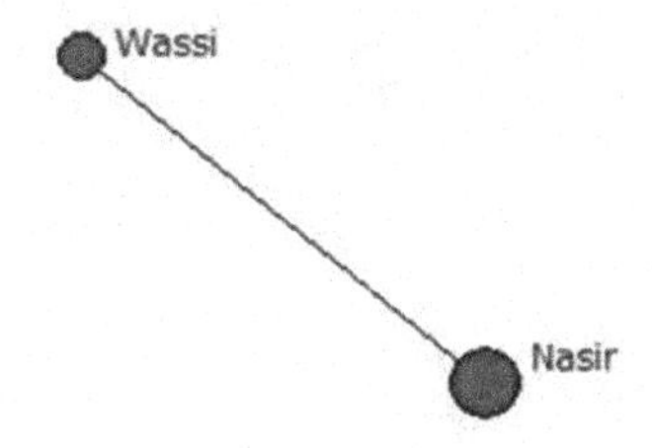

Fig. 13 Nasir acting as a broker

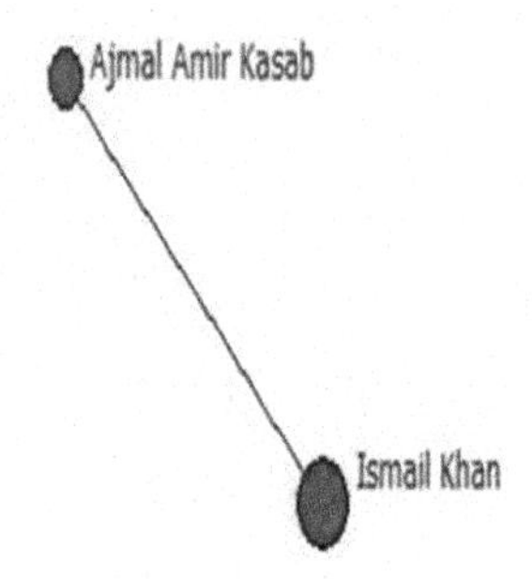

Fig. 14 Ismail Khan acting as a broker

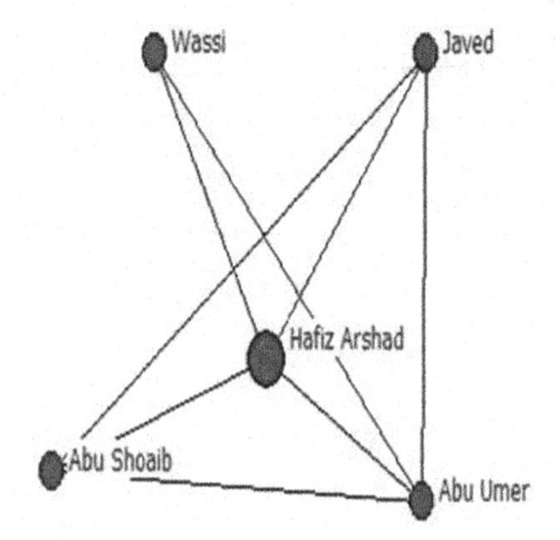

Fig. 15 Hafiz Arshad acting as a broker

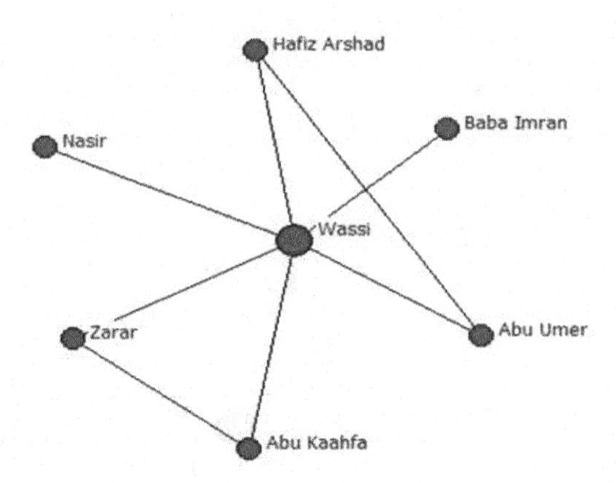

Fig. 16 Wassi acting as a broker

	Coordinat	Gatekeepe	Represent	Consultan	Liaison	Total
Abu Kaahfa	0	0	0	2	0	2
Javed	0	0	0	0	0	0
Fahadullah	0	0	0	1	0	1
Hafiz Arshad	1	2	2	0	0	5
Nasir	0	0	0	0	0	0
Abu Shoaib	0	0	0	0	0	0
Abu Umer	1	2	2	0	0	5
Ismail Khan	0	0	0	0	0	0
Baba Imran	0	0	0	0	0	0
Ajmal Amir Kasab	0	0	0	0	0	0
Wassi	0	4	4	18	0	26
Abdul Rehman	0	0	0	1	0	1
Zarar	0	1	0	1	0	2

Fig. 17 Brokerage Result

Wassi played number of roles where the gain of total count is also the highest for him. The results in addition reveal that Wassi contributed as the consultant maximally i.e. he has been an intermediator within the group leading to maximal activities performed through him. The graph (Fig. 19) shows the total count of different brokerage roles obtained from results in Fig. 18.

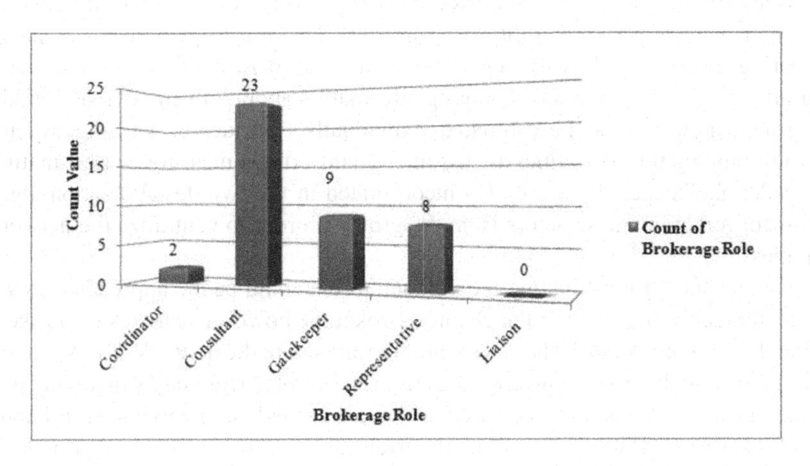

Fig. 18 Brokerage Role Count in 26/11 dataset

In accordance to the result obtained (in Fig. 18) the linking of these nodes may be visualized in Fig. 20.

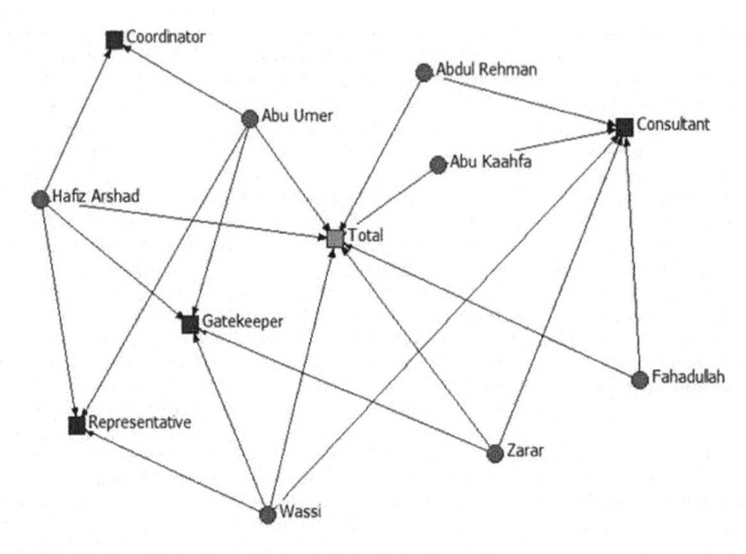

Fig. 19 Contribution of actors as brokerage roles

8 Conclusion and Future Work

The paper covered the very useful aspect of Social Network Analysis i.e. the centrality measures. As centrality measures: degree, betweenness, closeness and eigenvector helps the analyst to decide the various roles within a social network. In similar manner, SNA centrality measures aids law-enforcement agencies judge the various roles played by the terrorists within a terrorist network. Though these centrality measures contribute in gaining information about influential nodes within the group, they need to be estimated individually. The degree and eigenvector measures among the centralities are assumed as the crucial measures for estimating the leader/gatekeeper. They are also incorporated in the two destabilization algorithms for removing these actors from the group in order to neutralize the network activities.

Another aspect for estimating the influential role could be the approach of SNA i.e. brokerage as discussed in the chapter. Brokerage however searches for brokers within the network defining the roles a broker can play in the network. It is a general expectation that the maximum terrorist activities are not performed by involving just two actors. There are number of activities that are carried out involving a third node i.e. a broker node. Hence it would not be irrelevant to consider that broker node(s) serves as the influential node(s) within the network.

Moreover, the brokerage determines the five different influential roles as discussed in the chapter viz. coordinator, consultant, gatekeeper, representative and liaison that a broker can play for letting communication flow smoothly. Studying of these roles would be beneficial to gain knowledge about the dominating node(s), removal of which would affect the terrorist network activities and henceforth, neutralizing the network.

The future work could be to utilize the brokerage approach for the two destabilization algorithms (discussed in background section) which may replace the use of degree and eigenvector centrality measures. The work could also be done in the direction of making brokerage construct an effective algorithm for destabilization.

Appendix I

List of Equations (1) Similarity calculation among centroid vectors of the Typical -Terrorist-Behavior
(2) and (4) Adjacency matrix
(3) and (5) Symmetric property of adjacency matrix
(6) Degree Centrality
(7) Betweenness Centrality
(8) Closeness Centrality
(9) Eigenvector Centrality
(10) Matrix representation of Eigenvector Centrality

Appendix II

List of Tables (1) Adjacency Matrix
(2) 26/11 Attackers Names and their Location of operation
(3) Adjacency matrix for 26/11 Attackers
(4) Degree measure for 26/11 Attackers
(5) Betweenness measure for 26/11 Attackers
(6) Closeness measure for 26/11 Attackers
(7) Eigenvector measure for 26/11 Attackers

Appendix III

List of Figures (1) Learning of the terrorist behavior
(2) Detection of terrorist networks on web
(3) Network for example
(4) 26/11 terrorist network
(5) Abdul Rehman acting as a broker
(6) Abu Kaahfa acting as a broker
(7) Abu Shoaib acting as a broker
(8) Abu Umer acting as a broker
(9) Ajmal Amir Kasab acting as a broker
(10) Zarar acting as a broker
(11) Baba Imran acting as a broker
(12) Fahadullah acting as a broker
(13) Javed acting as a broker
(14) Nasir acting as a broker

(15) Ismail Khan acting as a broker
(16) Hafiz Arshad acting as a broker
(17) Wassi acting as a broker
(18) Brokerage Result
(19) Brokerage Role Count in 26/11 dataset
(20) Contribution of actors as brokerage roles

References

1. Krebs, V.E.: Uncloacking Terrorist Networks. First Monday 7, 4–1 (2002)
2. Borgatti, S.P., Everett, M.G., Freeman, L.C.: UCINET 6 for Windows, Analytic Technologies. Harvard University Press, Cambridge (2002)
3. Fellman, P.V., Wright, R.: Modeling Terrorist Networks - Complex Systems at the Mid-Range. In: Proceedings of Complexity, Ethics and Creativity Conference, LSE (2003)
4. Elovici, Y., Kandel, A., Last, M., Shapira, B., Zaafrany, O.: Using Data Mining Techniques for Detecting Terror-Related Activities on the Web. Proceedings of Journal of Information Warfare (2004)
5. Tube, V.G.: Measuring the Social Capital of Brokerage Roles, Connections. International Network for Social Network Analysis 36(1), 29–52 (2004)
6. Memon, N., Qureshi, A.R.: Investigative Data Mining and its Application inCounterterrorism. In: Proceedings of the 5th WSEAS International Conference on Applied Informatics and Communications, Malta, pp. 97–403 (2005)
7. Burt, R.S.: Brokerage and Closure: An Introduction to Social Capital. Oxford University Press (2005)
8. de Nooy, W., Mrvar, A., Batagelj, V.: Exploratory Social Network Analysis with Pajek. Cambridge University Press, New York (2005) ISBN: 0521841739
9. Han, J., Kamber, M.: Data Mining: Concepts and Techniques, 2nd edn. Morgan Kaufmann Publishers (2006)
10. Memon, N., Larsen, H.L.: Practical Approaches for Analysis, Visualization and Destabilizing Terrorist Networks. In: Proceedings of the First International Conference on Availability, Reliability and Security, ARES (2006)
11. Memon, N., Larsen, H.L.: Practical Algorithms And Mathematical Models For Destabilizing Terrorist Networks. In: Mehrotra, S., Zeng, D.D., Chen, H., Thuraisingham, B., Wang, F.-Y. (eds.) ISI 2006. LNCS, vol. 3975, pp. 389–400. Springer, Heidelberg (2006)
12. Memon, N., Larsen, H.L.: Structural Analysis and Destabilizing Terrorist Networks. In: The First International Conference on Availability, Reliability and Security, ARES 2006. IEEE (2006)
13. Shaikh, M.A., Jiaxin, W.: Investigative Data Mining: Identifying Key Nodes in Terrorist Networks. In: Multitopic Conference, INMIC 2006, pp. 201–207. IEEE (2007)
14. Memon, N., Kristoffersen, K.C., Hicks, D.L., Larsen, H.L.: Detecting Critical Regions in Covert Networks: A Case Study of 9/11 Terrorists Network. In: Second International Conference on Availability, Reliability and Security (ARES 2007). IEEE (2007)
15. Memon, N., Larsen, H.L., Hicks, D., Harkiolakis, N.: Detecting Hidden Hierarchy in Terrorist Networks: Some Case Studies. In: Yang, C.C., Chen, H., Chau, M., Chang, K., Lang, S.-D., Chen, P.S., Hsieh, R., Zeng, D., Wang, F.-Y., Carley, K.M., Mao, W., Zhan, J. (eds.) ISI Workshops 2008. LNCS, vol. 5075, pp. 477–489. Springer, Heidelberg (2008)

16. Memon, N., Qureshi, A.R., Will, U.K., Hicks, D.L.: Novel Algorithms for Subgroup Detection in Terrorist Networks. In: 2009 International Conference on Availability, Reliability and Security. IEEE (2009)
17. Bellotti, E.: Brokerage roles between cliques: a secondary clique analysis. Methodological Innovations Online 4, 53–73 (2009)
18. Wiil, U.K., Memon, N., Karampelas, P.: Detecting New Trends in Terrorist Networks. In: 2010 International Conference on Advances in Social Networks Analysis and Mining (2010)
19. Sozen, H.C., Sagsan, M.: The Brokerage Roles in the Organizational Networks and Manipulation of Information Flow. International Journal of eBusiness and eGovernment Studies 2(2) (2010) ISSN: 2146-0744
20. Azad, S., Gupta, A.: A Quantitative Assessment on 26/11 Mumbai Attack using Social Network Analysis. Journal of Terrorism Research (2011)
21. Kang, U., Papadimitriou, S., Sun, J., Tong, H.: Centralities in Large Networks: Algorithms and Observations. In: SIAM International Conference on Data Mining (SDM 2011), Phoenix, U.S.A. (2011)
22. Chaurasia, N., Dhakar, M., Tiwari, A., Gupta, R.K.: A Survey on Terrorist Network Mining: Current Trends and Opportunities? International Journal of Computer Science and Engineering Survey (IJCSES) 3(4), 59–66 (2012)
23. Abraham, A., Hassanien, A.-E.: Computational Social Networks: Tools, Perspectives and Applications, 4th edn. Series in Computer Communications and Networks. Springer, London (2012) ISBN 978-1-4471-4047-4
24. Abraham, A.: Computational Social Networks: Security and Privacy, 4th edn. Series in Computer Communications and Networks. Springer, London (2012) ISBN 978-1-4471-4047-4
25. Abraham, A.: Computational Social Networks: Mining and Visualization, Series in Computer Communications and Networks. Springer, London (2012) ISBN 978-1-4471-4047-4

Index

Printed in the United States
By Bookmasters